The Digitization of Cinematic Visual Effects

The Digitization of Cinematic Visual Effects

Hollywood's Coming of Age

Rama Venkatasawmy

LEXINGTON BOOKS
Lanham • Boulder • New York • Toronto • Plymouth, UK

Published by Lexington Books
A wholly owned subsidiary of The Rowman & Littlefield Publishing Group, Inc.
4501 Forbes Boulevard, Suite 200, Lanham, Maryland 20706
www.rowman.com

10 Thornbury Road, Plymouth PL6 7PP, United Kingdom

British Library Cataloguing in Publication Information Available

Library of Congress Cataloging-in-Publication Data

Venkatasawmy, Rama, 1972-
The digitization of cinematic visual effects : Hollywood's coming of age / Rama Venkatasawmy.
p. cm.
Includes bibliographical references.
ISBN 978-0-7391-7621-4 (cloth : alk. paper)—ISBN 978-0-7391-7622-1 (ebook)
1. Motion pictures—United States—History—20th century. 2. Cinematography—Special effects. 3. Digital cinematography. I. Title.
PN1993.5.U6V455 2013
791.430973—dc23
2012034920

The paper used in this publication meets the minimum requirements of American National Standard for Information Sciences Permanence of Paper for Printed Library Materials, ANSI/NISO Z39.48-1992.

Printed in the United States of America

Acknowledgements

I am grateful to Emeritus Professor Alec McHoul for his valuable contribution during the earlier stages of this work and to Professor Tom O'Regan who supported my interest in cinematic visual effects many years ago.

Thanks to Lenore Lautigar, Johnnie Simpson, Alison Northridge, and Kelly Shefferly at Lexington.

And special thanks to my parents L.L. & V. Venkatasawmy who let me watch on public television countless re-runs of low budget VFX movies during my childhood. . . .

Notes on the Text

With the kind permission of *TMC Academic Journal*, this book includes up to 40 percent of material originally published in the following article:

Venkatasawmy, R. 'The Evolution of VFX-Intensive Filmmaking in 20th Century Hollywood Cinema: An Historical Overview.' *TMC Academic Journal*, Vol. 6, Issue 2, March 2012, 17–31.

All translations from the French are my own unless otherwise indicated.

This book uses a variant of the MLA system of citations and references throughout. See:

Gibaldi, Joseph. *MLA Handbook for Writers of Research Papers*. New York: The Modern Language Association of America. 1999.

Contents

Introduction

It should be highlighted early on that, as the focus of this book, the 'Hollywood cinema industry' (and the motion pictures it produces) is fundamentally understood in this context as being constituted by all the dominant movie production studios or 'majors' that have existed or are currently in existence. These are: Columbia Pictures/Sony Pictures Entertainment, DreamWorks SKG, Metro-Goldwyn-Mayer Inc., Paramount Pictures, RKO Pictures, Twentieth Century-Fox, United Artists, Universal, Walt Disney Co., and Warner Bros. (independents and relatively smaller studios operating within or on the fringes of the industry are excluded here).

'Just as *Gone with the Wind* (Fleming 1939) was the pinnacle of the classical-era Hollywood system, so too is *Avatar* the pièce de résistance of the nascent digital age of Hollywood filmmaking.'[1] By the beginning of the second decade of the 21st century, *Avatar* had indeed become a turning point in the history of cinema in general and of Hollywood cinema in particular. But what led to such degree of technical advancement and sophistication in visual storytelling and spectacularity to be achieved in this film? To understand what enabled such a movie to be made the way it was at that particular point in time, it is productive to take into serious consideration that amidst the widely diverse nature of its filmic output, the Hollywood cinema industry has, for most of its 20th century history, been particularly prolific in its output of visual effects-intensive movies. "Visual effects" or VFX, as we shall refer to it, is briefly defined for now in terms of three main functions and objectives in application during the filmmaking process and in relation to film spectatorship: (i) filmic "fantasy" construction; (ii) imitative re-production, creative simulation and storytelling support; (iii) visual enhancement and correction. In Chapter three, the sole use of the term "visual effects" or VFX in this book will be explained in more details and filmic effects will be more extensively conceptualized and defined.

With more attention often being given to the construction and composition of elaborate spectacularity rather than to plot, dialogue or character development, VFX-intensive filmmaking is essentially interested in, as well as remembered for, identifying, formulating and projecting an aesthetic of visual excess predominantly meant to trigger the viewer's sensual gratification. Indeed, most of the biggest 20th century money makers at the American box-office have been VFX-intensive Hollywood movies:

The Exorcist, Jaws, Star Wars, Raiders of the Lost Ark, E.T. The Extra--Terrestrial, Return of the Jedi, Ghostbusters, Batman, Jurassic Park, Independence Day, Men in Black, Star Wars: Episode I—The Phantom Menace, to name a few.[2] Such motion pictures epitomize the Hollywood cinema industry's brand of unashamedly escapist commercial entertainment that appeals to the widest range of spectatorship and guarantees massive financial returns to investment—that is, attracting a maximum number of viewers to generate as much profit as possible.

Comprehensive VFX applications have not only motivated the expansion of filmmaking praxis: they have also influenced the evolution of viewing pleasures and spectatorship experiences. Following the digitization of their associated technologies, VFX have been responsible for multiplying the strategies of re-presentation and storytelling as well as extending the range of stories that can potentially be told on screen. By the same token, the visual standards of the Hollywood film's production and exhibition have been growing in sophistication. On the basis of displaying groundbreaking VFX—immaculately realized through the application of cutting-edge technologies and craftsmanship—and of projecting such a significant degree of visual innovation and originality, certain Hollywood movies have established techno-visual trends and industrial standards for subsequent filmmaking practice.

There was a significant shift during the 1990s in the production and exhibition mode of Hollywood motion pictures—as illustrated, for instance, by *Toy Story*'s entirely computer-generated images that, as such, do not originate from an actual pro-filmic event. This shift signaled the beginning of the most important transition in the evolution of (analog) cinema whose centenary was also being grandly celebrated during that period. It took a hundred years to see a feature-length motion picture whose *only* connection with acetate was in its theatrical exhibition: its actual making had little to do with celluloid-related technologies.[3] The year 1999 witnessed the first theatrical exhibition in digital (albeit in a handful of theaters only) of Hollywood feature films: *Star Wars: Episode I—The Phantom Menace* and the entirely computer-generated *Toy Story 2*. With the release of these two films, and the subsequent increase in low-budget digital video production, exhibition and distribution, celluloid film could only be on its way out, as confirmed by the gradual phasing out of celluloid film production by the manufacturers concerned. And at the Cannes film festival in 2000, Sony was launching its new F900 high-definition digital movie camera, capable of images comparable to 35mm film.

> The 1990s in particular can be considered as a decade in which the experience of film going, the economy of the industry and the ontology of film transmute into new formations. This claim rests on three aspects of change. First, Hollywood (the logo of a corporate empire) no longer

> primarily produces film. Its productive remit has extended horizontally and vertically into the production of related products, a versioning of narrative across diverse media, connected and supported by an industry of promotion and distribution. Second, the theatrical experience of film, the cinematic context, is no longer central to revenue production. . . . Cinema, economically, has become displaced as the critical focus, repositioned by ancillary objects and interests. Third, the materiality of film, the form of acetate ('celluloid' is a commercial label rather than a strictly descriptive term), is increasingly replaced by digital materials incorporated into various stages of production. Whilst film production in the present is characterized by a thoroughly hybrid mixing of digital and celluloid forms, Eastman Kodak's announcement that it intends to suspend celluloid production by 2010 provides a cut-off point for the period of transition.[4]

Digital VFX and filmmaking had indeed completely changed the very concept and practice of film production in Hollywood. Originally conceived in terms of making things technically easier for directors, cinematographers, actors and producers, digital VFX eventually evolved more in the direction of satisfying storytelling requirements, expanding storytelling possibilities and reducing production costs.

With filmmaking going almost entirely digital after the 1990s, the physical process of motion picture production in Hollywood had changed for good. The Hollywood film was getting made mostly during pre-production and post-production, via computer software, and less during principal photography on location or the studio soundstage. As expressed by David Rodowick:

> In the course of a single decade, the long privilege of the analog image and the technology of analog image production have been almost completely replaced by digital simulations and digital processes. The celluloid strip with its reassuring physical passage of visible images, the noisy and cumbersome cranking of the mechanical film projector or the Steenbeck editing table, the imposing bulk of the film canister are all disappearing one by one into a virtual space. . . .[5]

Just like so-called "traditional" film cutting and editing, many "traditional" VFX-related crafts such as mould making, casting, modeling, drawing, optical compositing and matte painting have also been fading away. With the production of digital VFX as well as the need to conceal their application becoming more dominant concerns in Hollywood filmmaking, compared to focusing on film style and aesthetics for instance, the role of computer programmers, graphic designers and animators has correspondingly become more crucial to filmmaking in comparison with that of other technicians and artists of the cinema.

From the late 1970s, digital VFX progressed from being a source of techno-visual novelty and problem solving in Hollywood to becoming the most defining characteristic of the Hollywood film and of its viewing experience. If the extent to which a Hollywood film is entertaining is established on the basis of its box-office success, then the movies that audiences found the most entertaining in the last two decades of the 20th century were all digital VFX-driven Hollywood spectacles involving large production budgets. The trend does not seem to have changed much during the first decade of the 21st century, as illustrated by two *Star Wars* movies, *The Lord of the Rings* trilogy, the *Harry Potter* films and *Avatar*.

In the mid-1980s, the application of digital VFX predominantly followed a "middle-of-the-road" approach by virtue of combining analog technologies and techniques (more than 50 years old but still reliable) and digital technologies (newly emerging and quite expensive). The period between the late 1980s and the mid-1990s was particularly productive, with many groundbreaking digital VFX styles and innovations emerging to set the pace and direction for the next decade. By the end of the 1990s, there were fewer analog and more digital technologies and techniques being applied in practically all areas of filmmaking in the Hollywood cinema industry, with digital VFX having proven their potential for enhancing drama and story, reducing production costs and increasing profits, as opposed to being a short-lived gimmick. Hollywood studios and producers had all been convinced that continuous investment in digital VFX and filmmaking would be the key to product differentiation and increased grosses in the future.

The originally expensive high-end digital VFX of *The Abyss*, *Terminator 2: Judgment Day* and *Jurassic Park* not only enabled the subsequent making of *Titanic*, *The Matrix* and *Star Wars: Episode I—The Phantom Menace* in the late 1990s, but they also further reinforced the normalization of digital VFX in Hollywood filmmaking as such. By the next decade, the same digital VFX had become affordable for extensive application throughout the industry, for medium- to low-budget independent productions, and for television movies, serials and programs. By the end of the first decade of the 21st century, digital filmmaking and VFX were enabling the relatively rapid production of profitable motion pictures on medium-to-low budgets.

The film studios were originally set up in the Los Angeles region because it had sea, mountains, desert, groves and fields nearby. These provided producers and directors with a wide range of settings and backgrounds while enabling endless narrative possibilities. In a context of increasingly internationalized and globalized productions, digital VFX and digital filmmaking have made almost redundant the centrality of Hollywood itself as well as previous concerns with location settings and backgrounds that would mostly be simulated in computer-generated virtual spaces from the late 1990s.[6]

VFX-intensive filmmaking has indeed significantly contributed to the Hollywood cinema industry's main motivation for selling thrills, emotions, dreams, nightmares and sensual pleasures: to generate financially profitable filmic spectacles—the motion picture business has, after all, always fundamentally been about making money. Originally more pertinent to the production of B-movies, double-bill 'fillers' and exploitation serials during the studio system era, VFX-intensive filmmaking progressively became a normalized and standardized aspect of A--movie production practices in the Hollywood cinema industry.

In the light of the above, the main aims of this book are to:

- investigate Hollywood cinema's techno-visual metamorphosis following the digitization of VFX-related tools and of filmmaking technologies commonly used in the industry;
- locate the origins, nature and ramifications of the most important techno-industrial transition for Hollywood cinema since the advent of sound, color and widescreen;
- examine the techno-industrial circumstances, motivations and outcomes of digitization for Hollywood cinema;
- identify the impact on the Hollywood digital cinemascape of continuous techno-industrial innovation, research and development in the VFX domain;
- analyze and evaluate the impact of digitization on VFX, on its practitioners, on filmmaking technologies, and on Hollywood film production practice generally.

As a way of "setting the scene" so-to-speak, Chapter one provides a general overview of the evolution of VFX-intensive filmmaking in the context of the Hollywood cinema industry during the 20th century.

Chapter two defines more extensively this book's object of enquiry, that is, VFX in 20th century Hollywood cinema from the perspective of film history and theory. This chapter also raises questions about the writing of film history and of the different strands that contribute to specific historical understandings of VFX.

Chapter three addresses the various ways of conceptualizing, defining and understanding VFX. In addition to exploring the importance, functions and objectives of VFX in Hollywood films, it addresses the impact of the digitization of VFX in the Hollywood cinema industry. This chapter also discusses issues of perception and cinematic re-presentation, especially in the light of the digitization of VFX and filmmaking in Hollywood.

Chapter four provides conceptual clarification about "digital" and "analog," in the process of discussing "science," "technology" and their intrinsic relationship with regard to the Hollywood cinema industry. This chapter also situates the digital conversion of images into "numbers," especially in terms of the emergence of "technoscience" as a specific model of 20th century industrial R&D. And it further argues that this digital conversion occurred in the Hollywood cinema industry in a way that was less a rupture with its tradition of filmmaking than a continuation and intensification of its long-term reliance on technological innovation.

Chapter five addresses the evolution of computer imaging, its application to and impact on VFX production. It also looks into how the Hollywood cinema industry progressively embraced the digital realm by providing a map of interlocking networks of specific creative individuals who were influential in the realization of digital filmmaking. The description of technological exchanges, transfers and synergies in relation to the Hollywood cinema industry is tied to how "technology," "industry," and "Hollywood cinema" are conceptually understood to interact in the first place. Theoretical and normative positions on these determine the range of technological exchanges, transfers and synergies in the industry that are consequently prioritized in its progressive digitization.

Chronicling the evolution and eventual digitization of VFX has its roots in, and is intertwined with, a number of other complex Hollywood cinema industry narratives:

- how and why certain specific VFX-intensive films were actually produced;
- the normalization of VFX-intensive filmmaking in the Hollywood cinema industry's A- and B-movie production practices during the studio system era;
- the studio system era's impact, its demise and aftermath;
- the constant upgrading of industrial standards through on-going techno-scientific innovation in filmmaking and media-related R&D both within and outside Hollywood;
- Hollywood's adaptation to internal and external changes through continuous updating of movie business practices (vertical integration, conglomeration, alliances, mergers and synergy).

André Bazin cautions that '[a]ny account of the cinema that was drawn merely from the technical inventions that made it possible would be a poor one indeed. On the contrary, an approximate and complicated visualisation of an idea invariably precedes the industrial discovery which alone can open the way to its practical use.'[7] Hence, to further understand the digitization of VFX in relation to the Hollywood film depends on understanding how and why: certain techno-scientific developments were encouraged and prioritized over others; certain movie business practices as well as alliances and individuals were favored over others in the Hollywood cinema industry; the convergence between corporate media, information technology and the entertainment industries eventually occurred, with digital becoming the techno-industrial standard across the industry by the end of the 1990s.

In the context of Hollywood as a global center for the production of VFX-intensive feature films, what is provided here is an integrated account of the digitization of VFX, while also addressing the digitization of Hollywood cinema as a whole albeit in general terms.[8] Digital post-production, digital capture and digital exhibition for instance are considered in discussing the digitization of cinema, but this discussion is not as comprehensive as to include detailed analysis of other related aspects such as the introduction of the DVD and its growing role in Hollywood revenue streams, debates around piracy, the role of the Internet in film distribution, or the attempt by Hollywood studios to establish standards for digital cinema through the formation of the Digital Cinema Initiative. Hence, this book is to be understood as an integrated account of the transition to digital of VFX in the context of the Hollywood cinema industry during the late 20th century—from predominantly techno-industrial and critico-analytical perspectives.

NOTES

1. Sickels, *American Film* 7.
2. See *Appendix B.*
3. Although entirely computer-generated, *Toy Story* had to be converted from digital to analog and transferred to celluloid film so that it could be theatrically exhibited in cinemas worldwide that were at the time all equipped with analog projectors only.
4. Harbord, *The Evolution of Film* 42.

In spite of having produced the first consumer digital still camera and shifted its focus to the digital realm, Eastman Kodak has ended up filing for bankruptcy in January 2012.

5. Rodowick, *The Virtual Life* 8.
6. Goldsmith & O'Regan nevertheless remind us that:

> The district of Hollywood is still more or less the geographic centre of a cluster of production facilities, soundstages, office buildings, and studio ranches, stretching from Culver City, Venice, and Santa Monica in the south to Glendale, Burbank, North Hollywood, and even the Simi Valley in the north. The dozen or so companies that control more than half of the world's entertainment have headquarters in Los Angeles, within a thirty-mile radius of Hollywood. The executives, agents, producers, actors, and directors are there. The meetings to decide what movies will be made are held there. At some point, every major figure in world entertainment has to come to Hollywood. . . .

Goldsmith & O'Regan *The Film Studio* 182.

7. Bazin, *What is Cinema? (vol. I)* 18.
8. As such, there is reference predominantly in this book to so-called "popular" Hollywood films. This approach may not necessarily sit well with many because 'film history has, in almost its entirety, been written without regard to and often with deliberate disregard of the box office. . . . [C]onventional film history omits the great majority of Hollywood's most commercially successful products . . . because, perhaps, no one wants to write the history of a cinema of complacency.'

Vasey, 'Hollywood Industry Paradigm' 299.

ONE

An Overview of VFX-Intensive Filmmaking in 20th Century Hollywood Cinema

The evolution of VFX-intensive movies and the digitization of VFX, in conjunction with Hollywood's entry into the digital realm, are irremediably linked to the industry's overall growth and global domination of filmed entertainment since the establishment of its principal motion picture production studios—especially in terms of how this growth and domination bears the permanent traces of certain key films, practices, events, producers, filmmakers and craftsmen. The contemporary big-budget VFX-intensive Hollywood movie has its roots in a "big picture" style and practice of filmmaking going as far back as the time when the American motion picture industry began to undergo "vertical integration" and actually started 'producing reasonably sophisticated narratives'—from about 1915 onward, as illustrated by the large-scale epic *Birth of a Nation.*[1] In addition, *Intolerance* (1916), *Cleopatra* (1917 and 1934), *The Iron Horse* (1924), *The Phantom of the Opera* (1925), *Ben-Hur: A Tale of Christ* (1925), *Wings* (1927), *Hell's Angels* (1930), *Alice in Wonderland* (1933), *The Hurricane* (1937), *Spawn of the North* (1938), *The Rains Came* (1939), *Union Pacific* (1939), and *Gone with the Wind* (1939) are also part of that select group of pre-World War II Hollywood films that became the models and established the standards for subsequent Hollywood "big picture" productions to follow. *Gone with the Wind,* for example, was

> the last big picture of the [1930s] decade and the greatest box-office hit of the sound era, epitomized to a remarkable degree almost every trend and taste of the 1930s. . . . *Gone with the Wind* cost more to make and had a longer running time than any previous American picture. *Its*

> *marketing made it the first modern "event movie" and created the blockbuster syndrome, which has dominated the industry's thinking to this day.*[2]

It can additionally be argued that the so-called 'event movie' and 'blockbuster syndrome' had in effect already emerged with *Birth of a Nation*, *Intolerance* and *Ben-Hur: A Tale of Christ*, hence many years before *Gone with the Wind*. In any case, blockbuster movies would essentially continue to increase in frequency, becoming permanent fixtures of Hollywood cinema, even after the collapse of the studio system and during later periods of crisis in the industry.[3]

After World War I, according to Kristin Thompson, the 'American film industry had become a group of multi-national corporations through their creation of distribution networks abroad: European corporations, the former imperialist powers, had to retreat and develop methods by which to respond to aggressive United States competition.'[4] And by the 1930s, the American motion picture industry had turned into

> a mature oligopoly. The merger movement had run its course, with the result that five companies dominated the screen in the U.S. Known as the Big Five, these companies were Warner Brothers, Loew's Inc. (the theater chain that owned Metro-Goldwyn-Mayer), Paramount Pictures, Twentieth Century-Fox, and RKO (Radio-Keith-Orpheum). All were fully integrated: they produced motion pictures, operated worldwide distribution outlets, and owned large theater chains where their pictures were guaranteed a showing. Operating in a sort of symbiotic relationship with the Big Five were the Little Three: Universal, Columbia and United Artists.[5]

A significant turning-point at the peak of the Hollywood studio system era was that of the "Paramount Decrees" which, although setting in motion the dismantling of the dominant studio system, triggered long-lasting industrial, structural, practical and stylistic changes with regard to the production strategy of Hollywood films. By forcing the separation of vested interests in exhibition from those in distribution and production, the "Paramount Decrees" weakened the oligopolistic control of the American film industry by a handful of studios and movie moguls. In 1942,

> the United States government reached a temporary settlement with the major studios in the so-called Paramount case, and, for the "big five" vertically-integrated majors at least, the twin practices of block booking and blind selling were either effectively curtailed or banned outright. For the first time all their products had to be sold to exhibitors individually, and this included B-films, which were even, occasionally, now subject to press reviews and trade shows.[6]

The Hollywood cinema industry had little choice but to implement the transformations that were being imposed if it did not want to lose a lucrative market and product. One of the consequences of the "Paramount Decrees" (finally issued in 1948) was that fairly 'rapidly, B-films were encouraged to become increasingly competitive, compulsorily and compulsively different, distinctive. What had previously, perhaps, been a rather static aesthetic and occupational hierarchy between Bs and As became suddenly more flexible.'[7] Second-bill low-budget films, serials and sequels—the B-movies—became as equally important as A-movies.[8]

From westerns to gangster movies, from alien invaders to mad scientists, from murder mysteries to slapstick comedies, from jungle adventures to spy thrillers, from detective stories to hospital dramas, from treasure hunting to medieval sword-play, from crime-fighting organizations to farcical families: during the entire studio system era, Hollywood studios (some more than others), as well as "independents" operating on the fringes, combined, used and abused practically every kind of generic vehicle possible in order to maintain a steady supply of lower grade B-movies to eager audiences in cinema theaters.[9] In existence since the mid-1910s, theatrical serials for example became much more prominent with regard to B-movie production, as illustrated by Universal's detective series *The Ace of Scotland Yard* (in ten episodes) which was, in 1929, the first talking serial of the sound era. Although having less resources than the "big five" Hollywood studios, Universal nevertheless gained financially nearly as much as they did during the rest of the "studio system" era, especially from its low- to medium-budget monster movies and their respective spin-offs [*Dracula* (1930), *Frankenstein* (1931), *The Mummy* (1932), *The Invisible Man* (1933), *The Wolf Man* (1941)] and from its slapstick comedies starring Bud Abbott and Lou Costello. For its part, Columbia relied for profit on low-budget comic strip-based serials [*Mandrake the Magician* (1939), *Terry and the Pirates* (1940), *The Shadow* (1940), *Captain Midnight* (1942), *Batman* (1943), *The Phantom* (1943), *Superman* (1948)].

Producing second-bill features cheaply, rapidly and in large quantity was essentially contingent upon: (i) paying very little or no attention to the high production-value requirements of A-movies; (ii) the ability to exploit "tired" stories and stretch thin narratives to make as many sequels or serial episodes as possible; (iii) continuous product diversification—basically the recycling of anything from footages, studio sets, props and costumes to themes, plots and characters belonging to other productions. As confirmed by Lea Jacobs: 'a B film is one made quickly on a low budget, without major stars or expensive literary properties, in a specialized production unit—such as Lee Marcus's unit at RKO—or by a B studio like Monogram or Republic.'[10] B-movies generated substantial box-office revenues for many years, sometimes to the extent of absorbing losses on big-budget flops. Hollywood studios and producers, as well as various "independents," enthusiastically prolonged such a strategy of low-grade mass-production filmmaking over nearly three decades for purely commercial reasons: it allowed easy money to be made continuously from the box-office with minimal investment and overheads. A less obvious but more significant contribution of mass-produced serials and B-movies of the Hollywood studio system era is to the development and diversification of countless themes, narrative formulas, plot devices, stories, characters, icons and generic vehicles for future filmmaking. Some of these have remained in use in Hollywood cinema many decades later—in the form of anything from subtle "references," "tributes," "remakes," to more flagrant "copying," "borrowing," "recycling" and "mixing."

In conjunction with the growth of blockbusters and B-movies in Hollywood, another kind of film was also beginning to surface: VFX-intensive movies—especially horror, fantasy, and science fiction. Notable VFX-intensive horror, fantasy and sci-fi movies produced during the studio system era include: *Dr Jekyll and Mr Hyde* (1920, 1931 and 1941), *The Thief of Bagdad* (1924 and 1940), *The Lost World* (1925), *London After Midnight* (1927), *Dracula* (1930), *Frankenstein* (1931), *Doctor X* (1932), *Island of Lost Souls* (1932), *The Mummy* (1932), *The Mask of Fu Manchu* (1932), *King Kong* (1933), *The Invisible Man* (1933), *The Son of Kong* (1933), *The Bride of Frankenstein* (1935), *Mark of the Vampire* (1935), *The Wizard of Oz* (1939), *Dr. Cyclops* (1940), *The Wolf Man* (1941), *Sinbad the Sailor* (1947) and *Mighty Joe Young* (1949).

Upon its appearance during the Hollywood studio system era, VFX-intensive filmmaking was often confined to the cheaper, less glamorous but very lucrative end of routine B-movie production in the industry: low-budget horror, fantasy and sci-fi serials, spin-offs and sequels. For such movies, high production values were usually deemed less important than the need for visual excitement and thrills. VFX were hence ideal inexpensive alternatives to replace or imitate anything needed on-screen that would have been too costly or time-consuming to record on location in real conditions, such as: dangerous or wild animals; "exotic" or immense background landscapes; fantastic beings (monsters, supernatural and mythological entities, extra-terrestrials) and settings; unmanageable occurrences in nature (hurricanes, thunderstorms, lightning); enormous or very fast moving vessels (airplanes, ships, rockets). Visual concepts and themes of relevance to the imaginary realms of the "horrific," the "fantastic" and the "futuristic" in particular—comic-book superheroes, destructive forces of nature, extra-terrestrial encounters, monstrous mutants, mythological creatures, prehistoric animals, and space adventures—were usually better conveyed on-screen as a result of a VFX-intensive approach to filmmaking, as illustrated for example by *The Mysterious Island* (1929), *The Vanishing Shadow* (1934), *Werewolf of London* (1935), *Flash Gordon* (1936), *The Devil-Doll* (1936), *The Invisible Ray* (1936), *Buck Rogers* (1939), *One Million BC* (1940), *Ali Baba and the Forty Thieves* (1944) and *Superman* (1948).[11]

While the business of motion pictures was definitely thriving in the immediate aftermath of World War II, it also became evident that the diverse domestic and international audiences of Hollywood cinema were undergoing significant changes in terms of consumption patterns, demographics and leisure interests. The U.S. box-office had been on the rise during World War II and in its immediate aftermath. By late 1947, it would begin a steady decline 'that would last for ten years and result in a 50 percent drop in attendance.'[12] If the commercial expansion of television and the massive proliferation of television sets in the United States from 1950 onwards were considered alarming enough at first, to the extent of additionally influencing the breakdown of the studio system, the Hollywood cinema industry as a whole nevertheless quickly regained its business composure and again adapted to change. It began to co-exist profitably with the burgeoning television industry: by producing programming for television networks increasingly needful of content to fill expanding air-time across America, by using television as a lucrative ancillary market for its movies after their first-run theatrical release, and eventually by branching into the low-cost production of "made-for-television" features.[13]

The glory days of double-bill "fillers" would be drawing to a close by the late 1940s while the studio system's demise was not too far off. 'After the televisualization and suburbanization of the 1940s, the B-movie was no longer a cheap commodity, and the studios focused on seeking hits rather than staples; thus, the economics of cinema shifted from seeking returns on all texts to a few.'. . .[14] Many sections of Hollywood cinema's audiences were increasingly expecting to see sophisticated and "quality" productions in theaters as opposed to low-budget material that they had become less interested in watching from the early 1950s onward. The Hollywood industry responded with a surge in the production of big-budget epics [*The Ten Commandments* (1956), *War and Peace* (1956), *Ben-Hur* (1959)], musicals [*An American in Paris* (1951), *Singin' in the Rain* (1952), *The King and I* (1956)], westerns [*Shane* (1953), *The Searchers* (1956)], and war films [*From Here to Eternity* (1953), *The Bridge on the River Kwai* (1957)]. Big- to medium-budget productions from Hollywood majors in the 1950s also included VFX-intensive horror, fantasy and sci-fi movies, such as: *The Day the Earth Stood Still* (1951), *When Worlds Collide* (1951), *The War of the Worlds* (1954), *Conquest of Space* (1955), *20,000 Leagues Under the Sea* (1955), *Forbidden Planet* (1956), *The Invisible Boy* (1957), *The Fly* (1958), *The Space Children* (1958), and *Journey to the Center of the Earth* (1959).[15] According to Melvin Matthews, *This Island Earth*

> along with *Forbidden Planet* (1956), seemed to signal a new chapter in Hollywood science fiction. . . . Yet, owing to the expense and intricate sets such films necessitated, these productions, while box-office hits, weren't sufficiently profitable for other studios to lavish the kind of money needed for similar movies. Subsequently, American science fiction films declined in quality. . . .[16]

But 1950s VFX-intensive films would end up having an enduring influence on Hollywood filmmaking for decades to come—as illustrated, for instance, by the remake of so many of them in recent times (especially in the wake of the late 20th century's boom in digital VFX.)

Responding and adapting to significant transformations in the consumption patterns, demographics and interests of its diverse audiences had become inevitable for the Hollywood cinema industry. Along with major socio-economic changes in the United States and the widespread diffusion of television, the Hollywood cinema industry was cornered into making a clear break from the decreasingly efficient studio system. By 1959, in the wake of the demise of the studio system, *Variety* was cynically pointing out that people 'no longer consider every film exciting because it moves on the screen.'[17] The Hollywood cinema industry promptly reacted to this kind of criticism by turning the motion picture into

> a special event, which had the effect of widening the gap between commercial winners and losers. When second-rate product may have barely scraped along before, it now did dismal business. On the other hand, pictures striking the public's fancy acted like magnets for the consumer dollar. Hollywood, as a result, concentrated its production efforts on fewer and more expensive pictures in its quest for profits.[18]

Hence, compared with the 1950s, the VFX-intensive movies generated by Hollywood majors often benefitted from larger production budgets during the 1960s than was previously the case—such as *Jack the Giant Killer* (1962), *The Birds* (1963), *7 Faces of Dr Lao* (1964), *Mary Poppins* (1964), *Robinson Crusoe On Mars* (1964), *Fantastic Voyage* (1966), *Doctor Dolittle* (1967), *Planet of the Apes* (1968), *2001: A Space Odyssey* (1968), *The Valley of Gwangi* (1969), and *Marooned* (1969).[19] Lured by expensively made blockbusters, audiences were returning to cinema theaters in large numbers, box-office figures were on the rise and the motion picture business enjoyed a healthy streak again albeit temporarily.

Following the demise of the studio system, the sustained re-organization of the Hollywood cinema industry during the 1960s witnessed the introduction of many innovative film production practices and management strategies (that have lasted well into present-day Hollywood cinema industry) as well as a greater acceptance of the big-budget philosophy.

> [At] Twentieth Century-Fox, the production policy under Darryl Zanuck had been to produce one key picture each season to keep the company in the black. . . . Zanuck launched three pictures in 1965–67: *The Sound of Music*, *The Agony and the Ecstasy*, and *The Bible*. *The Sound of Music* set the pace. Although it cost a hefty $10 million, the picture grossed over $100 million within two years. Other studios quickly embraced the larger risks of high cost films . . . and by 1968 six of the eight majors were contentedly producing blockbuster movies.[20]

But not all of these big-budget films were sufficiently successful at the box-office for the majors concerned to break even in the short run, not to mention the added burden of the reduction in fees offered by television companies to broadcast such films. Except for broadcast television, movie distribution and exhibition at that time were not yet extended to the more diverse ancillary markets which would open up later in the 1980s and 1990s. Because too many movies were not being profitable enough at the box-office to absorb very high production costs and to make up for short run losses resulting from big-budget flops, Hollywood majors were progressively facing serious financial problems by the end of the 1960s. Also during that decade, 'Hollywood went through its first series of corporate takeovers, when large diversified conglomerates acquired the major companies. This was a period of industrial instability for the industry, during which its existing profit-making strategies proved unreliable.'[21] In spite of novel industrial methods, innovative management strategies and policies, by 1969 the Hollywood majors at the time (Columbia, Disney, Metro-Goldwyn-Meyer, Paramount-Gulf+Western, Twentieth Century-Fox, United Artists-TransAmerica Corp., Universal-MCA, Warner Communications) had altogether lost some $200 million and the industry was going through a convulsion.[22]

The box-office successes of *Airport* (1970), *Patton* (1970), *Fiddler on the Roof* (1971), *The Godfather* (1972) and *The Poseidon Adventure* (1972) still did not prevent the slowing down of big-budget productions from late 1968 to 1972—a period of retrenchment for Hollywood as a result of pressure from banks. With stocks falling in value due to temporary recession in the movie business, companies resorted to drastic measures, such as cutting back on production and placing ceilings on production budgets, ranging from $1.5 million to $3 million.[23] As most majors were getting financially in the red by the end of the 1960s due to massive losses caused by big-budget flops, the Hollywood cinema industry began to witness the accommodation of "alternative" styles of low-budget filmmaking that actually did reasonably good business at the box-office. But as William Wolf points out:

> . . . the hope that *Easy Rider*'s triumph [in 1969], coupled with the financial plight of the major studios, would ensure a future for low-budget films was short-lived. The popularity of *Earthquake, The Poseidon Adventure, The Towering Inferno, The Exorcist,* and particularly, *Jaws,* switched the emphasis toward fewer, costlier films with huge profit possibilities. Producers, instead of calling for another *Easy Rider,* demanded: "Make me another *Jaws*!" Next, the magic words were *Star Wars.*[24]

Under pressure for even more structural re-organization, following the dismantling of the studio system, and hit by a recession culminating with all the majors losing millions of dollars by the end of the 1960s, the

Hollywood cinema industry again had to adapt to change in order to survive financially. Hence, it

> entered the age of conglomerates during the sixties as motion picture companies were either taken over by huge multi-faceted corporations, absorbed into burgeoning entertainment conglomerates, or became conglomerates through diversification. The transformation reflected changes in the economy as well as the vagaries of the motion picture market.[25]

The 1966 takeover of Paramount by Gulf+Western Industries Inc. signposted the first entry of a conglomerate in the American motion picture industry. This was rapidly followed by Transamerica Corp's acquisition of United Artists in 1967, and the purchase of Warner in 1969 by Kinney Services Inc. By the mid-1970s, conglomeration had allowed the Hollywood cinema industry to effectively adapt and upgrade itself to benefit from the kind of strategies and practices that were becoming increasingly dominant in American business at large.[26] The involvement of conglomerates in the Hollywood cinema industry would continue during the next two decades. Columbia was purchased by Coca-Cola Co. in 1982 and later acquired by Sony Corp in 1989. In the same year, Time Inc. would merge with Warner. Twentieth Century-Fox would be acquired in 1985 by News Corp. while Universal would be bought in 1991 by Matsushita Electric Industrial Co.[27]

By the mid-1970s, a new generation of "baby-boomer" filmmakers was rapidly rising to power in the Hollywood cinema industry: the so-called "movie-brats" (Francis Ford Coppola, George Lucas, Steven Spielberg, Martin Scorsese, Brian de Palma and John Milius). The "movie brats" intuitively understood the industry's need for change because, compared to many of their predecessors, they possessed a comprehensive knowledge of Hollywood cinema—much of it acquired from watching a lot of television and Saturday matinees while growing up. In spite of a few "surprises" once in a while, almost everything about the movies directed by George Lucas and Steven Spielberg in particular is often already "familiar." Their respective motion pictures tended to project a major sense of borrowing, recycling and mixing of so many characters, icons, narrative formulas and plot devices that originated from: (i) the B-grade movies, double-bill "fillers" and serials mass-produced during the Hollywood studio system era, and (ii) the VFX-intensive horror, fantasy and sci-fi movies of the 1950s—the most common themes being dangerous/mad science, treasure hunts, comic-book heroes, mutants, monsters, alien invasions, space adventures, and western frontier adventures that dominated the television and cinema theater screens of their childhood and youth. As echoed by Thomas Elsaesser:

> Typical of the New Hollywood is a self-conscious use of old mythologies, genre stereotypes, and the history of the cinema itself. But even

> more striking is the revival of genres which—in the 1950s—were regarded as 'B-movies': the sci-fi film, the 'creature-feature' or monster film, and the many other variations on the horror film.[28]

From past viewing experiences during their younger years, whatever George Lucas and Steven Spielberg had studied meticulously or had impressed them inevitably re-appeared in their respective movies.[29] For example, when they both discuss *Indiana Jones* in interviews, for them 'the four films have a simple ambition—to recreate the uncomplicated pleasures offered by the adventure stories of the 1930s and 1940s, the films of non-stop action built around the exploits of superheroes such as Zorro and Flash Gordon.'[30] Their filmmaking style and approach to storytelling that emerged during the 1970s would consequently influence Hollywood cinema's generic orientations in subsequent decades.

Hollywood cinema was, by the mid-1970s, also on the verge of yet another significant turning-point in its evolution: the techno-industrial transition into the digital realm. The new generation of technologically savvy film directors, especially George Lucas and Steven Spielberg, enthusiastically supported R&D in VFX, animation and filmmaking technologies. They also encouraged the synergistic relationship between film, computer technologies and their related service providers. Hence, they influenced techno-industrial product innovation as well as updated the production practices of big-budget blockbusters by the intermediary of their own respective films. According to Thomas Schatz,

> the mid-1970s ascent of the new Hollywood marks the studios' eventual coming-to-terms with an increasingly fragmented entertainment industry—with its demographics and target audiences, its diversified "multi-media" conglomerates, its global(ized) markets and new delivery systems. And equally fragmented, perhaps, are the movies themselves, especially the high-cost, high-tech, high-stakes blockbusters, those multi-purpose entertainment machines that breed music videos and soundtrack albums, TV series and videocassettes, video games and theme park rides, novelizations and comic books.[31]

The launching of commercial geo-synchronous satellites in the mid-1970s, for example, tremendously benefitted cable television, direct satellite broadcast and other "direct-to-home" delivery systems. With the information, communication and entertainment industries all converging, 'new delivery systems' would multiply and indeed expand the number of distribution and exhibition outlets available for Hollywood's films, long before the industry had gone digital.[32]

In conjunction with renewed interest in the big-budget philosophy, VFX-intensive filmmaking would become prominent again in Hollywood by the mid-1970s. Accordingly, generic vehicles that depended on fairly extensive VFX application were pushed to the fore and would become very profitable at the box-office: science fiction, fantasy, disaster and horror in particular.[33] Examples of these include: *The Towering Inferno* (1974), *Earthquake* (1974), *The Hindenburg* (1975), *Jaws* (1975), *Logan's Run* (1976), *The Exorcist II: The Heretic* (1977), *Close Encounters of the Third Kind* (1977), *Star Wars* (1977), *Alien* (1978), *The Black Hole* (1978), *Superman* (1978), and *Star Trek: The Motion Picture* (1979).

The popularity and commercial success of their 1970s films trumpeted the Hollywood supremacy of George Lucas and Steven Spielberg by the early 1980s. Their respective filmmaking practices and financial clout would end up permeating the Hollywood cinema industry throughout the 1980s and 1990s. Amongst other things, they would become influential producers, trendsetters and decision-makers in the motion picture industry. They would also be responsible for enabling first "breaks" for many new talents (directors, scriptwriters and actors), and they would have a strong influence on the techno-scientific progress of the industry and on the aesthetic evolution of the Hollywood film.[34]

In the case of George Lucas for example, huge earnings from the *Star Wars* franchise in particular would allow him to invest more and create additional divisions in his own production company Lucasfilm Inc. These included: Industrial Light & Magic, Lucas Licensing, Computer Research and Development Division, and Skywalker Sound. Although it was originally set up in 1975 solely for the purpose of VFX production for *Star Wars*, Industrial Light & Magic would subsequently expand and make significant contributions to the Hollywood cinema industry's VFX boom as well as to an additional techno-industrial characteristic of Hollywood movie production in the 1980s: digitization.[35] Industrial Light & Magic and Lucasfilm's Computer Division would subsequently deliver innovative digital VFX for such projects as *Star Trek II: The Wrath of Khan* (1982), *The Last Starfighter* (1984), *Young Sherlock Holmes* (1985), *Back to the Future* (1985), *Willow* (1988) and *The Abyss* (1989); while Lucasfilm would deliver innovative digital filmmaking services—such as the first digital non-linear editing systems for picture (EditDroid) and sound (SoundDroid). And Lucasfilm's lesser known Computer Division, originally set up in the late 1970s, would end up achieving one of the most important technological breakthroughs of the 1980s: the completion of an entirely computer-generated animation film, 1.8 minutes long, namely *The Adventures of André and Wally B.* in 1984. The Computer Division of Lucasfilm, later to become Pixar Animation Studios, would produce even more groundbreaking and award-winning computer-generated animation shorts during the rest of the 1980s. Lucasfilm and Industrial Light & Magic would keep evolving too, especially in terms of intensive motion picture technology R&D, to the point of being able to cover the whole spectrum of filmmaking services—digital VFX production, computer-generated 3D animation, digital sound engineering and digital film post-production.

The substantial box-office earnings of VFX-intensive movies of the late 1970s encouraged an increase in often expensive VFX application in Hollywood filmmaking during the early 1980s. Both released in 1982, Paramount's *Star Trek II: The Wrath of Khan* and Disney's *Tron* each opened the Pandora box of digital VFX wide enough to stimulate the imagination of producers and filmmakers alike.[36] The possibilities and profit potential associated with integrating digital media into the industry stopped being somewhat of a speculative "promise" from technocrats, corporate moguls and studio CEOs following the box-office success of *Star Trek II: The Wrath of Khan*. In spite of an unsurprising run-of-the-mill "Trekkie"-type of story line and narrative progression, it nevertheless contained the first continuous sequence of computer-generated images in a Hollywood feature film. Entitled "Genesis" and involving the visual transformation of a dead planet into a verdant world, this sequence was completely accomplished by means of 3D computer-generated graphics and animation produced at Lucasfilm's Computer Division.

Other prominent VFX-intensive Hollywood movies of the 1980s include: *Star Wars: The Empire Strikes Back* (1980), *Dragonslayer* (1981), *Raiders of the Lost Ark* (1981), *E.T. The Extra-Terrestrial* (1982), *Brainstorm* (1983), *Star Wars: Return of the Jedi* (1983), *Gremlins* (1984), *Ghostbusters* (1984), *Indiana Jones and the Temple of Doom* (1984), *Poltergeist* (1984), *Star Trek III: The Search for Spock* (1984), *The Last Starfighter* (1984), *Back to the Future* (1985), *Cocoon* (1985), *Aliens* (1986), *The Fly* (1986), *Poltergeist II: The Other Side* (1986), *Star Trek IV: The Voyage Home* (1986), *Innerspace* (1987), *Predator* (1987), *Cocoon: The Return* (1988), *Back to the Future II* (1989), *Batman* (1989), *Ghostbusters II* (1989), *Indiana Jones and the Last Crusade* (1989), *Leviathan* (1989), and *The Abyss* (1989).[37] *The Abyss* in particular had the merit of accelerating the evolution and acceptance of 3D computer graphics in the Hollywood cinema industry of the late 1980s. Its digital VFX "pseudopod" creature was visible proof that computer graphics had indeed become a feasible new tool for creating effective cinematic illusions that dazzled and attracted audiences to theaters.

Although their production costs in the 1980s had risen considerably, Hollywood motion pictures nevertheless generated a lot more revenue—from the box-office, merchandizing, related spin-offs and ancillary products—than in previous decades. In effect much more money was often being made from a Hollywood movie's related merchandizing and ancillary markets than from the movie itself, especially if the production company concerned had interests in more than just the motion picture business.[38] 'In the early 1980s, worldwide sales of *Star Wars* goods were estimated to be worth $1.5 billion a year, while *Batman* (1989) made $1 billion from merchandizing, four times its box-office earnings': more than just an entertaining movie, *Batman* was also a fast-food meal deal, a lunchbox, a soundtrack, a t-shirt, a toy, and so many other things.[39] Such a state of affairs was also an outcome of the increasing application of the "high concept" approach to filmmaking that was fast spreading in the 1970s to grow in strength during the 1980s. As explained by Justin Wyatt,

> the most overt qualities of high concept—the style and look of the films—function with the marketing and merchandizing opportunities structured into the projects. The result is a form of differentiated product adhering to the rules of "the look, the hook, and the book." . . . The high concept style, the integration with marketing, and the narrative which can support both of the preceding are the cornerstones of high concept filmmaking.[40]

VFX-intensive movies that best illustrate the growing predilection of Hollywood majors for "high concept" filmmaking during the 1980s include: *Batman, E.T. The Extra-Terrestrial, Ghostbusters,* the *Superman* series, and both the *Indiana Jones* and *Back to the Future* trilogies.

In spite of the stock market crash of 1987, the Hollywood cinema industry as a whole did not experience major setbacks during the 1980s.[41] In fact, '[o]verall industry revenues had increased hugely, from $4 billion in 1980 to $13.2 billion in 1990.'[42]

> Hollywood maintained its dominant position in the worldwide entertainment market by engaging in another round of business combinations beginning in the 1980s. The new urge to merge departed significantly from the merger movement of the 1960s, which ushered the American film industry into the age of conglomerates. . . . The merger movement of the 1980s was characterized in part by vertical integration, the desire to control the production of programming, the distribution of programming, and even the exhibition of programming.[43]

As a result, Columbia, Disney, MGM-United Artists, Paramount Communications, Time-Warner, Twentieth Century-Fox and Universal-MCA would become the leading "players" in the 1980s American motion picture business.[44]

The conglomerated (and progressively digitized) Hollywood cinema industry of the 1980s additionally saw scripts being increasingly developed with sequel potential, which quickly became one of the "preferred recipes" for long-term financial gain—through cable television and video, with or without a theatrical release. And audiences accordingly developed a fidelity to certain franchises as well as the habit of expecting sequels from most Hollywood movies.[45] The production of sequels more commonly applied to clear-cut VFX-intensive generic vehicles: horror (*Poltergeist, Friday the 13th*), action/adventure (*Indiana Jones, Back to the Future*), science fiction (*Star Trek, Alien*) and comic-book superheroes (*Superman, Batman*). Angela Ndalianis explains further that

> not only has the sequel become a phenomenon associated with contemporary Hollywood cinema, but also in recent decades audiences have been exposed to an increase in story and media crossovers, one that reflects a new aesthetic that complicates classical forms of narration.

> Batman, like Superman, began as a character in a comic-book serial in the late 1930s then was accompanied by media crossovers into films in the 1940s and television in the (in)famous 1960s series. More recently, however, the cross-media serialization of Batman has become more extreme. In addition to its multiple comic-book variations (*Batman, Detective, Shadow of the Bat, Legends of the Dark Knight, Gotham Nights,* and occasional crossovers into *Robin, Nightwing,* and *Catwoman*), the Batman story has also found a popular form of expression in four blockbuster films and numerous computer games. Other serial productions have followed suit.[46]

But generic constraints and requirements notwithstanding, the practice of sequel production would fast extend to practically any genre of Hollywood film: *The Karate Kid, Police Academy, Lethal Weapon, Beverly Hills Cop,* and so on.

The proliferation of sequels was also associated with a buzzword (and profitable commercial strategy) being increasingly applied in the Hollywood cinema industry: "synergy"—as bandied about in the 1970s by the new breed of corporate manager to describe the beneficial outcomes of conglomeration. In arguing the case that the multi-market company constituted the most appropriate structure in an era of rapid technological and social change, John Beckett defined "synergy" as

> the art of making two and two equal five. . . . We believe in the synergistic effects of running more than one business. By that I mean the sales effect, the cost control effect, the cost spreading effect. And as a result, we can run these businesses and make more money from them than can be done running them individually. What we're trying to do is get more utilization out of the same set of executives and the same computers and very much the same overhead and payroll. We think this is the way modern business is going.[47]

With more conglomerate entities becoming involved in the movie business by the early 1980s, the Hollywood cinema industry was indeed ripe for "synergy." According to Barry Langford, '[t]he model and inspiration for the synergistic remodeling of the film industry in the eighties was Warner Communications' successful exploitation of the first two *Superman* films, each the second-highest domestic earner in its year of release (1978, 1981).'[48]

The positive attitude towards "synergy" in the motion picture business would influence Hollywood cinema's techno-industrial transition to digital. The ease of circulation and re-disposition of technologies and of techno-industrial innovations across the spectrum of media production, communication, information and entertainment industries was especially facilitated by the digital medium becoming a standard technological platform. This was precisely what encouraged even more advanced R&D in the industry that would ultimately benefit all sectors of the movie production business.

> Thus there has been a trend toward "tight diversification" and "synergy" in the recent merger-and-acquisitions wave, bringing movie studios into direct play with television production companies, network and cable TV, music and recording companies, and book, magazine, and newspaper publishers, and possibly even with games, toys, theme parks, and electronics hardware manufacturers as well.[49]

This trend would only grow stronger as Hollywood "players" commercially diversified even more in future years. Hence, in the highly "synergized" and conglomerated Hollywood cinema industry of the late 1980s, a surge in feature film production was all but inevitable.

> Domestic feature film production jumped from around 350 pictures a year in 1983 to nearly 600 in 1988. . . . Rather than producing more pictures, the majors exploited a new feature film format, the *'ultra-high-budget'* film. . . . Contrary to common sense, pictures costing upwards of $75 million became conservative investments. Containing such elements as high concepts, big-name stars, and visual and special effects, such pictures reduced the risk of financing because: (1) they constituted media events; (2) they lent themselves to promotional tie-ins; (3) they became massive engines for profits in ancillary divisions like theme parks and video; (4) they stood to make a profit in foreign markets; and (5) they were easy to distribute.[50]

The "ultra-high-budget" production philosophy—incorporating "high concept," synergistic practices and VFX-intensive filmmaking—would quickly be adopted across the Hollywood industry to become common practice from the very early 1990s onward. *Back to the Future III* (1990), *Total Recall* (1990), *Terminator 2: Judgment Day* (1991)—the first motion picture with a $100 million production budget—and *Hook* (1991), for instance, all illustrate Hollywood's turn-of-the-decade inclination toward "ultra-high-budget" but extremely profitable VFX-intensive filmmaking.

The 1990s would be characterized by pervasive technological digitization across the converging information, communication and entertainment industries at large—in sync with the emergence of the World Wide Web and the proliferation of home computers and home theater appliances—to the point that entertainment, information technologies and communications became so inextricably intertwined as to be considered one activity. As Yannis Tzioumakis explains, 'Digital technology has been utilized as a means to an end, as the best available vehicle for the domination of the global entertainment market by a small number of giant corporations.'[51] Digital convergence would considerably extend the channels of distribution, circulation and exhibition of movies. By early 1997, the first DVDs would become commercially available. For example, the DVD for VFX-intensive *Twister* (1996), one of the very first Hollywood movies distributed on DVD after theatrical release, was characterized by sharper image resolution, better sound quality (2.1 Surround Sound) and durability compared to video home system (VHS) magnetic tapes. The sales of DVD players and of DVDs would surpass those of videocassette recorders and VHS tapes within a few years—as such, establishing the digital takeover of the home video market from analog technologies.[52] 'In addition to its influence on the home video market, the DVD significantly changed the video game market as well,' and with Sony releasing the PlayStation 2 in 2000 and Microsoft releasing the Xbox in 2001—each of them functioning as a gaming system *and* a DVD player—'the practices of the video game and home video industries converged in an important way, as both began producing and marketing the same object.'[53]

With digital convergence, digital VFX and digital filmmaking would simply become normal aspects of motion picture production by the late 1990s.[54] As predicted by filmmaker James Cameron in the mid-1990s:

> The lines will blur between visual effects, editing, color correction, and conventional image manipulations like flops, dissolves, and fades. It will all become one process, under the director's and the producer's control. Visual effects, composites, live action photography, animation, titles, dissolves, multi-image layouts, color and contrast effects . . . all of these will stream together electronically into a "final" image mix . . . a final, high-resolution online. The speed and flexibility of digital, non-linear editing has been and will continue to reduce postproduction time substantially. . . . There will be no generation loss between effects shots, titles, dissolves, and the rest of the film.[55]

Digital convergence within the Hollywood cinema industry would also encourage the more frequent production during the 1990s of so-called "event movies" and VFX-intensive blockbusters which, although financially risky, tended to fully exploit the kind of synergies associated with conglomeration. Simultaneously, Hollywood cinema's fast evolving visual aesthetic, as well as its production, distribution and exhibition practices would all become much more sophisticated. The digitization of most aspects of filmmaking (and of the VFX domain in particular) would enable the production of increasingly larger amounts of complex digital VFX. This can be witnessed in such "ultra-high-budget" VFX-intensive movies as: *Batman Returns* (1992), *Jurassic Park* (1993), *Apollo 13* (1995), *Batman Forever* (1995), *Waterworld* (1995), *Independence Day* (1996), *Twister* (1996), *Batman & Robin* (1997), *Jurassic Park II: The Lost World* (1997), *Men in Black* (1997), *Titanic* (1997), *Star Wars: Episode I—The Phantom Menace* (1999), *The Matrix* (1999), and *The Mummy* (1999). Other notable digital VFX-intensive Hollywood spectacles of the 1990s include: *Predator 2* (1990), *The Rocketeer* (1991), *Last Action Hero* (1993), *Star Trek VII: Generations* (1994), *Casper* (1995), *Jumanji* (1995), *Dragonheart* (1996), *Mars Attacks!* (1996), *Star Trek VIII: First Contact* (1996), *Contact* (1997), *Dante's Peak* (1997), *Event Horizon* (1997), *Volcano* (1997), *Deep Blue Sea* (1999), and *Wild Wild West* (1999).

As for George Lucas and Steven Spielberg who had spearheaded the "new Hollywood" in the 1970s, although doing less directing, their films as well as many of those they only produced or financially guaranteed would still end up generating massive profits at the box-office. They had achieved a status similar to that of the studio system era "moguls" by the 1990s, influencing much of what was going on in the Hollywood cinema industry. Spielberg would go so far as to co-set up a new Hollywood movie studio altogether. The outcome of a joint venture between Spielberg, ex-Disney CEO Jeffrey Katzenberg and record producer David Geffen, DreamWorks SKG was launched in October 1994 amidst extensive media coverage. Because of the kind of capital, network, influence and talent associated with these three very experienced contemporary "moguls" of the American popular entertainment industry, DreamWorks SKG did not confront the common barriers to entry experienced by new film production companies when they begin to compete with older, more affluent majors.[56] George Lucas had acquired by the mid-1990s the means to re-structure his California-based entertainment empire Lucasfilm Inc. hence taking even more advantage of the synergies enabled by digital convergence and conglomeration in the American motion picture, entertainment and communication industries at large.[57] More importantly, he was gearing up towards the production, marketing and distribution of his highly anticipated forthcoming *Star Wars* franchise pre-quel trilogy (also marking his return to active film directing): *The Phantom Menace, Attack of the Clones* and *Revenge of the Sith*.

All the "players" involved in Hollywood being simultaneously rivals, suppliers, distributors as well as customers of similar products, the more intense the competition became the greater was the need to partner up in the 1990s. With corporate activities and business operations continuously overlapping each other, monopolistic control of such products—from their initial development phase to the final stages of their distribution, exhibition and sale—hardly ever occurred and all involved usually benefitted in one way or another.[58] These products (movies, magazines, software, video games, music, and so on) are often initially configured to suit American consumption patterns first, then subsequently "re-formatted" according to the various specifications of targeted global markets, as illustrated by the international DVD and satellite TV distribution of Hollywood movies.

Corporate vertical integrations, acquisitions, takeovers, consolidations, mergers and conglomerations have indeed determined the recurrent formation of strategic partnerships and alliances between the film, television, broadcasting and communications industries, and so on. This strategic tendency in movie business operations received a boost from the Federal Communications Commission's deregulation of television—proposed during the mid-1970s but mostly implemented during the 1980s—when the 'U.S. government relaxed the rules that prohibited broadcast networks from having a financial interest in the shows they aired—and with that wall breached, the movie studios and their TV-production arms all had to have networks of their own.'[59] With the breakdown of the distribution and exhibition boundaries for the filmic medium, more "variations" in the production format of the Hollywood film cropped up. For example, if studios invested millions of dollars in deals with television writer-producers to maximize their return they would end up requesting various aesthetic concessions from the filmmakers concerned to satisfy the (technical) requirements of such lucrative ancillary distribution markets as cable and syndicated television from the late 1970s, rental/retail VHS video from the 1980s, rental/retail digital video and the Internet from the late 1990s, and videogames from the early 2000s onward (as exemplified by how film writers have been 'including some games in their screen narratives, as was attempted in the crossover between *The Matrix: Reloaded* movie . . . and the *Enter the Matrix* game, where some of the scenes in the film only made complete sense if the viewer had completed the videogame, and vice-versa').[60]

Conglomeratic trends and practices in the entertainment business in particular, as triggered by the increasing involvement of corporate giants, could only encourage the kinds of synergy that would subsequently influence the techno-industrial evolution of Hollywood. And it would especially influence how it would continue making its most profitable "staple": the VFX-intensive theatrical feature film.

> Popular films often initiate or continue an endless chain of other cultural products. A film concept or character often leads to a TV show, with possible spin-offs, video games, and records. Merchandizing efforts also include toys, games, t-shirts, trading cards, soap products, cereals, theme park rides, coloring books, magazines, how-the-movie-was-made books, etc. The major Hollywood corporations are transnational conglomerates, often involved in all of these activities. Thus it becomes increasingly more difficult to distinguish the film industry from other media or entertainment industries.[61]

Between the 1960s and 1990s, the major communication-, media-, leisure- and entertainment-related companies, conglomerates and corporations operating in the United States would simultaneously acquire substantial shares and assets (or at least develop considerable vested interests) in practically all areas of cultural, technological and media production—from filmmaking to publishing, from theme parks to online content services, from computer hardware/software development to graphic design, from music licensing to cable television programming, from consumer electronics to video retail, amongst many others. And what ultimately emerged by the end of the 20th century is a massive American entertainment industry that was not only thriving on returns to investment in multi-media forms but also on the very profitable dispersion of multi-media products to a global market.

Why have film studios always been so attractive to conglomerates to the extent of motivating so many partnerships, alliances, mergers and takeovers?

> The official explanation is that they serve as the idea factories that drive other, generally more profitable enterprises—TV production, theme parks, merchandizing. . . . The studios have been global all along. They were also vertically integrated, their Hollywood production facilities feeding their distribution apparatus and their nationwide exhibition circuits, until government trustbusters forced them to sell off their theaters in the early 1950s. But now they have been rolled into combines that are active in a half-dozen different media and all twenty-four time zones.[62]

If the big corporate money of banks and tycoons changed the rules of the game in Hollywood during the studio system era, even bigger conglomerate money would become necessary to change the rules of the game in "post-classical" Hollywood. As an outcome of the wave of conglomeration and mergers, the Hollywood cinema industry—and the American communication and entertainment industries at large—became more encouraging of technological transfers, exchanges and synergies. These were precisely what it depended on in order to maintain its constant dynamism and renewal. Hence, by the end of the 1990s, Hollywood film production had become but

> one component of the economic drive of the conglomerates that dominate the contemporary entertainment industry. The contemporary entertainment aesthetics that emerge from this drive support an industry that has multiple investment interests. In turn, the serial structure that manifests itself in entertainment narratives is supported by an economic infrastructure that has similarly expanded and adjusted its boundaries.[63]

The conglomerated and synergized Hollywood motion picture industry of the late 1980s and 1990s would be characterized by:

- significant expansion of film distribution and exhibition beyond traditional theatrical release (such as home video, cable and satellite television), thereby increasing demand for content;
- synergistic practices of movie development and production favoring sequels, spin-offs, franchises and merchandizing potential;
- a largely "high concept," "event movie" and VFX-intensive approach to filmmaking;
- the digitization of filmmaking;
- the beginnings of techno-digital standardization across the converging media, broadcasting, communication and entertainment industries at large.

By 1998, the Hollywood cinema industry's most prominent production studios and movie business "players" had become: Columbia (Sony Pictures Entertainment, Sony Pictures Classics, TriStar Pictures); DreamWorks SKG (DreamWorks Pictures); MGM-United Artists (MGM Pictures, United Artists Films, Orion Pictures); News Corporation (Twentieth Century-Fox Pictures, Fox Film Corporation); Paramount-Viacom (Paramount Pictures, Viacom Entertainment); Time-Warner-AOL (Warner Bros. Pictures, New Line Cinema); Universal-Seagram (Universal Pictures, Polygram Pictures); Walt Disney (Disney Pictures, Touchstone Pictures, Miramax Films). Hence, by the end of the 1990s, more or less the same handful of companies and brand names that had emerged and dominated the industry more than fifty years before still controlled the creation, production, distribution and exhibition of Hollywood movies worldwide as well as most of the profits generated. And by the early 2000s, six conglomerates—Viacom, Time Warner, NBC-Universal (owned by General Electric), Sony, Fox News Corp., and Disney—overwhelmingly dominated 'the production and distribution of entertainment products in the United States. Together they control 98 percent of the programs that carry commercial advertising during prime time television . . . and 96 percent of total U.S. film rentals.'[64] The Hollywood industry's structure had also changed some more, with its most prominent movie business 'players' being: Columbia-Sony, DreamWorks SKG, MGM-United Artists, Twentieth Century-Fox-News Corporation, Paramount--Viacom, Time-Warner-AOL, Universal-Seagram and Walt Disney. The first decade of the 21st century would continue to be characterized by 'the culmination of U.S. media industry control in the hands of a half-dozen global media superpowers,' namely Comcast Corporation, Fox News Corporation, Sony Corporation, The Walt Disney Company, Time Warner, Viacom.[65]

The Hollywood cinema industry would approach the end of the 20th century with a combination of synergistic business practices, digital VFX-intensive movie-making methods, "ultra-high-budget" and "high concept" philosophies firmly in place.[66] Tino Balio tells us that '[f]amiliar formulae in familiar production trends aided by increasingly sophisticated computer-generated imagery and attuned to changing pop-culture trends kept audiences entertained.'[67] For instance, even though disaster movie *Twister*'s digital VFX-intensive production carried a $92 million price tag, the sensual gratification it provided in thousands of cinema theaters was effective enough to bring in more than $240 million in American ticket sales alone.[68] And producing creature/sci-fi movie *Jurassic Park*'s visual thrills and excitement was more expensive than *Twister*'s but it eventually brought in more than $1 billion worldwide during many subsequent years of circulation. Then 1997 saw the release of the most profitable film at that point in Hollywood cinema history: James Cameron's "ultra-high-budget," "high concept" digital VFX-intensive *Titanic*—

fundamentally a disaster movie—which on its own generated about $1 billion in revenues from theatrical exhibition worldwide. By the end of 1999, the Hollywood cinema industry had earned close to $7.5 billion from the American box-office alone, a major increase from the $6.95 billion made in 1998. Amongst the American box-office's top ten films in the first week of December 1999 for example, the only three movies that had continuously been in exhibition for more than 25 weeks were all lavishly produced "ultra-high-budget," hi-tech, digital VFX-intensive movies: *Star Wars: Episode I—The Phantom Menace, The Matrix* and *The Mummy*. Interestingly enough, throughout the 1990s 'the majors released less than half the total number of films annually distributed in the USA, but studio releases consistently accounted for some 95 percent of domestic box office.'[69]

The 1990s would also witness the last major techno-industrial turning-point in 20th century cinema history—the first entirely computer-generated theatrical feature film: *Toy Story* in 1995 produced by Pixar Animation Studios in collaboration with Disney. *Toy Story* was the apotheosis of many years of continuous R&D in computer-generated 3D animation carried out via a string of short films produced by Pixar since 1984. *Toy Story* also inaugurated yet another phase in the development and creation of computer-generated moving pictures, animation and digital VFX while opening a gateway into a whole "new" universe of the film/digital synergy for filmmaking in the Hollywood cinema industry. Before the end of the 1990s, Pixar would have released two more computer-generated 3D animation feature films, namely *A Bug's Life* in 1998 and *Toy Story 2* in 1999.[70] Originally intended for "direct-to-video" release, *Toy Story 2* would rapidly become a box-office hit upon its theatrical release, making $81.1 million after just two weeks of its American exhibition. More importantly, in being an entirely computer-generated 3D animation feature, it would be the first feature film in motion picture history to be entirely created, mastered and exhibited digitally.[71] The (limited) theatrical exhibition in digital of *Toy Story 2* proved that a film could be entirely made, released, circulated and exhibited without the actual use of acetate. As such, this event in itself constituted a significant rupture with more than 100 years of analog cinema technology—as echoed by Stéphane Bouquet's crucial question 'Est-ce encore du cinéma?' in a special issue of *Cahiers du Cinéma* in 2000:

> Que devient le cinema? Il change. Ni progrés, ni décadence, pas trace d'un temps positiviste, simplement métamorphose, beaucoup de métamorphoses. On ne fabrique plus les images pareil, on ne les met plus en circulation pareil, on ne les regarde plus pareil. Cela oblige aussi à y penser autrement. Les vieilles définitions du cinema deviennent lentement caduques. Il y a du cinéma sans enregistrement, sans réel et sans empreinte, du cinéma sans projection, du cinéma hors de toute salle et de toute expérience collective. . . . Il faut souligner l'extraordinaire

> dilution du medium cinéma. On a envie de dire le cinéma est partout: dans les clips, dans les jeux vidéos, dans certaines séries télé héritières de la série B, dans les installations de plasticiens, sur internet, dans les casques des mondes virtuels, etc. Sauf que: *est-ce encore du cinéma*?[72]
>
> What is becoming of the cinema? It is changing. What the cinema is going through is not just a matter of progress or decline, nor does it relate to any anterior positivistic era, but is rather a matter of its significant metamorphosis. Images are neither manufactured, circulated nor watched in the same manner as they used to be. Such a state of things also mandates alternative ways of thinking about the cinema. "Traditional" approaches to its definition and conceptualization are slowly fading into obsolescence. The cinema can now exist without requiring any form of recording or physical exhibition, without being grounded in any kind of concrete reality and without being imprinted with any sense of "realness"; it can additionally exist independently of any public space of theatrical exhibition and of any collective experience. . . . The incredibly diluted nature of the cinematic medium should be highlighted. One is inclined to say that the cinema is everywhere: in music videos, in video games, in certain types of TV series that have become the direct descendants of the B-grade serial movie, in visual artists' installations, on the Internet, in virtual reality headsets, etc. But: *is this still cinema*?

Bouquet's commentary indeed pertinently describes as much as it problematizes the major transitional phase of metamorphosis that the art of motion pictures has undergone, via the Hollywood cinema industry, towards the end of its first hundred years of existence.

NOTES

1. Staiger, 'Introduction'—*The Studio System* 4.
2. Balio, 'Introduction'—*Grand Design* 1. Emphasis added.

The work of Tino Balio is authoritative and is relied upon throughout this chapter. It should be noted that the actual usage of the term 'blockbuster' dates to the early 1950s, more than a decade after *Gone With the Wind*'s release.

3. For a good overview of the development of the blockbuster, see Hall, 'Tall Revenue Features.'
4. Thompson, *Exporting Entertainment* 149.
5. Balio, *United Artists* 11–12.

Richard Jewell cautions that

> it is wrong to lump MGM, Paramount, Warner Brothers, Twentieth Century-Fox, RKO, and, oftentimes, Columbia, Universal, and United Artists together and treat them as if they were carbon copies of one another. Each of these companies had its own special characteristics, and each underwent significant changes during the studio system era. Each was a world unto itself with its own ways of making movies and making money.

Jewell, 'Howard Hawks' 48.

6. Kerr, 'Joseph H. Lewis' 52.

7. Kerr, 'Joseph H. Lewis' 52.
8. According to Tino Balio,

> [d]uring the war, dollars were plentiful, while commodities were not. Movies were the most readily available entertainment, which benefited Hollywood. Domestic film rentals for the eight majors jumped from $193 million in 1939 to $332 million in 1946. Every night was Saturday night at the movies. B-pictures, low-grade pictures, pictures featuring unknown players—all commanded an audience. Weekly attendance by the end of the war reached 90 million. . . . Predictions for the post-war era looked even better. On the domestic scene, returning servicemen were expected to boost attendance. Indicators of rising wages, shorter working hours, and greater leisure also implied prosperity. On the foreign scene, American film companies had a huge backlog of pictures ready for the re-opening of overseas markets.

Balio, *United Artists* 18–19.

9. According to Robin Cross, the B-movie originally 'was a direct response by Hollywood to the falling cinema audiences of the early Depression years. . . . In an attempt to lure the audiences back the double bill was introduced. Where previously audiences had paid to see a single feature supplemented with shorts and cartoons, they were now treated to two features, one of which was a low-budget supporting film.'

Cross, *The Big Book of B Movies* 43.

10. Jacobs, 'The B Film' 148.
11. Made in fifteen chapters on a low-budget deal with Columbia by quickie producer Sam Katzman in 1948, *Superman* was the highest grossing Hollywood serial of all times.
12. Balio, *United Artists* 24.
13. Releasing vintage films to television networks at first, the Hollywood cinema industry would eventually start supplying newer products at rental levels starting at $200,000 in 1960 and approaching $1 million per picture by the end of the decade.

Balio, *United Artists* 89.

14. Miller *et al*, *Global Hollywood* 92.
15. As a result of the combinatory re-invention of the science fiction and horror genres in particular, fuelled by Cold War anxieties and paranoia, independent studios (especially Universal International Pictures) were even more prolific in their output of low-budget VFX-intensive horror, fantasy and sci-fi movies. Examples are: *The Thing from Another World* (1951), *Invaders from Mars* (1953), *It Came from Outer Space* (1953), *The Beast from 20,000 Fathoms* (1953), *Creature from The Black Lagoon* (1954), *It Came from Beneath the Sea* (1955), *Tarantula* (1955), *This Island Earth* (1955), *20 Million Miles to Earth* (1957), *The Incredible Shrinking Man* (1957), *The Monolith Monsters* (1957), *From the Earth to the Moon* (1958), *It! The Terror from Beyond Space* (1958), and *The Blob* (1958).
16. Matthews, *Hostile Aliens* 61.
17. Quoted in Balio, *United Artists* 88.
18. Balio, *United Artists* 89.
19. There would nevertheless be during the 1960s a substantial output of medium- to low-budget VFX-intensive horror, fantasy and sci-fi movies, like those produced by independents such as American International Pictures and Roger Corman Productions.
20. Balio, *United Artists* 315–16.
21. Vasey, 'Hollywood Industry Paradigm' 293.

The "conglomerate" is understood here 'as a diversified company with major interests in several unrelated fields.' See Balio, *United Artists* 303.

22. See Balio, *United Artists* 316–17.
23. Balio, *United Artists* 316–17.
24. Wolf, *Landmark Films* 332.

25. Balio, *United Artists* 303.
Tino Balio further explicates that

> conglomerates from outside the motion picture industry were attracted to motion picture companies for several reasons: (i) film stocks were undervalued during the sixties as a result of erratic earnings records; (ii) studios owned strategically located real estate and other valuable assets such as music publishing houses and theaters in foreign countries; and (iii) film libraries had the potential of being exploited for cable and pay television.

Balio, *United Artists* 304.

26. Further readings that can provide a more comprehensive overview of conglomerization in the American film industry include: Bruck, *Master of the Game*; Gomery, 'The American Film Industry of the 1970s'; Guback, 'Theatrical Film'; Mayer, *The Film Industries*; Monaco, *American Film Now*; Phillips, 'Film Conglomerate Blockbusters.'

27. See Appendix A.

28. Elsaesser, 'Spectacularity' 195.

29. For instance, *Raiders of the Lost Ark* and subsequent *Indiana Jones* movies have their sources in "treasure hunt adventures" like Universal's 13-episode serial *Lost City of the Jungle* (1946) and United Artists's *The Lone Ranger and the Lost City of Gold* (1958), and in "pulp action adventures" like MGM's *The Mask of Fu Manchu* (1932). And dueling Jedi sequences in *Star Wars* films "borrow" from, at least, the Errol Flynn swashbucklers and "medieval swordplay" serials, such as Columbia's *The Adventures of Sir Galahad* (1949), made during the studio system era.

30. Sergi & Lovell, *Cinema Entertainment* 66.

31. Schatz, 'The New Hollywood' 9–10.

32. See Appendix A.

33. Robert Kapsis's explanation should also be noted here:

> During the boom phase of a film cycle, dozens of films of a particular genre are released or re-released each year while dozens more are either in pre-production, production, or post-production. One factor that might affect the duration of a film cycle is the perception of how markets (audiences) might respond to the genre. . . . [F]or a genre boom to emerge and persist, there must be strong indications that the genre will thrive in at least two of the three major theatrical markets—domestic, foreign, and ancillary (such as sales to network and cable television and videotape playback).

Kapsis, 'Hollywood Genres' 5.

34. See Appendix A.

35. "Digitization," in this context, means the process of transformation from an analog to a digital state; the term "digitalisation" or "digitalization" has alternatively been used by some to also mean precisely that (for instance, see Harbord, 'Digital Film'). But "digitalisation" or "digitalization" is in fact a term more commonly used to refer to a form of medical treatment, namely the administering of digitalis (a drug preparation) to treat certain heart conditions.

36. While the release of *Tron* conferred a "digital pioneer" status upon Disney, its commercial failure (having gone over budget and making only $33 million at the box-office) not only caused its important achievements in computer-generated imaging to receive far less attention than those of the more popular *Star Trek II: The Wrath of Khan*—*Tron* would not even be nominated at the Academy Awards in the VFX category—but also forced Disney and the film's director Steve Lisberger out of the burgeoning domain of computer-generated digital VFX in the Hollywood cinema industry.

37. Many of these films have ended up in plenty of Hollywood cinema lists of most profitable movies of all times. *E.T. The Extra-Terrestrial, Star Wars: Return of the Jedi, Batman, Star Wars: The Empire Strikes Back, Raiders of the Lost Ark, Ghostbusters* and *Back to the Future*, for instance, had each respectively generated by the end of the 1990s cumulated profits in excess of $200 million. See Appendix B.

38. See Biskind, 'Blockbuster.'
39. Maltby, 'Nobody knows everything' 24.
40. Wyatt, *High Concept* 188.
41. The worst disaster in the 1980s for the Hollywood cinema industry was the bankruptcy of United Artists caused by Michael Cimino's *Heaven's Gate* (1980): it made only $1.5 million at the American box-office and lost at least $40 million for the studio. Tino Balio attributes the responsibility for the demise of United Artists to Transamerica Corporation, a San Francisco-based insurance and financial services conglomerate, which was formed in 1928 as a corporate umbrella for the expanding financial enterprises of the Bank of America. It took over United Artists in 1967. 'Transamerica Corporation's belief that sophisticated business systems could replace leadership had a disastrous effect on United Artists, culminating in the *Heaven's Gate* fiasco'; consequently, Transamerica Corporation left the movie business and sold United Artists to MGM in 1981.
See Balio, *United Artists* 5–6.
42. Langford, *Post-Classical Hollywood* 198.
43. Balio, 'A major presence' 67.
44. Most Hollywood majors would experience some box-office flops during the 1980s that would lose large amounts of studio money, but not so bad as to cause bankruptcy: MGM-United Artists' Korean War epic *Inchon* (1981) made only $5 million from a $50 million budget; Time-Warner's *Revolution* (1985) made less than a million from a $28 million budget; Universal-MCA's *Howard the Duck* (1986) only grossed about $15 million from a $30 million budget; Columbia's *Ishtar* (1987) made about $14 million from a $55 million budget while *The Adventures of Baron Munchausen* (1988) grossed only $8 million from a budget of about $46 million.
45. It should be noted that the franchising of medium- to big-budget films in Hollywood was not an entirely new phenomenon as such in the 1980s, considering the James Bond, *Planet of the Apes*, and *Pink Panther* series of films released during the 1960s and 1970s.
46. Ndalianis, *Neo-Baroque Aesthetics* 34.
47. Quoted in Balio, *United Artists* 310.
John Beckett, 'The Case for the Multi-Market Company'—a lecture presented at the Financial Analysts Seminar, sponsored by the Financial Analysts Federation in association with the University of Chicago, Rockford, Illinois, 22nd August 1969. John Beckett was at the time president of Transamerica Corporation, Paramount studio's conglomerate "partner."
48. Langford, *Post-Classical Hollywood* 193.
Barry Langford explains further that:

> Beyond the direct earnings of the films at home and overseas, WCI also profited from sales of both special editions and regular runs of *Superman* comic books published by DC (owned by Warner since 1971); nine different products, ranging from a novelization to a *Superman* encyclopedia and calendar, published by Warner Books; dozens of new spin-off Superman toys and novelties authorized by Warner's licensing division; a soundtrack album on Warner Records; and even an arcade pinball table from Warner-owned games manufacturer Atari.

49. Schatz, 'The New Hollywood' 31.
50. Balio, 'A major presence' 59. Emphasis added.
51. Tzioumakis, 'From the Business' 11.
52. In the United States, by 2001 DVD players had reached a 21 percent household penetration rate and by 2006, 75 percent of consumer households owned a DVD player, according to Consumer Electronics Association reports. See Brookey, *Hollywood Gamers* 9.
53. Brookey, *Hollywood Gamers* 11.

54. Just a few years later, George Lucas's *Star Wars: Episode II—Attack of the Clones* (2002) would be the first major Hollywood motion picture to be entirely filmed in digital video (at 24 frames-per-second equivalent).

55. Quoted in Ohanian & Phillips, *Digital Filmmaking* 242.

56. After a somewhat "slow" start—its first feature-length movie, *The Peacemaker*, only coming out in 1997—DreamWorks SKG quickly implemented creative management and synergistic practices quite similar to those which had contributed to the prosperity of its main competitors. It would successfully expand its activities, diversify its range of products and services as well as increase returns on investment. *Antz* and *The Prince of Egypt*, both in 1998, would also establish DreamWorks SKG as a serious competitor in the domain of 2D and 3D animation features, usually monopolized by Disney and Pixar Animation Studios. In 2006, as a result of growing financial difficulties however, its live-action studio would be sold to Viacom's Paramount Pictures and, by 2008 DreamWorks SKG was under the control of Indian investment firm Reliance ADA Group.

57. In preparation for the handling of the forthcoming massive *Star Wars* prequel franchise (hence to retain most of the generated revenues), Lucasfilm Inc. would expand further to comprise more divisions and commercial subsidiaries: Industrial Light & Magic, Lucas Digital Ltd., Lucas Arts, Lucas Learning, Lucas Licensing Ltd., Industrial Light & Magic Commercial Productions and Skywalker Sound. These would provide state-of-the-art filmmaking and cinema services—digital sound mixing, computer-generated animation, digital VFX, theatrical sound engineering and so on—while simultaneously conducting film technology R&D and catering to anything else related to the filmed entertainment business (gaming, merchandizing, licensing, and so on).

58. It should be pointed out that digital convergence has also disrupted synergies as much as it has sustained them, as illustrated for instance by the proliferation of digital consumer products resulting from intensive developments that occurred within a short period of time especially with regards to computer hardware and software. See Appendix A for an overview of the rapid evolution of computer chips and digital consumer products.

59. Rose, 'There's No Business' 55.

60. Brown & Krzywinska, 'Movie-Games' 90.

Although finally becoming a multi-billion dollar industry in its own right during the 2000s, videogame movie tie-ins have in effect existed since Atari Inc.'s *Raiders of the Lost Ark* (released in November 1982) and *E.T. The Extra-Terrestrial* (released in December 1982) video games of the early 1980s.

61. Wasko, *Hollywood in the Age* 94.

62. Rose, 'There's No Business' 58.

63. Ndalianis, *Neo-Baroque Aesthetics* 39.

64. Christopherson, 'Behind the Scenes' 192.

65. Schatz, 'New Hollywood, New Millenium' 20.

66. It should be noted that Hollywood majors additionally created subsidiaries or purchased independent companies to also produce low- to medium-budget films for distribution in niche markets (such as art, foreign language and "indie" cinemas).

67. Balio, 'Hollywood Production Trends' 181.

68. See Appendix B.

69. Langford, *Post-Classical Hollywood* 197.

70. The release of *Toy Story* would get the attention of other Hollywood cinema industry "players" who obviously also wanted to partake of the boom in computer-generated 3D animation feature filmmaking. Within the next ten years after *Toy Story*, DreamWorks SKG (in collaboration with Pacific Data Images which specializes in the production of digital VFX, animation and graphics) would have also produced a number of fully computer-generated 3D animation feature films such as *Antz* in 1998, *Shrek* in 2001 and *Shrek 2* in 2004, while Twentieth Century-Fox Animation made *Titan A.E* in 2000, *Ice Age* in 2002 and *Robots* in 2005.

71. *Toy Story 2* would have six digital "playdates": the AMC Pleasure Island in Florida, the Cinemark Legacy in Texas, and in California: the El Capitan Theater, the Edwards Irvine Spectrum, the AMC Burbank and the AMC Van Ness.

72. Bouquet, 'Est-ce encore du cinéma?' 20. Emphasis added.

TWO

Defining the Object of Enquiry

Writing Film History

Why are certain histories and conceptualizations of cinema considered more "plausible" than others? What allows certain forms of judgment and commentary about cinema to be considered more legitimate than others? What makes certain discourses more valid than others in the discussion, theorization, analysis and critical assessment of Hollywood cinema?

Historical and institutional contexts inform not only the perspective and mode of address to be adopted but also the choice of films and topics to be commented upon. Dudley Andrew relevantly points out that

> events in the field of film occur as an interplay of novelty and stability. The complexity of this interplay has a name: history. Not a sequence of events, history is rather the revaluation by which events are singled out and understood in successive eras. When formalized, the concepts of stability and novelty can be defined as institution and change. *Institutions are the sites of history because only within or against institutions does the process of revaluation run its course.*[1]

Certain preoccupations and problems with Hollywood cinema indeed predominate more in specific eras than in others: visual aesthetics, representation, narrative styles, techno-industrial standards, production methods, business practices, and so on. As such, at any point in time the 'project of film historical research—what topics receive scrutiny, which questions get asked, which approaches are taken—is conditioned by (1) the history of film history and film studies as an academic discipline, (2) the perceived cultural status of film as an art form and industry, and (3) the particular research problems presented by the nature of film technology and economics.'[2] And, as Robert Allen and Douglas Gomery remind us, ultimately:

> There is no one correct approach to film history, no one "superhistory" that could be written if only this or that "correct" perspective were taken and all the "facts" of film history uncovered. All models and approaches . . . have their advantages and disadvantages. All make certain assumptions and are based on certain notions as to what makes history "work." . . . Doing history requires judgment, not merely the transmission of facts.[3]

While film history writing is definitely the main agenda and motivation in this book, it may not seem clear what specific type of film history is being dealt with, considering that there are many possible discursive orientations from which to write in historical terms about Hollywood cinema. These include:

- a techno-industrial discourse, commonly used by those professionally involved in Hollywood filmmaking to talk specifically about film production-related technological standards, work practices and tools of the trade;
- a "passionate" cinéphilic discourse, normally used by "film buffs," fans and aficionados outside institutional bounds in the mostly informal artistic and aesthetic appreciation of cinema, films and their makers;
- an economic discourse, commonly used by those in business and financial milieux to analyze cinema as "industry," films as "products" and their theatrical exhibition as "markets";
- a critico-analytical discourse, which formally addresses cinema and films as art.

All of these discursive orientations, as well as their permutations and combinations, are equally—independently or interactively—applicable to the kind of investigation and analysis that informs any history writing pertinent to Hollywood cinema and its films. Each allows for a certain transformation to occur to the raw material of facts, figures, comments and judgments, by the intermediary of specific means and agents to create a determinate outcome. As summarized by Toby Miller's 'tripartite line of reasoning: (a) filmmakers use imagination and practical knowledge; (b) filmgoers work with common sense; and (c) film theorists unravel the magic and escape of cinema.'[4]

A techno-industrial discourse about Hollywood cinema evolved through the interaction of its technical practitioners—as illustrated for instance by industry-oriented publications from the Society of Motion Picture and Television Engineers (est. 1916) or from the Academy of Motion Picture Arts and Sciences (est. 1927). *American Cinematographer* (est. 1920), for example, conveys the techno-industrial discourse associated with cinematography practitioners in particular. It is officially endorsed by the American Society of Cinematographers (est. 1919) which describes itself as 'not a labor union or a guild, but an educational, cultural and professional organization. Membership is by invitation only to those who are actively engaged as directors of photography and have demonstrated outstanding ability.'[5] Such publications allow those in the profession to share techniques, to become aware of new projects and to generally keep in touch with the most recent techno-industrial R&D, innovations and practices.

A "passionate" cinéphilic discourse was developed by "film buffs," fans and aficionados interested in expressing their love of film and cinema more than anything else, as illustrated by the early film commentaries of Ricciotto Canudo or Vachel Lindsay, for instance, during the silent era of cinema.[6] More contemporary examples of that kind of cinéphilic discourse about (Hollywood) movies can be found online, such as at Jasmine Park's *cinephilia.com* blog or Harry Knowles's *Ain't-It-Cool-News* website. It has usually been Hollywood cinema's regular output of VFX-intensive movies in particular that originally triggered the emergence of numerous fan-oriented magazines and bulletins ("fanzines")—as exemplified by such publications (some still in existence and some defunct) like: *Famous Films* (1964), *Famous Monsters of Filmland* (1958–1983), *Fangoria* (1979–), *Gore Creatures* (1963–1976), *Gore Gazette* (1981–1991), *Monster Times* (1972–1976), *Monster!* (1988–1992), *Screem* (1993–), *SFX* (1992), and *Starlog* (1976–2009), amongst many others. Michele Pierson pertinently points out, however, that

> while many different types of writing about special effects—academic criticism as well as entertainment journalism and history—regularly reach into the popular archives for firsthand accounts of the business, technology, and craft of special effects production, the significance of these publications for the cultural reception of special effects has received no more than passing comment in even the few scholarly accounts of special effects that have specifically been concerned with theorizing their reception.[7]

A more sophisticated and advanced form of cinéphilic discourse directed not only at "film buffs," fans and aficionados but also of use to professional film historians and critics can be identified in "*The Making Of . . .*"-type publications that became more common from the 1970s onward, such as Jerome Agel's *The Making of Kubrick's 2001* (1970), Edith Blake's *On Location on Martha's Vineyard: The Making of the Movie Jaws* (1975) or Alan Arnold's *Once Upon a Galaxy: A Journal of the Making of The Empire Strikes Back* (1980). And such discourses additionally exist in video form and online.

With corporate big money becoming increasingly involved and at stake in the production of Hollywood films, it was inevitable that the industry's activities would be closely observed, analyzed, monitored and speculated about in economic terms. An economic discourse about Hollywood cinema emerged since the transformation in the 1930s of 'the American film industry into a modern business enterprise. No longer run by their founders as family businesses, motion picture companies were managed by hierarchies of salaried executives who rationalized operations to ensure long-term stability and profits.'[8] Trade magazines and periodicals like *Variety* (1905–), *The Hollywood Reporter* (1930–), *Film Journal International* (1934–) and *Boxoffice Magazine* (1920–) have been conveying information and analysis pertaining to the state, evolution and predicted growth of the Hollywood cinema industry as a business enterprise since the time when made-in-Hollywood movies had begun to clearly prove 'their potential as big business, and the great Wall Street and La Salle Street investment houses vied for the underwriting of new stock issues for capital expansion, which they sold to the public.'[9]

It should be noted that the construction of sound studios in the district of Hollywood in Los Angeles involved extensive financing by the Morgan and Rockefeller banking groups. This allowed investment houses to assign their own representatives to the boards of directors of film studios and production companies in Hollywood, 'where they worked hand in hand with top management to oversee fiscal matters.'[10] By the mid-1940s it was already clear to Mae Huettig, a "pioneer" analyst of the Hollywood motion picture business, that the 'production of films by the major companies is not really an end in itself, on the success or failure of which the company's existence depends; it is an instrument directed toward the accomplishment of a larger end, i.e. domination of the theater market.'[11] In addition to trade magazines and periodicals, more advanced and comprehensive forms of economic discourse about the Hollywood cinema industry would also become available from the 1930s onward, as illustrated by such publications as: Klingender and Legg's *Money Behind the Screen* (1937), Huettig's *Economic Control of the Motion Picture Industry* (1944), Conant's *Antitrust in the Motion Picture Industry: Economic and Social Analysis* (1960), Kindem's *The American Movie Industry: The Business of Motion Pictures* (1982), Wasko's *Movies and Money: Financing the American Film Industry* (1982), Vogel's *Entertainment Industry Economics: A Guide for Financial Analysis* (1986), Daniels et al.'s *Movie Money: Understanding Hollywood's (Creative) Accounting Practices* (1998), de Vany's *Hollywood Economics: How Uncertainty Shapes the Film Industry* (2004), and Epstein's *The Hollywood Economist: The Hidden Financial Reality Behind the Movies* (2010), amongst others.

A critico-analytical discourse about the cinema was developed by those not directly and professionally involved in filmmaking as an industry and as a business. Often from academic and journalistic circles, they comment and pass judgment on films and also address (Hollywood) cinema as *art* and as an aesthetic object of formal study and enquiry. Films of the silent era—a period that witnessed the emergence of and experimentation with many of the fundamental aesthetic intricacies that would shape the future evolution of cinema as an art form—were addressed in predominantly formalist terms by the likes of Sergei Eisenstein, Rudolf Arnheim, Bela Balàzs, Hugo Münsterberg and Siegfried Kracauer.[12] By identifying and discussing the inherent artistic value of film, these pioneer critics and theorists intended to formulate an aesthetic of cinema and, ultimately, to validate cinema as a legitimate art form, as opposed to simply being a techno-visual process for re-presenting, re-producing, recording, and re-projecting nature and "reality." Hence, for Rudolf Arnheim in 1932: 'People who contemptuously refer to the camera as an automatic recording machine must be made to realize that even in the simplest photographic reproduction of a perfectly simple object, a feeling for its nature is required which is beyond any mechanical operation.'[13] And a few years earlier, in 1929, Sergei Eisenstein had suggested that

> film, freed from traditional limitations, will achieve direct forms for thoughts, systems and concepts without any transitions or paraphrases. And which can therefore become a synthesis of art and science. That will become the really new watchword for our epoch in the field of art. And really justify Lenin's statement that 'of all the arts . . . cinema is the most important.'[14]

Adopting for his part a predominantly psychological perspective, Hugo Münsterberg recognized as early as 1916 that the "photoplay" provided

> a view of dramatic events which was completely shaped by the inner movements of the mind. To be sure, the events in the photoplay happen in the real space with its depth. But the spectator feels that they are not presented in the three dimensions of the outer world, that they are flat pictures which only the mind moulds into plastic things. . . . The events are seen in continuous movement; and yet the pictures break up the movement into a rapid succession of instantaneous impressions. We do not see the objective reality, but a product of our own mind which binds the pictures together.[15]

The writings of pioneer film critics and theorists seemed inclined to suggest that the cinema was doing much more than simply providing "exotic" sensual pleasure and entertainment, that the art of cinema fulfils more than just a superficial sense of wonder and gratification: 'Erwin Panofsky, Siegfried Kracauer, and André Bazin, amongst others, argued that film was not an art in opposition to nature but somehow, in a rich paradox, an art *of* nature.'[16]

The end of the pioneer phase in the evolution of a critico-analytical discourse about cinema coincided with the collapse of Hollywood cinema's studio system, as well as with the occurrence of critical global sociopolitical events in the aftermath of World War II: the outbreak of the Cold War and the Korean War, the rise of McCarthyism and the Civil Rights movement in the United States, the Vietnam War escalation, political assassinations, civil protests and riots, and the emergence and disintegration of counter-cultures of resistance. For Dudley Andrew, 'film theory came to life by burying work that was representative of all earlier film theory: 1964 was the date of both the publication of Jean Mitry's *Esthétique et Psychologie du Cinéma* and the appearance of Christian Metz's writings [such as his famous essay "Le cinéma: langue ou langage"].'[17] The independent, alternative and experimental wave of filmmaking that passed through practically all cinematic apparatuses in Europe and in the United States from the late 1950s to the early 1970s triggered "radical" modes of critical and analytical enquiry meant to question, oppose and problematize earlier pioneering critico-analytical discourses about cinema.[18] And such modes of enquiry indeed fitted with the prominent counter-cultures of resistance, rebellion and change characteristic of the 1960s.

With the arrival and rapid growth of "post-pioneering" critico-analytical discourses about cinema, what was

> new was the fact that theory no longer sought accommodation with the existing criticism and aesthetics, and was not presented as an improvement or refinement of current critical practice, but was avowedly bent on its overthrow. Not evolution but revolution was on the agenda; the allegiance of the new theory was to a radical Left Politics totally opposed to existing régimes, both social and intellectual.[19]

This politically "leftist" attitude to the art of cinema eventually came to dominate critico-analytical discourses during the period, causing film theory to turn into a kind of exclusive discourse between theorists, about and with films. More than ever before, critico-analytical discourses about cinema became institutionally and historically bound ways of discussing, commenting on and judging films, framed by and in accordance with standards and rules established by each succeeding wave of film theorists and critics, and by the respective institutions they were associated with.

In the case of commercially successful VFX-intensive films for instance—which have always been an integral component of Hollywood cinema—these tended to become ideal "scapegoats" for critico-analytical discourses to illustrate the unproductiveness of a profit-oriented cinema.[20] Hollywood cinema and its films regularly resuscitated some of the negative attitudes and arguments pertaining to commercial predicaments. Hollywood cinema's evolution in the direction of pure sensual gratification and escapist entertainment, as projected by the extensive application of VFX for instance, reinforced the views of Hollywood filmmaking's detractors pertaining to how it emphasized a superficial retrograde art form that dulled the intellect. This kind of attitude had its origins in various Marxist-oriented debates—generated by the *Cahiers du Cinéma* (1951–), *Positif* (1952–) and *Screen* (1959–) collectives in particular—more concerned with the cinema's intellectual and socio-political obligations than with the sensual pleasures and gratification inherent to the cinema spectator's viewing experience.

During their "post-pioneer" and "counter-culture" phase, critico-analytical discourses evolved much faster than previously as a consequence of the art of cinema—as an aesthetic object worthy of formal study and enquiry—acquiring respectability in American and European universities, as well as having numerous scholarly publications entirely devoted to it, such as *Film Culture* (1955–1999), *Cinema* (Switzerland, 1961–), *Film Comment* (1962–), *Cinema Journal* (1966–), *Cinéaste* (1967–), *Monogram* (1971–1975), and *Jump Cut* (1974–). Challenging earlier formalist and realist perspectives that privileged the artistry of film, this particular phase of critico-analytical discourse often favored particularly erudite approaches to the art of cinema, especially through conceptual "borrowing" from academic disciplines as diverse as anthropology, linguistics, sociology and psychology. This tendency was conducive to the centralizing of discursive tools more pertinent to structuralism, psychoanalysis and semiology, amongst others, in influencing a predominantly institutionalized evolution of critico-analytical discourses about cinema towards greater epistemological sophistication and complexity.[21] Although it did not start as such, over the years critico-analytical discourses about cinema became, according to Dudley Andrew, 'an accumulation of concepts, or rather, of ideas and attitudes clustered around concepts.'[22]

Whenever the cinema underwent significant techno-industrial transitions in the course of its evolution, how it would consequently be defined, conceptualized and theorized would always be re-adjusted so that the meaning of the "cinematic" could then be expanded accordingly. This sort of "conceptual re-adjustment" exercise—fundamentally involving the identification and assessment of the cinema's additional possibilities of operation, expression and purpose, as enabled by the specificities of the changes in question—has always been a routine event across the cinema's associated critico-analytical discursive landscapes. As David Rodowick pertinently comments:

> Periods of intense technological change are always extremely interesting for film theory because the films themselves tend to stage its primary question: *What is cinema?* The emergent digital era poses this question in a new and interesting way because for the first time in the history of film theory the photographic process is challenged as the basis of cinematic representation. If the discipline of cinema studies is anchored to a specific material object a real conundrum emerges with the arrival of digital technologies as a dominant aesthetic and social force.[23]

Indeed, most critico-analytical conceptualizations of the "cinema" prior to the 1990s, in essentially attempting, amongst other things, to further legitimate film as art, were mainly informed by and derived from the analog techno-visual properties, possibilities and limitations of the medium. These properties could arguably be deemed not to have undergone drastic changes since the late 19th century. They had not progressed considerably from the original intentions of its many "inventors"—in relation to which André Bazin boldly comments that the 'cinema was born . . . out of a myth, the myth of total cinema.'[24] For Bazin, the cinema's true pioneers (such as Étienne-Jules Marey, Eadweard Muybridge, the Lumière brothers and Georges Méliès) imagined

> the cinema as a total and complete representation of reality; they saw in a trice the reconstruction of a perfect illusion of the outside world in sound, color, and relief. . . . There are numberless writings, all of them more or less wildly enthusiastic, in which inventors conjure up nothing less than a *total cinema* that is to provide that complete illusion of life which is still a long way away.[25]

Regular techno-industrial innovations and transformations have indeed continuously nurtured the prevalence of this quest for the cinema's state of perfection as "total cinema"—which, from Bazin's perspective, instead remained a myth and became increasingly more elusive rather than a practically feasible goal, not even in the aftermath of such technological transformations in the cinema as pivotal as: the conversion from orthochromatic to panchromatic film stock during the 1920s; the introduction of sound in the late 1920s; the implementation of the three-color film process in the mid-1930s; and the standardization of the widescreen process and color negative film in the 1950s.[26] Bazin further observes that

> it would be absurd to take the silent film as a state of primal perfection which has gradually been forsaken by the realism of sound and color. . . . Every new development added to the cinema must, paradoxically, take it nearer and nearer to its origins. In short, cinema has not yet been invented![27]

Like many of its prior transformations, the cinema's digital metamorphosis has been an outcome of techno-visual innovation and novel filmmaking practices originating from the Hollywood cinema industry in particular. With digitization progressively implemented from the 1980s onwards, the "total cinema" imagined by the motion picture's pioneering "forefathers" and later explicated by André Bazin may have arguably been finally achieved through the digitization of VFX and of filmmaking technologies, a century after the so-called "birth" of cinema.

In its digital form, cinema may have finally turned into a "total cinema"—in the sense of being in a position at last to achieve perfectly what it had only approximated or promised to achieve in its analog state for the major part of a century. Via the digitization of the tools of filmmaking, the cinema has been techno-visually motivated into following a specific evolutionary pathway by means of which it arrived, more than ever before, at its closest to a state of perfection of operation, expression and purpose. On the basis of its metamorphosis into a "total cinema," contemporary cinema may have become arguably more "cinematic" than ever—contrapuntally to the Bazinian point-of-view of the impossibility of such an occurrence—and, hence, needs to be addressed as such from a historical perspective here. But, in relation to how analog celluloid-based cinema has been commonly conceptualized, categorized, theorized and understood over many decades, to reiterate Stéphane Bouquet's crucial question: 'is this still cinema?'[28]

Discursivizing VFX

Most available critico-analytical discourses have traditionally perceived VFX as a *technical* component of the cinematic apparatus along with everything else techno-industrial in nature. As pertinently commented by Peter Nicholls for instance:

> Film snobs often talk as if special effects were somehow vulgar—at best, the icing on an otherwise realistic cake. It probably makes more sense to regard special effects as completely fundamental to film, the cake itself. After all, the language and grammar of film is nearly all "special effects," but most of the tricks are now so familiar that we pay them no more attention than people just talking pay to nouns, verbs and adjectives.[29]

And Lev Manovich further explains that 'optical effects and other techniques which allowed filmmakers to construct and alter moving images . . . were pushed to cinema's periphery by its practitioners, historians and critics. Today, with the shift to digital media, these marginalized techniques move to the centre.'[30] Although having evolved in parallel with the art of cinema, VFX have rarely been addressed as art or in dynamic aesthetic terms, but more according to techno-industrial and cinéphilic discourses. The increasing output of VFX-intensive Hollywood movies (especially horror, science fiction and fantasy) since the "studio system" era witnessed the parallel appearance of specialized magazines dedicated to VFX practitioners, their on-screen creations and, sometimes, to related technologies of VFX production. Cases in point: *L'Ecran Fantastique* (1969–), *Cinéfantastique* (1970–2002), *FXRH* (1971–1974), *Reel Fantasy* (1978), *Cinemagic* (1979–1987), *Cinefex* (1980–), *SF Movie Land* (1985–), *Cinescape* (1994–2004), and *Sci-Fi & Fantasty Models* (1994–2000 & 2006–). Involving professional practitioners and enthusiastic aficionados of VFX-intensive films in particular—usually amateurs and semi-pro practitioners on the fringe of the filmmaking industry—these publications tend to project a combination of techno-industrial and cinéphilic discourses that cater to informed readers already working in the VFX industry, to VFX fans and film buffs. *Cinefex*, for instance, has been endorsed by VFX supervisor Richard Edlund (*Ghostbusters, Star Wars: Return of the Jedi, Raiders of the Lost Ark*) as

> *the* journal for the practitioners of special visual effects—we all read it and learn from it. Always beautifully printed with color production shots and rare behind-the-scenes material, *Cinefex* presents detailed and poignant information of value to fellow effects artists, interested fans of the art, and even producers who need visual effects in their films.[31]

Professional publications on the subject of VFX, other than specialized magazines, include: Bulleid's *Special Effects in Cinematography* (1954); Clark's *Special Effects in Motion Pictures: Some Methods for Producing Mechanical Special Effects* (1966); Fielding's *The Technique of Special Effects Cinematography* (1972); Wilkie's *Creating Special Effects for TV and Film* (1977); Perisic's *Special Optical Effects in Film* (1980); and Hayes's *Trick Cinematography: The Oscar Special-Effects Movies* (1986).

As a result of the Hollywood cinema industry's entry into the digital realm, from the 1980s onwards techno-industrial discourses about filmmaking would be increasingly informed by non-cinematic discourses that seem more related to computer science, software engineering, cybernetics and information technology than to film. This state-of-affairs can be illustrated by Hayward and Wollen's *Future Visions: New Technologies of the Screen* (1993); Ohanian and Phillips's *Digital Filmmaking: The Changing Art and Craft of Making Motion Pictures* (1996); and McQuire's *Crossing the Digital Threshold* (1997). And digitization in the Hollywood cinema industry also influenced techno-industrial discourses pertaining specifically to VFX to become more technically complex than ever.

Techno-industrial discourses about VFX are sometimes "translated" or "mainstreamed" by industry practitioners, "experts," and "veterans" to make them more understandable for a larger popular readership; although these tend to be techno-centric nonetheless in terms of involving more talk about the "mechanical" achievements and potential of the technologies and methods used during production. Instances are: Fry and Fourzon's *The Saga of Special Effects* (1977); Culhane's *Special Effects in the Movies: How They Do It* (1981); Abbott's *Special Effects: Wire, Tape and Rubber Band Style* (1984); Finch's *Special Effects: Creating Movie Magic* (1984); Smith's *Industrial Light & Magic: The Art of Special Effects* (1986); Clise's *Special Effects* (1986); Hutchison's *Film Magic: The Art and Science of Special Effects* (1987); Scott's *Movie Magic: Behind the Scenes with Special Effects* (1995); and McClean's *So What's This All About Then?—A Non-Users Guide to Digital Effects in Filmmaking* (1998). Another form of discursive "mainstreaming" exercise in relation to VFX that additionally made some reference to their aesthetic features and their artistic contribution to the filmic medium also emerged: McKenzie's *Hollywood Tricks of the Trade* (1986); Millar's *Cinema Secrets: Special Effects* (1990); Vaz's *Industrial Light*

& Magic: Into the Digital Realm (1996); Bizony's *2001: Filming the Future* (1994); and Hall's *Pause: 59 Minutes of Motion Graphics* (2000).

Conceptualizing Hollywood as "Film," as "Cinema" and as "Industry"

While there was no doubt for Edward Buscombe in 1975 that cinema is simultaneously art as well as industry, he pointed out that 'proof that the mutual exclusion of art and industry operates at a level too deep to be affected by mere common sense can be found not only in the dominant critical attitudes but in the organization of social institutions.'[32] And more than thirty years later, Ruth Vasey would also be reasoning about how to interrogate Hollywood:

> Where should one begin—with hard-nosed financial imperatives of the "business" or with the intangible attractions of the "show"? . . . [I]t is impossible satisfactorily to account for the business of the movies without taking into account their aesthetic characteristics, and vice versa. To understand Hollywood is to understand the many interrelated strategies—some industrial and some aesthetic—that it has used to find acceptance amongst its diverse audiences.[33]

The definitions of Hollywood as "film," "cinema" and "industry" are as multiple as they are variable. 'A key distinction within "apparatus theory" of the 1970s and 1980s, *cinema* is regarded as the overall institution of *film*; not merely the primary site for *film* viewing but also the main intersection between *industry* and ideology.'[34] While the *cinema* implies film-making as a form of art whose associated creative practices are oriented towards aesthetic pursuits, the *film industry*, on the other hand,

> describes an economic system, a way (or ways) of organizing the structure of production, distribution, and consumption. . . . It has passed from the primitive stage of small-scale entrepreneurial activity to the formation of large-scale monopolies, securing their position by vertical integration, spreading from production into distribution and exhibition. Since the war, the industry has, like other forms of business, developed towards diversification and the formation of multinational corporations. . . . In film criticism, the term *film industry* implies a way of looking at film which minimizes its differences from other forms of economic activity; a way which is of course predominantly that of those who actually own the industry.[35]

The magnitude and complexity of the task of defining Hollywood in particular as "film," as "cinema" and as "industry"—in economic, techno-industrial, and aesthetic terms—can be illustrated, for instance, via the introduction to Bordwell, Staiger and Thompson's seminal 1985 work *The Classical Hollywood Cinema: Film Style and Mode of Production to 1960*. This lists some of the "familiar" descriptions of and (mis)conceptions about the Hollywood cinematic apparatus (according to the more prominent ideologies and critical preoccupations associated with particular eras): 'the dream factory'; 'an arm of the culture industry'; 'celluloid imperialism'; 'escape'; 'nostalgia'; 'imaginary landscape.'[36] The 'Hollywood mode of film practice constitutes an integral system, including persons and groups but also rules, films, machinery, documents, institutions, work processes, and theoretical concepts.'[37] It is indeed this totality that needs to be taken into account when addressing Hollywood cinema in historical terms.

Common terms recurrently used to study, qualify and make sense (from a historical perspective) of Hollywood cinema's mode of film practice have included: "Fordist," "classical," "post-classical," and "new Hollywood." The use of such terms is usually linked to specific historical phases in Hollywood cinema's techno-industrial, economic and aesthetic evolution. The definition of these terms is generally shaped by the different ways of conceptualizing the nexus between filmmaking style, industrial operations, technological standards and business practices inherent to the industry at particular points in time. And these conceptualizations are derived from the identification and articulation of relationships between temporally framed filmmaking strategies, industrial practices, aesthetic motivations, technological standards and limitations in the production of the Hollywood film.

While the concept of a "classical" American cinema had been circulating for decades, it became a focus of theoretical attention predominantly in the 1970s, particularly in such journals as *Monogram* and *Screen*. According to Murray Smith, 'Classicism may refer to certain narrative and aesthetic features (the stability of a system of genres, or of continuity principles, for example); or, it may refer to the studio system as a mode of production.'[38] From the 1920s to the 1950s, the Hollywood cinema industry was characterized by an efficient industrial approach applied by all major studios for the cost-effective mass production of motion picture entertainment: the breakdown of a film's production into discrete components, each being handled by separate departments all operating within a particular studio. This particular form of industrial division of labor, similar to that applied in Ford's automobile factories, is what became better known as the "studio system."[39] Most accounts and analyses of the Hollywood cinema industry's "classical" phase invoke a combination of institutional control, efficient industrial division of labor, technological innovation and commercial acumen to explain the success of the studio system—as instigated and maintained by a handful of movie "moguls," especially: Samuel Goldwyn, Louis B. Meyer, Jack Warner, William Fox, David O. Selznick, Darryl Zanuck, Adolph Zukor, Irving Thalberg and Harry Cohn.

The dominant artistic and aesthetic characteristics of "classical" Hollywood cinema were established through increasing domination of exhibition markets worldwide during the silent era, especially after WWI. During that period, a large number of films screened in most countries were of one type, described by Kristin Thompson as: 'the classical Hollywood narrative film in continuity style.'[40] With the studio system quickly structuring Hollywood cinema into an efficient business during that period, the foundations were being laid for a very profitable commercially motivated cinema of quality that would consequently prevail for decades, especially by the intermediary of the industry's continuous aesthetic, techno-industrial and economic upgrading.

In accordance with the studio system's business philosophy, 'Hollywood executives, like television programmers of today, rarely thought in terms of a single work; each film was one component of a total schedule, each production a fraction of a yearly budget.'[41] This kind of film production strategy inherent to Hollywood's studio system era has been labeled "Fordist," as we have seen, on the basis of the analogy established between the Hollywood cinema industry and Ford's automobile factories.[42] The two

> foundations of Fordism are abstraction and homogenization. Products and tasks are standardized thereby making mechanical parts as well as workers interchangeable elements in the assembly-line process. The shift from a craft economy to an economy based on mass production required the development of mass-marketing techniques. . . . Planning is always centralized to create a hierarchy of management extending from producers to consumers.[43]

In such terms, the Hollywood studio system seems to structurally mirror Ford's automobile assembly lines. This by virtue of the film studio's separate departments and units organized according to a complex division of labor and meant to increase efficiency of output and cost-effectiveness, as well as maintain quality standards with regard to film production. It was believed that, by maintaining consistent techno-visual standards for its specific product, namely the Hollywood feature film, the studio system could guarantee continued audience loyalty and hence profitability. And the motion picture business was indeed booming during the studio system era.[44] The large output of films—from what appeared to be a smoothly running filmmaking operation, based on division of labor, aesthetic and techno-industrial standardization, as well as the tight management style by studio heads of actors, personnel, physical resources and finances—seems to have been confirmation enough for addressing the Hollywood cinema industry's studio system era in so-called "Fordist" terms.

On the other hand, mainly on the basis of his study of the production of Howard Hawks's *Bringing Up Baby* (1938), Richard Jewell has argued that

> the time has come to dispense with the assembly-line analogy for studio production. Although the moguls no doubt wished their operations

> could be as efficient and predictable as those of a Ford plant, their product mitigated against standardization. . . . The departmental structures and operating methods of studios never turned filmmaking into a conveyor-belt business. Most pictures presented special problems which could not have been solved by inflexible, factory-inspired methods.[45]

Further confirming the problematic of a too clear-cut "factory assembly-line" analogy for Hollywood film production during the studio system era is Thomas Schatz's identification of some of the "unpredictable" factors in the Hollywood movie business. These include: 'the talents and personalities of key creative personnel, the growing power of labor unions and "guilds," and the intervention of federal and state government via the courts.'[46] In addition, we should consider the unwavering inclination of the industry for product innovation and differentiation.

What clearly transpires however is the Hollywood cinema industry's recognized ability to adapt productively and repeatedly to both short-term and long-term problems and changes, and to find solutions quickly in times of crisis. Common changes, problems and crisis included spectatorship and business interest, production methodology and distribution strategy, technology and filmmaking practice, re-presentation and aesthetics, socio-political and ideological configuration, amongst others. Leo Rosten in 1941, at a time when Hollywood cinema's studio system was peaking, noted that

> movie making is not a systematized process in which ordered routine can prevail, or in which costs can be absolute and controlled. Too many things can and do go awry, every day, every hour, during the manufacture of a movie. Movies are made by ideas and egos, not from blueprints and not with machines. Every story offers fresh and exasperating problems; every actor, director, writer carries with him curious preferences and needs. . . . *The movie business moves with relentless speed, change is of the essence, and Hollywood must respond to change with short-spanned flexibility.*[47]

Conceptualizing and defining Hollywood cinema—in economic, techno-industrial, and aesthetic terms—fundamentally always already seems to involve historicizing its willingness to change and seeing how productively it reacts and adapts to change—as projected by the very films it produces in any era. This adaptability and willingness to change will later become central to understanding the digitization of VFX and of filmmaking generally.

Following the demise of the studio system, the re-organization of the Hollywood cinema industry during the 1950s and 1960s witnessed innovative business strategies such as those of 'Lew Wasserman (at MCA-Universal) and Steven J. Ross (at Warner Communications) [who] both initiated practices which were widely emulated by other corporations, acting as the catalysts for transformations in the industry as a whole.'[48] For instance, Wasserman's novel methods included the independent marketing of stars, the promotion of made-for-television movies, and the saturation release and intensive use of prime-time television advertising for high-budget films, amongst other things.[49] Arthur Krim and Robert Benjamin are also credited with contributing significantly to Hollywood's adaptation to change. They turned United Artists into a modern motion picture company as the studio system was beginning to fall apart. Hence, under Krim's and Benjamin's impetus, United Artists' 'brand of independent production moved production closer to the concept of one man-one film by granting filmmakers autonomy and creative freedom over the making of their pictures and by rewarding talent with a share of profits.'[50] Originally an immediate response to the collapse of the studio system, innovative film production practices and management strategies (introduced by Wasserman, Ross, Benjamin and Krim, amongst others) have lasted well into the present-day Hollywood cinema industry. And the re-organization of and changes inherent to the Hollywood cinema industry from that period onward has warranted its characterization as "post-classical" and sometimes as "post-Fordist."[51]

As indicated earlier, by the 1970s, the Hollywood cinema industry was experiencing the firm entry and rapid rise to power of a new generation of cinema-literate and technology-savvy filmmakers. And they would constitute the driving force of the so-called "new Hollywood." Thomas Schatz explains that the

> post-1975 era best warrants the term "the New Hollywood," and for many of the same reasons associated with the "classical" era. Both terms connote not only specific historical periods, but also characteristic qualities of the movie industry at the time—particularly its economic and institutional structure, its mode of production, and its system of narrative conventions.[52]

The term "neo-classical" has sometimes additionally been associated with the "new Hollywood" especially by virtue of the massive revival of big-budget VFX-intensive filmmaking, previously one of the most important components of the Hollywood cinema industry during the studio system era.

The emerging "neo-classical" Hollywood cinema of the 1970s had many structural affinities with its "classical" ancestor, on the basis of its revamping of some aspects of the Fordist approach to film production and the system of narrative conventions inherent to the studio system. This enabled the transition from the age of mechanical production to the age of electronic and subsequently digital production. As such, it could be argued that the industrial structure of Hollywood cinema never changed instantly and in radical terms: it evolved and adapted continuously over time with minor "disruptions" (such as the very short-lived phase of "alternative" low-budget filmmaking). "Neo-classical" Hollywood cinema, arguably a revamped continuation (by the "movie brats" in particular) of "classical" Hollywood cinema, lasted well into the 1990s. Hollywood cinema's early "neo-classical" period in the 1970s was in many ways similar to the formative years of its early studio system-oriented "classical" period.

From the Order of Facts to the Order of Ideas to the Order of Objects

One possible conclusion, titrated from the massive body of works attempting to conceptualize Hollywood cinema in one way or another, is that there does not seem to be any unanimous single concept or term that can ever adequately encapsulate all of what Hollywood cinema really is. 'The contradictory accounts of Hollywood, its periodization and its (in)consistency, mark the failure of consensus and are, perhaps, a sign of the discipline's [film studies'] maturity.'[53] Any conceptualization and application of the term "Hollywood" as "film," as "cinema" and as "industry" can never be definitive, remaining instead constantly open-ended and in the making.

No history of Hollywood cinema or of a particular aspect of Hollywood cinema can claim "universality." None can comprehensively display all available information or address absolutely everything that relates to it in one way or another. History writing of any particular era or aspect of Hollywood cinema implies an inevitable shift from the order of facts to the order of ideas and models, so as to conceptualize and make sense of such facts. The order of facts here implies a chronological historicizing of VFX, in terms of identifying the progressive digitization of their means of production and, consequently, of filmmaking technologies, in the context of the Hollywood cinema industry. This particular order of facts is established—using specific Hollywood films as markers—on the basis of: groundbreaking digital VFX; visual originality and techno-industrial innovation; the escalating prioritization of the digital medium over the analog due to the technological convergence of the information, communication and entertainment industries in general; synergies and privileged R&D partnerships interconnecting corporations, independent companies, movie studios, technology providers and particular individuals in the Hollywood cinema industry landscape. But the compilation of facts and events pertaining to digital VFX in Hollywood and to their related technologies should not monopolize any consequent theorization and discussion. 'To investigate a medium is to analyze and synthesize the historical nature of the material mediations that characterize a period in time.'[54] The material data at hand should serve to support theoretical claims critically and productively; it should not obstruct nor excessively influence the shape and direction of theorization and discussion.

To address digital VFX productively in the Hollywood cinema industry context requires a discourse derived from the "marriage" of technology, science, artistry and skilled craftsmanship—and which also combines techno-industrial and critico-analytical discourses about the Hollywood cinema industry, its films in general, and digital VFX production and application in particular. This "customized" discourse to address digital VFX is ultimately a matter of subjective judgment in that it favors particular orientations and angles, standards and coordinates which can be referred to and modified when the need arises. This should help to bridge the "gap" between the techno-industrial "face" of digital VFX production and its critico-analytical consideration. This is about theorizing the practice as well as the form of digital VFX. The aim is to obtain, in the end, a coherent conceptualization and understanding of digital VFX in the Hollywood cinema context.

Why there has been limited scholarly writing on special effects might well have to do 'with the intellectual traditions that have shaped contemporary thinking about film spectatorship, on the one hand, and analysis of the culture industries, on the other.'[55] Although four modes of analysis were identified earlier, the account of the digitization of VFX provided in this book operates mostly within the register of the techno-industrial and of the critico-analytical, as such downplaying more aesthetic and economic considerations.[56] Hence, what is provided here is an integrated historical account of the digitization VFX in the Hollywood cinema industry from two perspectives predominantly, namely: *techno-industrial film history* which investigates the origins and evolution of technologies and industrial practices that enable the production and exhibition of movies; and *critico-analytical analysis* which addresses cinema as a form of art from an institutional perspective.

The Time-Line

The construction and assembly of "A Time-Line of landmark (VFX-intensive) films, techno-scientific innovations, financial and industrial events that have contributed to the evolution of digital VFX in the Hollywood cinema industry up to the 1990s"—see Appendix A—has been tremendously productive for this entire enquiry in terms of generating, amongst other things, a constant preoccupation with looking for and establishing connections between (often seemingly unrelated) events, discoveries, inventions, business deals and alliances. Doing so has enabled an expanded understanding of the kind of role they subsequently played, both in the short and long run, with regard to the gradual metamorphosis of the Hollywood cinema industry and its films from an analog to a digital techno-industrial state. Film and various media forms are inevitably inter-related and inter-dependent in terms of historical narratives, as such emphasizing the shared history of all media, information, communication and entertainment industries. As Sean Cubitt reminds us: 'The historian must be alert to difference as much as to similarity, and the materialist historian has also the ethical duty to watch out for contradiction and alternatives.'[57] The time-line attempts as much as possible to include significant milestones in the mutual digital convergence of motion picture and computer technologies, of Hollywood cinema and the American entertainment industries, of VFX production, filmmaking and general techno-scientific R&D: relevant events that have been seminal in the techno-industrial, economic and aesthetic (re-)configuration of Hollywood cinema throughout the 20th century.

It is instinctive to conceptualize so-called "new" technologies and media as emerging and evolving in temporal linearity, as progressing logically and un-problematically from one state to another in accordance with formally acknowledged R&D, invention and innovation at specific points in time and specific stages of a shared techno-scientific "mythology of evolution." Accordingly, a linear topography initially appeared to be the best approach to adopt in mapping the techno-scientific and visual evolution of Hollywood cinema.

However, establishing a cinema-related chronology of techno-scientific, economic and visual evolution is inevitably problematic, especially when dealing with overlapping innovations, discoveries, events, research and development. This implies a consideration of various filmic and non-filmic technologies and media that have progressively surfaced and built the "family-tree" and "mythology of evolution" of filmmaking and digital VFX production in the Hollywood context. The linear chronological approach to this kind of time-line reveals a fundamental weakness: un-problematized "cause-and-effect" progressions and "neat" openings and closures in the historical narrative of techno-scientific innovation across overlapping innovations, discoveries, events, research and development. In being tied down to institutionally and consensually accepted dates of the conceptualization, invention or diffusion of techno-industrial devices and media, this time-line often reproduces a recurrent biased prioritization of certain logically related events and occurrences over others (which might have been of equal importance but remain unacknowledged due to lack of legitimate information or adequate confirmation).

Most of the resources consulted in constructing the time-line have shown a tendency to focus on dates of "invention" or "release" that are institutionally and consensually agreed upon, with little or no questioning of alternative factual possibilities nor consideration of influential "parentage" or "lineage" connections, so-to-speak, between most of the techno-scientific innovations, devices and events mentioned. Brian Winston's commentary about the introduction of the transistor, for example, confirms this observation:

> The transistor, even more than the computer, the laser or television, seems to have burst upon society. It is without infancy, without past history, born fully formed in the Bell Labs on 23 December 1947. . . . The impression of instantaneity is created at the cost of an amnesia profounder than that affecting our memories of the development of other devices.[58]

Winston's approach to techno-scientific progress is driven by the conviction that anything technological that has ever emerged is essentially characterized and governed by stages such as "conceptualization," "announcement," "building," "patenting" and "commercialization."[59] In Winston's terms, these stages rarely follow *any* set linear order or pattern. They digress and overlap so profusely that it is not always possible to establish unproblematically which device or product came first, which came later, what might be the connections in between and what was thought of but never surfaced. Is it the "announcement" of a particular technological device or product that marks its "origin"? Is it "conceptualization" or "commercial availability" that matters more in the historical narrative of its development? In any case, institutional and popular chronological accounts of techno-scientific evolution tend to employ such words as "invented," "introduced," "released" or "produced" indiscriminately and interchangeably, often showing no interest whatsoever in the prior stages that led to the emergence of particular innovations or devices. Brian Winston's *Misunderstanding Media* is one of the few versions of techno-scientific developmental history in which inventions do not unproblematically come into existence and grow in sophistication but go through complex, lengthy phases of evolution and perfection. His chosen approach to techno-scientific development and concurrent research methodology led him to establish, for example, how

> electronic television was described in 1908, partially demonstrated in 1911, patented in 1923, perfected (or *invented*) a decade later and how it took a further quarter of a century for it to be diffused. Video recording was proposed in 1927 and demonstrated in 1951. Its potential use as a home recording device was noted in 1953. The first cassette appeared in 1969. Diffusion was to take a further decade and a half. . . . The electric telegraph: proposed first in the seventeenth century, but re-proposed in 1820, demonstrated in 1825, perfected in 1837 and diffused over the following three decades. The electric telephone: proposed in 1854, demonstrated in 1876, *invented* in 1877 and widely available by 1914. Radio: suggested in 1872, demonstrated in 1879, *invented* in 1895 and diffused over the next 40 years. And the introduction of the laser . . . suggests that the pattern still holds good: ideation—1947, *invention* 1960, and slow diffusion from then on.[60]

During the development of the time-line presented here, however, it has not always been possible to follow Winston's lead nor to avoid the seemingly unproblematic perspective to linear chronology altogether.

Why should the time-line start at 1869 with the invention of celluloid, even though the cinema was then not quite ready to emerge? Why has the 19th century history of the photographic apparatus, for example, as one of the most important influences in the development of motion picture technology and aesthetics, not been included? Why has the advancement of mathematics or physics—by the intermediary of the discovery of electricity by Benjamin Franklin in 1780 or of Charles Babbage's designs of the "difference engine" in 1833, amongst other things—not been addressed in the time-line in relation to the later emergence of computer technology? These are, after all, crucial components in the "family tree" of 20th century techno-scientific progress and industrial innovation that enabled the Hollywood cinema industry to emerge and grow in the first place; but they have had to be excluded. To gather and include information relating to every significant event in techno-scientific history is an encyclopedic exercise that would have eventually turned into a digressive task considering the principal intentions of this book. In any case, the more detailed chronological treatment of techno-scientific progress—for cinema and other media—has already been carried out comprehensively elsewhere.[61]

The time-line is designed to present a chronological account of pertinent events that have triggered and molded, in one way or another, the digital convergence of cinema, media, communication, and information technologies in the course of the 20th century. 'Friedrich Kittler's research years ago initiated interest in ways technologies develop out of each other, with new ones piggybacking, as it were, on old ones, so that traces of the old remain and are carried forward in new modes.'[62] The ultimate purpose of this time-line is to make sense of how the convergence between corporate media, information and communication technologies and the entertainment industries eventually occurred, with the normalization of the digital medium across the Hollywood cinema industry by the end of the 1990s. This was an outcome of many decades of research and development, discoveries, inventions, business deals, alliances and setbacks. "Techno-scientific innovations, economic and industrial events" are meant to signpost significant stages—wherein digital VFX in particular were developed, tested, applied, commercialized and normalized—of Hollywood cinema's coming of digital age in particular. Any digital VFX breakthrough can usually be traced back to a prior innovation in the domains of filmmaking, computing and VFX production. After identifying their respective techno-scientific "family trees," what is of interest are the intersections and overlaps between the three.

Specific films mentioned in the time-line constitute "landmarks" of the Hollywood cinema industry, on the basis of introducing original visual innovations obtained through the application of groundbreaking technologies and techniques. These "landmark" films have guided the constant "rejuvenation" of Hollywood cinema—as an *art of moving pic-*

tures, as an *entertainment industry* and as a *business enterprise.* Influential stylistic and aesthetic visual innovations in these movies were enabled by continuous development in the domain of VFX in particular. These films have proposed innovative, or at least "alternative," forms of, and approaches to, visual re-presentation and aesthetic expression in the cinema. They have also irreversibly transformed the film viewing experience by broadening the spectator's range of sensual pleasures and aesthetic appreciation. Addressing, let alone mentioning, every single Hollywood motion picture that involves the extensive use of VFX would be an encyclopedic task indeed. It has been more productive to concentrate on a smaller number of visually, aesthetically and techno-industrially groundbreaking films in their use of VFX to support the claims made here.

Certain Hollywood films (and their associated technologies and techniques of production) are considered "landmarks" not necessarily by virtue of commercial success at the box-office but because they projected new possibilities with regard to the subsequent re-presentational, aesthetic and techno-visual evolution of Hollywood cinema. Some films that were box-office flops during the same period have also been mentioned. While their aesthetic and techno-industrial innovations might have received less attention at the time, they should certainly not be ignored simply because of commercial success or failure. And most of the Hollywood films mentioned have one thing in common: they introduced groundbreaking visuals and techno-industrial applications in one way or another. The time-line also includes movies that might not be considered techno-industrially "innovative" nor aesthetically "original" upon release, but they function as markers of continuity and on-going activity in between the emergence of "landmark" films (and their associated technologies and techniques of production). In other words, these films provide an indication of the pace at which transformation was occurring in Hollywood: that is from the "landmark" status of techno-visual innovation to that of normalization—meaning that the state of "innovation" ceases to be through repetition and reproduction, and becomes "normal" by virtue of saturated industrial application, usage and exhibition.

NOTES

1. Andrew, *Concepts* 128. Emphasis added.
2. Allen & Gomery, *Film History* 25.
3. Allen & Gomery, *Film History* iv.
4. Miller, 'Introduction'—*A Companion to Film Theory* 3.
5. 'Editorial Note'—*American Cinematographer* 6.
6. The filmic commentary of very early cinéphiles—who were often simultaneously theorists, critics and passionate fans of the new art of cinema—would eventually influence later critico-theoretical discourses about the cinema. See: Ricciotto Canudo (1877–1923), *L'Usine Aux Images;* Vachel Lindsay (1879–1931), *The Art of the Moving*

Picture. See also: Barthes, 'Upon Leaving the Movie Theatre'; de Baecque, 'Canudo'; Elsaesser, 'Cinéphilia'; Keathley, *Cinéphilia*; Rosenbaum & Martin, *Movie Mutations*.

7. Pierson, *Special Effects* 2.
8. Balio, 'Introduction'—*Grand Design* 8.
9. Balio, 'Surviving the Great Depression' 21.
10. Balio, 'Surviving the Great Depression' 23.

Ruth Vasey reminds us that 'the financial headquarters of the companies remained in New York City in order to facilitate movie distribution to both the large population centers of the East, after 1918 to Europe.'
Vasey, 'Hollywood Industry Paradigm' 292.

11. Huettig, *Economic Control* 295.
12. See Rudolf Arnheim (1904–2007), *Film als Kunst*; Béla Balàzs (1884–1949), *Theory of Film*; Sergei Eisenstein (1898–1948), *Selected Works*; Siegfried Kracauer (1889–1966), *Theory of Film*; Hugo Münsterberg (1863–1916), *The Film*.
13. Arnheim, *Film as Art* 55.
14. Eisenstein, *Selected Works* 67.
15. Münsterberg, *The Film* 45.
16. Braudy & Cohen, 'Preface' xv.
17. Andrew, *Concepts* 10–11.

Jean Mitry's *Esthétique et Psychologie du Cinéma* essentially addressed the two experiences offered by every film to its spectators: that of recognizing something they can identify with (which corresponds to the realist trend in film theory and criticism) and that of constructing something worth identifying with (which corresponds to the formalist trend in film theory and criticism). Christian Metz, on the other hand, introduced an entirely new theoretical paradigm in film studies, namely the semiology of film.

18. See Baudry, 'Ideological Effects'; Brakhage, 'Metaphors on Vision'; Henderson, 'Toward a Non-Bourgeois'; Metz, *Essais sur la signification*.
19. Lapsley & Westlake, *Film Theory* vii.
20. A notable exception was when *Cahiers du Cinéma* writers in the late 1950s—François Truffaut, Jean-Luc Godard, Jacques Rivette and others—analyzed and celebrated Hollywood films and filmmakers of the studio system era in developing "la politique des auteurs," better known as "auteur theory," for instance.
21. For an overview of "post-1968" tendencies in critico-theoretical discourses about the cinema, see for instance: Baudry, 'The Apparatus'; Carroll, *Toward a Structural Psychology*; Haskell, *From Reverence to Rape*; Mast, *Film/Cinema/Movie*; Metz, *Film Language*; Metz, *The Imaginary Signifier*; Mulvey, 'Visual Pleasure'; Wollen, *Signs and Meaning*.
22. Andrew, *Concepts* 3.
23. Rodowick, *The Virtual Life* 9.
24. Bazin, *What Is Cinema? (vol. I)* 22.
25. Bazin, *What Is Cinema (vol. I)* 20. Emphasis added.
26. See Appendix A.
27. Bazin, *What Is Cinema? (vol. I)* 21.
28. Bouquet, 'Est-ce encore du cinéma?' 20.
29. Nicholls, *Fantastic Cinema* 12.
30. Manovich, 'What Is Digital Cinema?' (online).
31. 'Advertisement for *Cinefex* subscription'—*American Cinematographer* 25.
32. Buscombe, 'Notes on Columbia Pictures' 18.
33. Vasey, 'Hollywood Industry Paradigm' 287.
34. Keane, *CineTech* 4.
35. Buscombe, 'Notes on Columbia Pictures' 18.
36. See Bordwell, Staiger & Thompson, *The Classical Hollywood Cinema* xiii.
37. Bordwell, Staiger & Thompson, *The Classical Hollywood Cinema* xiii.
38. Smith, 'Theses on the Philosophy' 3.
39. See Gomery, *The Hollywood Studio System*.

40. As a result, most other styles—whether on a national commercial level (for example, German Expressionism), on the level of the 'personal commercial (for example, Yasujiro Ozu) or on an experimental independent level (for example, Surrealism, New American Cinema, Michael Snow)—have generally been seen as *alternatives* to this style.'
Thompson, *Exporting Entertainment* ix.

41. Allen, 'William Fox Presents' 128.

42. To better understand the Fordist conceptualization of the Hollywood cinema industry, see: Bohn & Stromgren, *Light and Shadows*; Bordwell & Thompson, *Film Art*; Cook, *A History of Narrative Film*; Gianetti, *Masters of the American Cinema*; Jacobs, *The Rise of the American Film*; Hampton, *History of the American Film*; Thomson, *America in the Dark*; Mast, *A Short History*.

43. Taylor & Saarinen, *Imagologies* 61.

44. Balio, 'Introduction'—*Grand Design* 7–8.
Tino Balio tells us that

> no one studio had the capacity to produce a year's supply of pictures for its subsequent-run theaters, especially after the *double-feature vogue*. A subsequent-run theater typically showed double bills that changed two or three times a week. Because these theaters required as many as *three hundred pictures a year*, the majors needed supplemental products, particularly inexpensive class-B pictures to fill the bottom half of the double bill.

Balio, 'Introduction'—*Grand Design* 7–8. Emphasis added.

45. Jewell, 'Howard Hawks' 46–47.

46. Schatz, 'A Triumph of Bitchery' 74.

47. Rosten, *Hollywood, the Movie Colony* 255. Emphasis added.

48. Neale & Smith, 'Introduction'—*Contemporary Hollywood Cinema* xvi.

49. Neale & Smith, 'Introduction'—*Contemporary Hollywood Cinema* xvi.

50. Balio, *United Artists* 3.

51. See Kramer, 'Post-classical Hollywood.'

52. Schatz, 'The New Hollywood' 9.

53. Harbord, *The Evolution of Film* 61.

54. Cubitt, *The Cinema Effect* 2.

55. Pierson, *Special Effects* 7.
Notable instances of critico-analytical analysis and scholarly writing pertinent to digital VFX and digital cinema (some of which are addressed here) include: Bouquet, 'Est-ce encore du cinéma?'; Buckland, 'A Close Encounter'; Chanan, 'The Treats of Trickery'; Craig, 'Establishing New Boundaries'; Cubitt, 'Introduction: *le réel c'est l'impossible*'; Cubitt, *Digital Aesthetics*; Cubitt, *The Cinema Effect*; Dahan, '*Dark City*: l'écran noir'; Dahan, '*Matrix*: le règne'; Flanagan, 'Mobile Identities'; Gunning, 'Now You See It'; Hamus, 'Astérix'; Hamus, 'Retour vers le passé'; Hayward, 'Industrial Light and Magic'; Jones, 'Hollywood et la saga du numérique'; Keane, *CineTech*; Klein, *The Vatican to Vegas*; Lounas, '*Men In Black*'; Lunenfeld, *The Digital Dialectic*; Manovich, 'What is Digital Cinema?'; McAlister, *The Language of Visual Effects*; Ndalianis, *Neo-Baroque Aesthetics*; Neale, 'Hollywood Strikes Back'; Pierson, 'CGI Effects in Hollywood'; Pierson, 'No Longer State-of-the-Art'; Pierson, *Special Effects*; Robins, *Into the Image*; Prince, 'True Lies'; Sickels, *American Film*; Spielmann, 'Aesthetic Features in Digital Imaging'; Telotte, 'Film and/as Technology'; Tesson, 'Éloge de l'impureté'; Turim, 'Artisanal Prefigurations of the Digital'; Tryon, *Reinventing Cinema*; Wark, 'Lost in Space.'

56. The economic rationale of VFX films is also addressed although in very limited terms, mainly because of the unreliability of film industry accounting wherein some costs (such as marketing and distribution) are often manipulated to hide real profits and losses.

57. Cubitt, *The Cinema Effect* 5.

58. Winston, *Misunderstanding Media* 183.

59. See Winston, *Misunderstanding Media*; see also Winston, *Media, Technology and Society*.
60. Winston, *Misunderstanding Media* 183.
61. See: 'A Chronology of Computer History' (online); Barrett, *The State of Cybernation*; de Sola Pool, *Technologies of Freedom*; Dummer, *Electronic Inventions*; Hafner & Lyon, *Where Wizards Stay Up Late*; Handel, *The Electronic Revolution*; Neale, *Cinema and Technology*; Pennings, *History of Information Technology* (online); Ryan, *A History of Motion Picture*; Salt, *Film Style*; Spufford & Uglow, *Cultural Babbage*; 'Timeline for Inventing Entertainment' (online); Wiener, *The Human Use*; Winston, *Misunderstanding Media*; Winston, *Media Technology and Society*; Wulforst, *Breakthrough*.
62. Kaplan, 'The State of the Field' 86. See also Kittler, *Gramophone, Film*.

THREE

"Effects" in Hollywood Cinema

By 1902, Georges Méliès's *Le Voyage Dans La Lune* had already articulated a pivotal function for VFX in the cinema: to enable the visual realization of concepts and ideas that would otherwise have been, in practical and logistical terms, too risky, expensive or plain impossible to capture, re-present and reproduce on film according to so-called "conventional" motion picture recording techniques and devices. Since then, VFX—in conjunction with their respective techno-visual means of re-production—have gradually become utterly indispensable to the array of practices, techniques and tools commonly used in filmmaking as such.

In enabling moving pictures, the cinema at the beginning of the 20th century was fundamentally conceptualized in terms of "special effects" before their actual realization, especially by virtue of the fact that it transcended the limitations and expanded the capabilities of its precursor, photography. 'It is in fact essential to know that cinema in its entirety is, in a sense, a vast *trucage*, and that the position of *trucage*, with respect to the whole of the text, is very different in cinema than it is in photography.'[1] Once cinema had materialized as the exhibition on a white screen of recordings on celluloid, it ceased to be considered—at least from a filmmaking perspective—in terms of "special effects." Instead, any kind of direct intervention that in any way visually altered what was literally recorded by the camera was considered a "special effect." Hence, in the very early days of cinema, "special effects" would seem to have been conceptualized as an intrinsic, rather than as a separate, component of cinematography and filmmaking. "Special effects" were conceived as the application of visual trickery during cinematography to achieve illusions on screen, to create whatever was deemed impossible in physical terms to re-present and re-produce visually or beyond the imaginative capabilities of most people. Long before the availability of optical printers, all visual tricks were executed entirely inside the camera, before the filmstrip was chemically processed. Some of these in-camera effects involved multiple passes of the film strip, while others, such as "glass shots," were completed in one exposure. As used by Georges Méliès (1861–1938)—who was after all a professional magician before making films—at the beginning of the 20th century, "special effects" were commonly described as "trick photography," referred to as *trucages* in the French context, and later described as "in-camera visual effects."

For Christian Metz, *trucage* 'exists only when there is *deceit*' and he establishes a distinction between 'profilmic *trucages*'—which involve 'a small machination which has been previously integrated into the action or into objects in front of which a camera has been placed. It is before shooting that something has been "tricked"'—and 'non-profilmic *trucages*'—which 'come into play at another point in the production of film. They belong to the filming, not to the filmed.'[2] He further differentiates between ways of interpreting profilmic and non-profilmic *trucages*, from the perspective of film "reading":

> Some *trucages* are imperceptible, while others are on the contrary meant to be discernible (accelerated motion, slow motion, etc.). *Imperceptible trucages*, moreover must not be confused with *invisible trucages*. Resorting to a stunt man is an imperceptible *trucage*. [If] successful, the spectator will not notice that there has been a *trucage*. . . . Invisible *trucage* is another matter. The spectator could not explain how it was produced nor at exactly which point in the filmic text it intervenes. It is invisible because we do not *see* it. . . . But it is perceptible, because we perceive its presence, because we "sense" it. . . . The spectator who is accustomed to cinema, and who knows the rules of the game, has at his disposal three apprehensible systems which correspond respectively, in the film, to imperceptible *trucages*, to visible *trucages*, and to perceptible but invisible *trucages*.[3]

"Special effects" in the early days of cinema were understood from a "filmmaking perspective" to be visual manipulation by means of cinematographic cheating or "trick photography"; while, from a film "reading perspective," they were understood as being either "profilmic" or "non-profilmic" and in terms of their degree of (in)visibility and (im)perceptibility on screen. How the production of "special effects" and their "reading" evolved over time is linked to the intertwined development of filmmaking, of cinema language and of film viewing practices.

With the subsequent evolution of filmmaking technologies, *trucages* or "trick photography" became better known as *effets spéciaux* or "special effects." Once visual manipulation could be separately added to film in post-production, as opposed to being created during cinematography, "special effects" started to become a distinct component of filmmaking by virtue of how they could then transcend the limitations and expand the capabilities of their precursor "trick cinematography." Numerous methods, devices and practices continuously emerged with regard to how "special effects" could be devised: controlled explosions, miniature models, mechanical armatures, stop-motion animation, multiple exposure, matte paintings, optical printing, and so on. As both 'profilmic' and 'non-profilmic *trucages*,' these "special effects" can be conceived as "process" or "technique," on the basis of human skills, that is, transformative tools and technologies required in their creation and application. Within thirty years of their inception at the beginning of the 20th century, filmic effects gradually became the norm rather than the exception in Hollywood cinema: distinct from, yet intrinsic to, filmmaking practice.

Because "special effects" in the analog era could *only* be created and produced in close association with filmmaking technologies and were most commonly associated with the production of VFX-intensive films in particular, "special effects" practitioners were few and most of them usually "belonged" to film production studios rather than being independent. As explained by Michele Pierson,

> the industrial organization of the special effects industry during the middle decades of the last century, when the major Hollywood studios formed their own special effects departments and the rationalization of labor, technology, and skills that followed eventually resulted in the development of specialist sub-disciplines.[4]

"Special effects" production indeed turned into a highly specialized industrial domain by virtue of the increasing complexity of filmic effects required and demanded by directors and producers, leading to the development of additional technologies parallel to those of filmmaking, and to the emergence of specialists and experts, often as freelance effects supervisors and producers, performing the sole function of creating filmic effects. As such, the production of "special effects" not only constitutes a practical application for re-presenting and re-producing some desired visual phenomenon but also a *bona fide* process of artistic creation and aesthetic transformation in the cinema.

Towards the end of the 20th century, the progressive digitization of filmic effects further modified the conceptualization of "special effects." Digitization further expanded the range of creation, manipulation and illusion possible. It also increased the number of tools and technologies available for (re-)producing filmic effects. Digitization had the impact of making filmic effects producible outside the soundstages of movie studios since digital effects do not necessarily require the direct use of filmmaking equipment or of physical sets in their creation and production. They are often created and produced in the virtual domain independently of the filmmaking domain before being inserted into film during post-production, provided that the analog film is first digitized. CGI and digital "special effects," created within the entirely non-filmic realm of computer imaging and cyberspace, have in turn transcended the limitations and expanded the capabilities of their analog "special effects" precursor. As such, digital "special effects" first became a distinct component of the predominantly analog filmic effects domain, as well as a distinct component of filmmaking. With the digitization of the filmic effects domain influencing the digitization of filmmaking at large, digital filmic effects became the norm rather than the exception in Hollywood cinema. Accordingly, "special effects" in their digital form should hence again be conceived as intrinsic to, rather than separate from, filmmaking—as was the case originally during the early days of cinema at the beginning of the 20th century. In many ways, the conceptualization of digital "special effects" follows a fundamentally similar evolution to that of its analog precursor.

"Special Effects"? "Special Visual Effects"? "Visual Effects"?

The term "special effects" can refer to different things depending on its use according to particular discursive orientations. What makes "special effects" 'special'? Does the term automatically imply all the possibilities and potential of its associated technologies? Can it be subsumed under the sum total of its associated techniques as applied in the pursuit of better "realism" and storytelling in the cinema? Or are "special effects" just visual trickery to deceive an audience? Terms used to designate different forms of filmic effects in Hollywood have been abundant over the years. The term "effects," as used in the Hollywood cinema industry, has been conjoined with a number of other terms, all of which were not automatically intended to refer to exactly the same thing. The most common ones are: "*special* effects," "*special visual* effects" and "*visual* effects."

Although the practical use of filmic effects in American motion pictures precedes the embryonic stages of Hollywood's constitution as a substantial cinema industry, it was only in 1939 that a special award would be given to *Spawn of the North* for 'outstanding achievement in creating *special photographic* and sound effects,' at the 11th Academy Awards ceremony.[5] And it was at the 12th Oscars ceremony in 1940 that an Academy Award for "*Special* Effects" was officially created and awarded for the first time to *The Rains Came*. This particular category would prevail at every ceremony until the 36th in 1964.[6] In 1965, the award's appellation for the filmic effects category would be stretched to "*Special Visual* Effects" and presented annually as such until 1972. From 1973 to 1977, no film was nominated in that category. Instead, "Special Achievement Awards in *Visual* Effects" were bestowed by the Academy to specific films (except in 1974 when there was no effects-related award of any kind): *The Poseidon Adventure* in 1973, *Earthquake* in 1975, *The Hindenburg* in 1976, *King Kong* and *Logan's Run* in 1977—all "big picture" movies with large budgets. When the effects category re-appeared as "*Visual* Effects" in 1978, *Star Wars* collected the award against its only rival nominee *Close Encounters of the Third Kind*. This turn of events highlighted and inaugurated the forthcoming domination by filmic effects provider Industrial Light & Magic of this Academy Award category and of filmic effects production in the Hollywood cinema industry over the next two decades. Since 1978, the "*visual* effects" category has prevailed at every Academy Awards ceremony, with the exception of the four years when "Special Achievement Awards in *Visual* Effects" were again awarded instead to specific films: *Superman* in 1979, *Star Wars: The Empire Strikes Back* in 1981, *Star Wars: Return of the Jedi* in 1984 and *Total Recall* in 1991.[7]

As the organizer of the annual Academy Awards ceremony, amongst other things, the Academy of Motion Picture Arts and Sciences or AMPAS establishes, maintains and publicizes professional standards of all kinds that are associated with the Hollywood cinema industry, as shaped by the combined artistic, techno-industrial and economic requirements of commercial filmmaking. Over many decades, AMPAS has officially employed a number of terms for designating filmic effects, each time setting a "standard" for filmmakers in the industry to follow.

As part of its online resources for school teachers, AMPAS indicates that the term "special effects" first appeared in screen credits in the 1926 Fox film *What Price Glory?* and further explains that there are

> two major types of special effects. Visual effects include all types of image manipulation, whether they take place during principal photography or in post-production. Physical effects, also called practical effects, are performed live using "real world" elements. These include explosions, weather effects and stunts.[8]

In a similar vein, for Steve Blandford et al., "*special* effects" are

> artificially contrived effects designed to create the illusion of real (or imagined) events, whether through special photographic effects or created, and recorded normally by a camera. . . . The term is usually reserved for three main classes of effects: mechanical effects such as simulated explosions, fires, floods, storms; illusion created by so-called 'trick photography' either in-camera, during shooting, or via the optical printer . . . ; and increasingly, the major use of the term, effects achieved on the photographic image by digital means, from electronically programmed motion control to computer graphics and animation.[9]

While David Bordwell and Kristin Thompson simply consider "*special* effects" as a 'general term for various photographic manipulations that create fictitious spatial relations in the shot,' on the other hand, James Monaco explains that "*special* effects" is 'a rather dull label for a wide variety of activities, each of which has direct creative potential. The craft of special effects rests on three premises: (1) film need not be shot continuously, each frame can be photographed separately; (2) drawings, paintings, and models can be photographed in such a way that they pass for reality; (3) images can be combined.'[10]

Techno-industrial discourses tend to be somewhat more "precise" with regard to the term "*special* effect"—as illustrated by Thomas Smith (ex-VFX supervisor at Industrial Light & Magic) who further fragments and specifies the definition of the "*special* effect" as 'any shot that uses optical or mechanical devices to create an illusion on film. In location filming, *mechanical* effects are sometimes called practical or physical effects. *Optical* effects are achieved using the optical printer during post-production,' while *digital* effects are obtained virtually through the application of specific computer software and programming.[11] And he considers the "*visual* effect" shot to be 'any visual manipulation of motion picture frames, whether accomplished in cameras, projectors, optical printers, aerial image printers, front- and rear-screen processes, and so on.'[12] From a techno-industrial point of view then, the main concern has to do with the *process* of performing a visual manipulation that will have a specific, intended visual *effect* or impact on the movie as well as on its viewer.

While "*special* effects" have traditionally been understood as occurring during production, either as 'mechanical effects' physically created and recorded "live" (such as mechanized props or scale models shot on the studio soundstage) or as 'optical effects' obtained purely in-camera (such as with multiple exposure or slit-scan photography), "*visual* effects" are the result of the direct frame-by-frame post-production transformation of already recorded moving images—by the intermediary of analog and digital film and image processing technologies (such as optical printing or compositing). But as Renee Dunlop et al. indicate, digital

> visual effects technologies blur the line between pre-production, production and post, almost the exact opposites as it was in the 1930s. This new reality leads to a far more active and collaborative role for VFX supervisors and artists across every phase of production. . . . In short, visual effects are no longer limited to post-production. The options . . . now allow for anything to be created anywhere at any time, across all time zones and all phases of production.[13]

Indeed, associating "*special* effects" and "*visual* effects" with the production and post-production phases respectively is a technical distinction that has progressively become obsolete because of techno-industrial digital convergence in filmmaking—as illustrated by the substantial application of computer programming, software and other image processing technologies that used to be solely dedicated to filmic effects production but now permeate all phases of film production operations in the Hollywood cinema industry.

Except in official film credits and on some websites, as well as in certain publications authored by effects industry professionals, the subtle techno-industrial distinctions between "*special* effects," "*visual* effects" and "*special visual* effects" are rarely upheld. Cinéphilic and critico-theoretical discourses about Hollywood cinema often use "*special* effects," "*visual* effects" and "*special visual* effects" interchangeably to refer to the same thing, or else they use one particular term—predominantly "*special* effects"—to refer to any form of film-related visual trickery or illusion. And techno-industrial discourses tend to be more specific in defining filmic effects in terms of techniques applied and visual outcomes achieved. While attempting to delineate the different ways in which the terms "*special* effects," "*visual* effects" and "*special visual* effects" are understood, differ from or relate to one another, it should be acknowledged that their etymological essence lies not in their separate individual conceptualizations and definitions but in the very combination of these. It is nevertheless practical to "separate" these to obtain a better sense of what is at stake: crucially, individual conceptualizations and definitions of these three terms tend to be overlapping rather than singularly distinctive.

Instead of filtering down the variety of jargon or attempting tedious etymological titrations of the plethora of terms used in various discourses when referring to filmic effects in the Hollywood cinema industry, and to minimize terminological confusion between the blurred categories and techniques of analog and digital filmic effects, the term "*visual effects*" or VFX has been chosen to solely designate the object of enquiry here. "*Visual* effects" is understood in this context as a conceptualization of *all* forms and types of filmic effects combined into one specific grouping of analog and digital filmic components and filmmaking tools relating to a purely *visual* aspect of cinema –"*visual* effects" as filmic manipulation of any kind to create visual illusions in their final on-screen materialization perceived and judged *visually* by the spectator. And "*visual* effects" is of more contemporary currency in the Hollywood cinema industry—as used by AMPAS and at the Academy Awards for instance—as opposed to the more archaic usage of "*special* effects" and "*special visual* effects" to refer to filmic effects in Hollywood movies. These two terms are considered "archaic" here because the sense of the "special" in "*special* effects" requires exclusion primarily as a result of effects becoming a *normal,* as opposed to *special,* aspect of filmmaking and motion picture production in the Hollywood cinema industry.

The Impact of the Digitization of VFX

Before the Hollywood cinema industry's entry into the digital realm, films were usually made according to the mechanical/physical possibilities and limitations of the filmic medium and available analog technologies. Hence VFX from the pre-digital era, although showing signs of visual "aging" when viewed within a contemporary hi-tech digital mediascape, sometime seem cinematically more "realistic" and "credible." *Jaws* and *Close Encounters of the Third Kind*, for instance, employed only mechanical and physical effects, which explains why they look neither too "artificial" nor too "real" but oscillate between these two poles. Had they been released in the hi-tech digital mediascape of the late 1990s, these two movies might arguably not have been as commercially successful as they originally were in the 1970s.

With its importance in Hollywood cinema becoming increasingly obvious and inevitable by the late 1980s, the advent of techno-industrial digitization was beginning to be accepted fairly rapidly in the VFX production industry as a logical updating and upgrading of analog VFX. CGI in particular revolutionized in every possible way VFX production as well as the overall movie business.[14] Some companies feared that CGI and digitization would mean the end of the then analog-oriented VFX industry.

> Many people regard Hollywood studios as aging dinosaurs looking for a place to fall down, kept alive by clinching their younger and less experienced opponents until the latter give up the ghost, or until they consent to be permanent sparing partners. On the other hand, computer animation companies, such as RGA, Digital Domain, Metrolight, or DreamQuest Images, which are considered part of the genre of New Media, have all grown up in a hurry as a result of the hard and frequent beatings each have been forced to endure.[15]

The smarter companies learned to work with the new technology, combining it with traditional analog VFX techniques to create even more remarkable results. Even if functions and desired objectives might have subsequently appeared to remain fundamentally similar, technique and execution became faster in the digital realm with end results often looking much more sophisticated. Analog VFX technologies and techniques were gradually converted to the digital medium, to the point of informing and regulating all the subsequent processes and stages of VFX production: the ways in which illusions should be re-presented and re-produced, how the "realistic" should be creatively simulated; what direction VFX would follow in terms of their future evolution. The introduction and "assimilation" of increasingly more sophisticated digital technologies re-formatted and re-standardized what VFX *should look like* at a particular point in time. Digitization led to a re-definition of what counts as acceptable visual standards for VFX in the Hollywood cinema industry, noticeably improving not only the "look" of VFX but also the "look" of the Hollywood motion picture in general.

Creating and producing in digital, compared to analog, allows VFX to be designed and conceived independently in the virtual domain before being inserted into film during post-production. Hence, digital VFX imply the creation and (re-)production of cinematic effects in purely visual, non-material terms within the essentially non-pro-filmic digital realm of computer interfaces. Computer-generated landscapes, objects, "creatures" and characters can be added, deleted and infinitely manipulated frame by frame for narrative, aesthetic or commercial reasons. By the end of the Hollywood cinema industry's "embryonic" phase of digitization in the early 1990s, the digital VFX production domain could be said to have reached a techno-industrial peak with regard to R&D and filmic applications, although high-grade digital VFX were still considered so costly that only blockbuster movies with substantial budgets could actually afford them. But the digitization of VFX had consequences not just for big budget Hollywood productions but also for the averagely budgeted "run-of-the-mill" Hollywood movie. Filmmaker and digital consultant Van Ling explains that

> as the stakes get higher and the technology improves, both the big-budget studio films and the smaller independent films are going to benefit. . . . The technology that is being used by the big guys and many of the new tools, especially the digital ones, are now available to everybody. . . . Whereas the big guys are moving into major paintbox or high-end digital manipulation, the older analog effects mixers and basic digital effects are becoming more affordable and accessible to the smaller guys.[16]

Cheaper and faster production of low-grade digital VFX indeed became quite commonplace at the lower-end of the audio-visual spectrum (in television, music videos, advertising, video games, low-budget filmmaking, and so on) by the early 1990s.

Digital VFX best conjure the feeling of awe and "magic" because of the Hollywood cinema industry's "quest for perfection," usually leaving the viewer dazzled with the "how-did-they-do it?" feeling. As confirmed by Michele Pierson:

> At the beginning of the twenty-first century, computer-generated visual effects are not only a major attraction of Hollywood blockbuster cinema but one that, despite being produced within an industrial context that is more highly rationalized than ever before, continues to be presented to contemporary audiences as magic. The artists, designers, and engineers working in this area of special effects production are revered as wizards and illusionists in the many different forms of media that now report on the current state of the art.[17]

In the very early days of cinema, the sense of amazement and wonder associated with the sight of moving pictures seemed sufficient to attract spectators to theaters. But relying mainly on this kind of techno-visually-driven sense of amazement and wonder to attract audiences could not last long. As those early 20th-century-cinema spectators became used to the sight of moving pictures and became familiar with the filmmaking technologies behind their creation, movies had to offer more and better "pleasures" rather than just being pictures in motion. A similar sort of situation occurred by the end of the 20th century with regard to cinema audiences and digital VFX-intensive Hollywood movies.

The big-budget Hollywood blockbuster movie had been continuously riding waves of groundbreaking technologies being developed for both the VFX production and filmmaking industries, and generating healthy box-office figures, from the late 1980s onwards. 'The early to mid-1990s . . . were the wonder years, a period in which CGI effects became the focus of intense speculation not only for cinema audiences but also for the special effects industry itself.'[18] After the initial sense of amazement and wonder associated with large "doses" of digital VFX in many Hollywood films within a short period of time, a slump in audience interest was predictable. This was the case around the mid-1990s when many digital VFX-intensive Hollywood movies did not actually fare very well commercially. Digital VFX-intensive Hollywood filmmaking was running out of steam, so to speak, due to too many high-budget productions paying more attention to visual gimmickry than to theme, storytelling, characterization and narrative. While digital VFX undoubtedly reinvigorated the tradition of spectacle and display in the cinema, for Barry Purves

> computer technology and image manipulation has now removed the boundaries of what is possible, and when anything is possible, I am no longer thrilled and excited. Knowing there are no limits, my brain is dulled and no longer surprised, let alone involved. I personally am far more in awe of what Georges Méliès achieved a hundred years ago, with his enormous technical restrictions than some of the special effects laden epics today.[19]

For many sections of the Hollywood cinema audience, VFX-intensive movies had indeed caused a film-viewing culture of "surprise" to be superseded by one of over-expectation and over-anticipation of visual spectacle and display.

The massive box-office profits generated worldwide by *Titanic* (1997) brought re-assurance to the VFX production industry and the major corporate players in Hollywood with regards to digital VFX-intensive filmmaking. After *Titanic*, the box-office success of many digital VFX-intensive Hollywood films contributed to a resurgence of hi-tech, complex and expensive digital VFX applications—such as *Contact* (1997), *Men in Black* (1997), *Godzilla* (1998), and *Armageddon* (1998)—an apogee being reached with *The Matrix* and *Star Wars: Episode I—The Phantom Menace* in 1999. These films confirm how the digital has indeed 'furthered a culture of spectacle and immediacy through the creation of fantastical worlds and the effects of superhuman efforts.'[20]

"Dream it and we'll make it" or "anything you want is possible" are indeed applicable mottos to describe the creative—and lucrative—direction taken in Hollywood in the 1990s with regard to the apparently unlimited potential of digital VFX. As Steven Spielberg puts it: 'It is [the creative spirit of Industrial Light & Magic] that has allowed my vision to fly freely, knowing that if I can imagine it, the brilliantly creative minds at ILM will get it on screen.'[21] And this is the kind of attitude that would enable his *Jurassic Park* to be massively successful worldwide. While prehistoric dinosaurs on film have a long history in the cinema, they looked more convincingly "realistic" than ever in *Jurassic Park*. This was essentially the result of various technical innovations achieved during VFX production at Industrial Light & Magic. Innovations included the new Viewpaint program that revolutionized how CG artists created surface-texture maps by allowing the surface of CG dinosaurs to be painted as if it were a real sculpture and the new in-house tool called Enveloping, which allowed animators to make computer-generated flesh movements with all the realism of actual organic skin moving against muscle and bone.[22] The progressive techno-industrial digitization of VFX production in the Hollywood cinema industry has contributed more than anything else to the pursuit of perfection in (re-)producing visual illusion and spectacularity. For instance, *Terminator 2: Judgment Day*'s visually innovative "liquid mercury" T-1000 android was the outcome of Morphing (short for metamorphosis), a new software and technique especially developed to meld film and computer-generated images through digital compositing.[23] Morphing enabled the T-1000 character to change very smoothly from one form to another, the effective continuity making it impossible to detect the boundary of where one form ends and another starts.[24] And in *Total Recall*, one of the most visually exciting illusions is the skeleton security screen in the spaceport sequence on Mars where an X-ray detection device projects the skeleton of Douglas Quaid (played by Arnold Schwarzenegger) onto a black glass shield as he walks along a 30-foot ramp; the half-dozen aqua blue skeletons seen walking across the charcoal screen in this sequence were separately computer-generated by CGI company Metrolight in association with digital VFX provider DreamQuest Images, and later inserted in the film during post-production.[25]

Resorting to expensive digital VFX is not always the best nor the only solution for all Hollywood films. VFX shots could still be delivered efficiently in the "digital era" without the use of expensive "space-age" wizardry. Hence Eric Brevig, VFX supervisor on *Total Recall* and *Men in Black*, points out that

> the more elaborate, high-tech and state-of-the-art your equipment is, the more you'll get sunk by something as simple as a three-color separation that doesn't register! . . . No matter how high falutin' the techniques we come up with are, and some of them are scary in their complexity, a shot like the pullback with all the people is still solved using technology that's been around since 1920![26]

There are many advantages and disadvantages in choosing to work exclusively within either the analog or digital medium to produce VFX in the Hollywood cinema industry, but the balance between the two is ultimately maintained by the reaction of VFX practitioners to the production studio's budget allocation and box-office expectations and to the director's vision and aesthetic requirements: to produce the "realistic" and to create spectacular, attention-grabbing VFX while using less expensive means and as little of the allocated film budget as possible.

Filmmakers and VFX practitioners regard digital technologies as *tools* that *contribute* to a film, but by themselves cannot "make" a film. The best VFX are often achieved through conventional analog techniques and tools combined with emerging digital software and technologies. There are gains and losses either way but the balance between the two approaches is always more or less maintained, ultimately with the benefit of creating the best VFX for specific moments in a film. Janet Wasko, for instance, argues that

> [s]ome of the techniques used by the earliest filmmakers, such as double exposures and miniaturized models are still employed. *King Kong* [1933] represented a landmark in special effects and incorporated many of the same techniques used by today's special effects teams: models, matte paintings for foregrounds and backgrounds, rear projections, miniature or enlarged props and miniaturized sets, combined with live action.[27]

It should be noted that digital VFX are often no more than digitized versions of analog VFX for which software programs have been developed so that they do not need to be produced mechanically or physically in real space but in cyberspace via a computer interface: CG sculpting and modeling, CG matte-painting, digital make-up and so on.[28] While most analog VFX methods have simply been converted into software programming inside digital film workstations, they are nevertheless still applied within the digitized techno-industrial landscape of the Hollywood cinema industry.

Digital VFX have inevitably transformed the nature of how stories can be told onscreen as well as enhanced the element of spectacle in the Hollywood film, as illustrated, for instance, by Ridley's Scott *Gladiator* (2000) and Wolfgang Petersen's *Troy* (2004). Through extensive digital VFX application, these two "sword-and-sandal" epics essentially brought back to life on the screen the long gone fighting arenas of the Roman Empire and the battlefields of ancient Greece respectively. Increasing reliance on digital VFX and computer-generated characters has also influenced significant transformations to the actor's body on screen. As illustrated by David Rodowick for instance,

> technological transformations of the film actor's body in contemporary cinema are indicative of a sea of change that is now nearly complete.

> One could say that the body of the film actor has always been reworked technologically through the use of special makeup, lighting, filters, editing and so on. Contemporary cinema, however, is taking this process to new levels. One of the many fascinating elements of digital cinema is not just the thematic idea of cyborg fusions of technology and the body. Digital processes are increasingly used actually to efface and in some cases entirely to rewrite the actor's body. . . . In *The Mummy* [1999], Im-Ho-Tep is a constantly mutating digital construction that has more screen time than Arnold Vosloo, the actor who plays him in the live-action sequences.[29]

Furthermore, practices associated with digital VFX production—such as blue/green screen, motion-capture or morphing—very often lead actors to perform in empty spaces against physically absent characters or creatures which are later added to the filmic frame during post-production: this becomes particularly obvious when mismatched eye-lines are noticeable between real actors and computer-generated characters, as illustrated, for instance, by scenes involving interaction between computer-generated character Jar Jar Binks and real human actors in *Star Wars: Episode I—The Phantom Menace*. The application of certain digital VFX-related techniques, therefore, led to the emergence of alternative ways of acting in response to a more digital technology-intensive filmmaking environment. But this is not quite a radically new state of affairs: actors similarly had to adapt in the past to innovative techno-industrial and filmmaking practices in the Hollywood cinema industry, for instance when sound recording technology (involving microphone placement constraints) was introduced or when the rear-screen projection technique began to be applied during the studio system era.

Digital VFX do not necessarily create or project anything "new" but, while maintaining an essence of their existing identity, they do multiply the possibilities of the "realistic" in the re-presentation and re-production of "real" objects and characters on screen. For Stephen Keane:

> Looking at aspects of form allows us to look much more specifically at film technique, the ways in which special effects sequences may or may not fit into the design and shape of the film as a whole. Similarly, the aesthetic extends to the perceptual properties of film viewing, principally whether we can see the 'join' or not. The problem with the strictly narrative approach is that any action sequence can be regarded as an interruption. Consideration of form brings us much closer to the ways in which particular aspects of special effects work are combined with existing film techniques: that is the overall fit and feel of effects sequences.[30]

The overwhelming presence of digital VFX in Hollywood films has practically "forced" the cinema spectator's acceptance of other ways to perceive, to see, and to make sense of what is on screen—ways which are visually "appealing" and enable a more complete film viewing experience.

Digitization has transformed visual re-presentation in the Hollywood film by enabling it to bypass the characteristics of the so-called "real" in favor of those of the "realistic"—as literally demonstrated in *Forrest Gump* (1994) for instance. The proliferation of computer-generated digital VFX in Hollywood cinema has indeed nurtured an innate preference for virtual re-presentation and re-production of the "realistic" in terms of illusion. It is worth highlighting here Sean Cubitt's comment that 'distinctions between realism and illusion make no sense in an epoch when it was neither the illusion of life nor the illusion of illusion that fascinated but rather the spectacle of their making.'[31] One could compare the modes of re-presentation in a number of movies with very similar characterization, themes or settings made during very different techno-visual phases in the Hollywood cinema industry: Neame's *Meteor* (1979) and Leder's *Deep Impact* (1998); Browning's *Dracula* (1931), Badham's *Dracula* (1979) and Coppola's *Bram Stoker's Dracula* (1992); Seaton's *Airport* (1970) and Harlin's *Die Hard II: Die Harder* (1990); Negulesco's *Titanic* (1953) and Cameron's *Titanic* (1997); Spielberg's *Jaws* (1976) and Harlin's *Deep Blue Sea* (1999); Wise's *The Day the Earth Stood Still* (1951) and Derrickson's *The Day the Earth Stood Still* (2008).

Digital filmmaking technologies have caused a loss of distinction between the "real" and the "realistic" as a result of all processes of re-presentation and re-production being unilaterally absorbed into the virtual realm predominantly favored by a techno-industrially digitized Hollywood cinema. Via technological mediation, Hollywood cinema's "realistic" virtual visions of natural disasters and calamities (hurricanes, volcanoes, avalanches, collapsing bridges, crashing planes or sinking ships) for instance—all belonging to the immediacy of the physical world of contact—become mere virtual simulacra in the age of digital re-production.

The narrowing gap between the human perception of "reality" and its digital re-presentation, re-production, recording as well as exhibition through the Hollywood film has changed the nature of the relationship between actors (real and synthetic), filmmaking technologies and the film viewing experience, inevitably modifying the liminal connection between the cinema spectator and the filmic text. The ever-increasing techno-visual complexity and sophistication associated with processes of re-presentation and re-production—as influenced by digital filmmaking and digital VFX application in particular—has indeed problematized cinematic analysis. Walter Benjamin has remarked about cinema that 'the equipment-free aspect of reality has become the height of artifice; the sight of immediate reality has become an orchid in the land of technology.'[32] His comment can only acquire more poignancy in the context of the Hollywood cinema industry in the age of digital re-production. Hollywood cinema's visual re-presentations on screen, as transformed by digital VFX, have become increasingly removed from the human perception of "reality" which itself has become less of a source of reference to its cinematic re-presentation and re-production, especially with mediating technological vehicles substituting for this function. Digital VFX hence further reinforce the unreliability of the fundamental indexical relationship between the re-presented and its re-presentation in the Hollywood film.

Re-Presenting and Re-Producing the Ordinary in Extraordinary Ways

Motion pictures as optically formed re-productions of moving objects, such as those formed by a lens or mirror, rely on various technological vehicles in order to be processed and to make the transition from one state to another—the celluloid-based camera, for instance, enables chemical emulsions to capture moving objects as visual re-presentations that are re-produced on screen at projection. The projected moving picture is the outcome of a "live" event or object being processed by the capabilities of the camera (as one of the main technological vehicles of mediation) and of the projector. The cinematographer and the projectionist act as "facilitators" who enable the two elements to merge and physically materialize into a series of images that move on the cinema theater screen. The

> simple notion of the visual image (in film, painting, or other media) is ruled by a plethora of terms that name the types of relation between visual re-presentation and the experience of "reality." Some of these are: designation, denotation, resemblance, expression, exemplification, depiction, representation, signification, imitation, and reproduction.[33]

These terms suggest that the production of the visual image, especially in the cinema, inevitably involves a breakdown of the experience of "reality" in order to better re-compose it on the screen.

Hollywood cinema has unequivocally been driven by the quest for perfection in the manufacture of the "realistic" as filmic optical illusion. This mostly depends upon the possibilities and outcomes associated with blurring the distinction between "real," re-presented and imagined worlds. The techno-visual evolution of Hollywood cinema has subsequently been motivated by the constant objective of blurring "real," "re-presented" and imagined worlds and by the search for the least amount of optical difference identifiable in the illusion produced: by making illusions "realistic."

The continuous demand by film viewers and the Hollywood cinema industry alike for the moving picture's better proximity to the "real" world has influenced the techno-visual evolution of filmmaking technologies toward emulating how human senses react, and operate in relation, to so-called "real" environments.

> The desire of contemporary masses to bring things "closer" spatially and humanly . . . is just as ardent as their bent toward overcoming the uniqueness of every reality by accepting its reproduction. Every day the urge grows stronger to get hold of an object at very close range by way of its likeness, its reproduction.[34]

The cinema as a mode of visualizing essentially implies, to use Dudley Andrew's term, "seeing as."[35] There are two paradigms to be taken into account with regard to "seeing as": (1) the perception of a cinematic representation of "reality" that is technology-dependent in creating the right conditions for a specific viewing experience, and (2) its source of reference, namely, real human sense perception and experience as founded in everyday "reality." Hence, film functions to re-present and re-produce as accurately as possible, through a techno-visually shaped cinematic "reality."[36]

The cinema's ways of "seeing as" have fundamentally been implemented through two broad approaches by most filmmakers, as identified by André Bazin (1918–1958) in his examination of the cinema between 1920 and 1940:

> Those directors who put their faith in the image and those who put their faith in reality. By "image" I here mean, very broadly speaking, everything that the representation on the screen adds to the object there represented. This is a complex inheritance but it can be reduced essentially to two categories: those that relate to the plastics of the image and those that relate to the resources of montage.[37]

From its inception, the cinema's progress—coterminous with the creation of alternative ways of perceiving, of seeing—has been predominantly techno-scientifically-driven: 'the full machinery of cinema, the cinema as an invention of popular science, ensures that we can see anew, see more but also see in the same way.'[38] But filmmaking technologies have always been fundamentally very limited and constricted with regard to how much they can in fact re-present and re-produce. The evolution of representation and re-production on the cinema screen could then only occur through techno-scientific progress in combination with economic and aesthetic pursuits, creative flair and artistic ingenuity. Since its emergence, the cinema thus evolved from being a mere recording of "reality," in the technicist Lumière sense, towards the creative re-presentation and re-production of "reality"—through 'film-*making*, make-believe, an aesthetic situation'—as implemented by Méliès for instance.[39]

The Hollywood film, as a manufactured and profit-oriented product, is essentially meant to give the average Cineplex film viewer his or her "money's worth" by dazzling, exciting and entertaining him or her in one way or another for a couple of hours. Techno-visual developments in the Hollywood cinema industry have hence not only been oriented towards simply making the real look more colorful, more clearly defined and more "realistic," but have involved a quest for more audiences to recover capital investment. This meant a continuous removal of the barriers between the "real" and the re-presented through constant improvements in the re-production of the "realistic."

Techno-visual developments in Hollywood have been driven as much by a "realist" project as by an illusionary one. The illusion of the spectacular always needed to be more convincingly "realistic." Bazin notes that 'studio reconstructions reveal a mastery of trick work and studio imitation—but to what purpose? To imitate the inimitable, to reconstruct that which of its very nature can occur only once, namely risk, adventure, death.'[40] Bazin's comments are particularly pertinent to the Hollywood cinema industry's fixation with re-presenting and re-producing the illusion of the "realistic" through the use of studio-based VFX, long before digitization. This is confirmed by RKO's optical printer work supervisor Linwood Dunn in 1943 referring to the extensive use of VFX during the making of *Citizen Kane* (1941): 'The picture was about 50% optically duped, some reels consisting of 80% to 90% of optically-printed footage. Many normal-looking scenes were optical composites of units photographed separately.'[41] In relation to the "realistic" worlds of Hollywood cinema, both filmmaker and spectator are fully aware that what ends up on screen is illusion and that its purported "reality" is ingrained in the imaginary of its make-believe worlds. 'The point is not whether we can achieve a certain distance and detachment from the fearful principles of reality, but whether we can ever become reconnected to a world that we no longer take for real, a world whose reality has been progressively screened out.'[42] The Hollywood film is ultimately the corollary of cheating optically to produce the illusion of the "realistic" and of the viewer's complicity in being optically cheated during a film's screening.

Moving pictures in general (in cinema, video, television) now saturate and govern the flow of human communication to such an extent that the crux of (post-)modernity can be deemed to have become principally visual: 'New image technologies are facilitating greater detachment and disengagement from the world. Vision is becoming separated from experience, and the world is fast assuming a de-realized quality.'[43] But if so-called "normal" 'three-dimensional space is already debased and impoverished by virtue of an excess of means (all that is real, or wants to be real, constitutes a debasement of this type),' as pointed out by Jean Baudrillard, attempting to flawlessly re-present, re-produce and project three-dimensional "real" worlds onto two-dimensional visual media can only remain inherently hypothetical or ideal.[44] And for every techno-visual enhancement, there is often also a reduction in other features of the film viewing experience. 'To magnify some observed object, optically, is to bring it forth from a background into a foreground and make it present to the observer, but it is also to reduce the former field in which it fit, and—due to foreshortening—to reduce visual depth and background.'[45]

Other than striving to look less "synthetic" in re-producing the "realistic," Hollywood films have additionally moved towards overwhelming the viewer with spectacularity, by displaying the ordinary in extraordinary ways. The quest for ever-more spectacle has been the driving force of a century of continuous innovation and techno-visual sophistication in the industry. It is in this respect that digital VFX in particular have made one of the biggest contributions to both the aesthetic and techno-visual evolution of the Hollywood film: perfecting the process of creatively re-presenting and re-producing the ordinary in extraordinary ways. The application of VFX in Hollywood filmmaking reflects a number of techno-industrial filmic "quests" for perfection: the perfect illusion, the perfect motion, the perfect shape, the perfect figure, the perfect color, the perfect texture, the perfect visual and so on. Digital VFX constitute a form of 'perceptual deviation' that, in the cinema, 'makes possible the conferring of value on this or that aspect of perception, the filtering here and enlarging there which makes a representation significant.'[46]

Verisimilitude is an important component of the shaping of the Hollywood film's "realistic" character. Considered as a 'complex construction of signs, not a privileged mode of knowledge connected to the nature of things' by Dudley Andrew, "verisimilitude" for Steve Neale 'means "probable," "plausible" or "likely." In addition, it entails notions of propriety, of what is appropriate and *therefore* probable (or probable and therefore appropriate).'[47] According to Tzvetan Todorov,

> a work is said to have verisimilitude in relation to two chief kinds of norms. The first is what we call rules of the genre: for a work to be said to have verisimilitude [in re-presenting "reality"], it must conform to these rules. . . . But there exists another verisimilitude, which has been taken even more frequently for a relation with reality. . . . The relation is here established between the work and a scattered discourse that in part belongs to each of the individuals of a society but of which none may claim ownership; in other words to public opinion. The latter is of course not "reality" but merely a further discourse, independent of the work.[48]

The state of verisimilitude, in shaping the "realistic" in Hollywood cinema, has been tremendously influenced by popular understandings and demands for visual spectacularity—as well as by the requirements of industrial economic pursuits to meet those popular demands—with the technological digitization of filmmaking and of VFX enhancing the industry's particular ability to simultaneously polish, extend and transform the "realistic" into visual spectacles for on-screen display.

The credibility ascribed by the viewer to the "realistic" in the Hollywood film depends upon the success of its internal logic in resolving ambiguities presented by its "true-to-the-probable" stories to consequently make them believable in terms of verisimilitude. The "true-to-the-probable" approach to storytelling and visual re-presentation in VFX-intensive Hollywood films in particular indicates 'an aesthetics of "studio production" in which the story space becomes more a matter of possible worlds, fantasy worlds, jumbled worlds made of bits and pieces of existing worlds put together in the studio and the computer.'[49] In addressing their own artificial and fictive narrative spaces in accordance with a studio-based production aesthetic, the "true-to-the-probable" worlds of VFX-intensive Hollywood films develop an internal logic that turns any ambiguity felt by the viewer into an elegantly appropriate solution by creatively re-presenting and re-producing the ordinary in extraordinary ways. Spectacularity in VFX-intensive films in particular remains credible insofar as Hollywood cinema's "realistic" worlds are anchored in some version of ordinary everyday "reality," as can be identified in many of Steven Spielberg's VFX-intensive films for instance. What his films often have in common are: (i) the theme of how an ordinary person transcends limitations that he or she expects to find and becomes a hero, a martyr, or an adventurer; (ii) the context of American middle-class suburbia. There are indeed more similarities than differences between the suburban settings, main and background characters of *Jaw*'s Amity in New England, of *Close Encounters of the Third Kind*'s Muncie in Indiana, and of *E.T. The Extra-Terrestrial*'s un-named Northern California town. These three films are 'at once a glorification of suburbia and a pointer to escape routes from that class.'[50] With regards to *E.T. The Extra-Terrestrial*, Robert Conn confirms how it is

> set in typical American suburbia, an environment that Spielberg knows well. It is where he grew up and where he started to tell his stories on film. He seems to enjoy injecting the cosmos into people's backyards, juxtaposing the ordinary with the extraordinary and exploring the way people would deal with it.[51]

And Spielberg's interest in American suburbia would still be cropping up again more than twenty years later in his *War of the Worlds* (2005) remake. This grounding in the ordinary, everyday "reality" of American middle-class suburbia is part of the explanation to why many of VFX-intensive Spielberg movies tend to have so extraordinarily wide an appeal, especially with domestic U.S. audiences. In the case of *Close Encounters of the Third Kind* and *E.T. The Extra-Terrestrial* in particular, they were also grounded in the pro–science fiction atmosphere that was getting stronger by the late 1970s to early 1980s, at least in the United States—as indicated for instance by James Berardinelli in his review of *Close Encounters of the Third Kind*:

> Although UFOs have been a popular subject for speculation, rumination, and investigation for more than 50 years, at no time was the phenomenon more popular than during the 1970s. . . . There were plenty of skeptics, but many people, including those who had never purported

> to see anything out-of-the-ordinary, wanted to believe. Maybe it had something to do with the world order being so bleak (racial tension, Vietnam, the Cold War), but more and more men and women looked to the stars to find hope.[52]

Hence, *Jaws*, *Close Encounters of the Third Kind* and *E.T. The Extra-Terrestrial* all illustrate how the Hollywood film tells stories that are "true-to-the-probable" by projecting "realistic" worlds founded upon, or supplemented by, fragments and components of existing, "real" worlds as a kind of concession to being "true-to-the-actual."

The Vital Necessity of VFX

A stage magician's trick is successful only when the illusion is sustained through a deliberate lack of information provided to the spectator. If the spectator does not see everything that happens during a particular trick, this lack of information prevents him or her from questioning the act in progress. Lack of information allows the trick to pass as "magic." If any particular trick is performed 300 times in a row, it is most likely that the spectator will acquire all the necessary information to work out how the trick is done, so that the feeling of illusion and "magic" eventually disappears to be replaced by the sense of "sleight-of-hand."

Whatever genre that a Hollywood film might be considered to belong to, VFX shapes desired visual re-presentations that are essentially meant to "trick" the film viewer. VFX supervisor Richard Edlund says, 'My job is to trick an audience that is very astute—they have seen millions of feet of film and when you try to achieve an effect that is the least bit funny, you lose the audience and you've failed as a visual effects artist.'[53] An expert technique at the service of imagination in the art of filmmaking, the production of VFX suggests not just the virtuosity of the VFX practitioner in creating credible illusions on screen but also the inexorability of his or her having to hide the process of producing these illusions. This process of creation does not constitute a random procedure of creative coincidences, but entails a process of high precision deeply rooted in techno-scientific knowledge, artistry and craftsmanship. The creation and production of VFX hence constitute a discipline that combines art and science, vision and pragmatism, experience and skill, creativity and know-how, techno-scientific expertise and competence. This enables the legitimate characterization of the VFX practitioner as "artist-engineer-scientist" in the tradition of Georges Méliès 'as a designer of *trucages* (trick pictures) for the stage [who] already in the eighteen-eighties occupied the key position of technologist-engineer-artist which has become central to digital cultures.'[54]

Every Hollywood motion picture has its own unique set of requirements with regard to what kind of VFX would most appropriately fit its producer's and director's agenda. Also coming into play are the practicalities associated with script requirements for any chosen story line, topic or theme. The availability and effectiveness of technologies and techniques applied also influence strongly the decision to include particular VFX in a film as well as their predicted final form and "look." Hence, as David Butler informs us:

> CGI enabled the vast landscapes in the *Lord of the Rings* to be convincingly populated. . . . The trilogy pioneered the use of Massive (Multiple Agent Simulation System in Virtual Environment), a 3D animation system developed by Stephen Regelous specifically for [VFX provider] Weta Digital to provide the trilogy with large-scale battles involving hundreds of thousands of digitally-animated characters, each of which moves and interacts in relation to the characters around them.[55]

For the *Lord of the Rings* trilogy and many other films, the creation of VFX draws on particular modes of operation, sets of conventions and practices habitually implemented by particular production studios and directors, as well as on particular filmmaking strategies routinely associated with certain kinds of motion pictures (blockbusters, B-movies, and so on). But 'beyond conflicts within product conventions and production practices also exist conflicts between the norms of a good movie and the practical restrictions on achieving that outcome.'[56] VFX ultimately satisfy purely visual concerns as informed by narrative requirements, aesthetic preoccupations and economic pursuits: to tell entertaining stories, trigger emotions, re-present and re-produce places, themes and concepts in movies that are in one way or another visually "realistic," narratively consistent, aesthetically appealing, and commercially profitable.

Although principally motivated by the vision of the film director, VFX production in the Hollywood movie is also meant to satisfy the needs of its producer(s); that is, to secure return on investment from the movie generating box-office profits. The extent of VFX application in any Hollywood film is both qualitatively and quantitatively determined not just by directorial vision and narrative requirements, but also by the size of the production funds allocated in proportion to subsequent commercial exploitation expectations. Small- to medium-sized budgets often imply moderate profit expectations that, in filmmaking terms, translate to either altering the narrative in order to use fewer VFX or resorting to cheaper, low-grade VFX. An example of this is the all-digital 1995 film *Rainbow* directed by Bob Hoskins. Shot entirely with high-definition television cameras, *Rainbow* had a budget of only $10 million which, according to its producer Robert Sidaway,

> is not very much for a major effects film such as this, so we had to find a way to do the special effects. . . . The bottom line is that using high-def allowed us to complete this film. Freddie [Francis, DOP] was able to shoot an enormous amount of exteriors and the digital compositing that we had was live, and in five days we accomplished what would

> have taken two-and-a-half weeks on film. . . . We shot forty-six days total, including underwater green screen shots—to do that work would have taken ten days on the stage, but we did it in two days. It was the key to bringing this film in on budget.[57]

Substantial budgets imply higher profit expectations, thereby enabling intensive, high-grade VFX application for the creation of grand cinematic spectacles believed to appeal to larger audience segments, as illustrated by the commercially successful *The Hunt for Red October* in 1990. Its high-budget production involved the innovative use of digital technology in the shape of a particle-generating system that could imitate natural, random movements such as those of stars, fire, or dust particles, developed during the production of its VFX at Industrial Light & Magic.[58]

There is also what can be described as a "gratuitous" use of VFX—the "effects for effects' sake" approach. Grand cinematic spectacles of visual excess obtained through "gratuitous" use of expensive VFX often seem to appeal to large audience segments, as such cashing in at the box-office, when nothing else in the film might be thought to attract people to watch it in theaters—as confirmed by some of the 1990s Hollywood "blockbusters" for example. Janet Staiger observed that 'movies that end up costing a lot of money seem to be the ones that attract the interest and pleasure of massive numbers of people.'[59] And novelist David Foster Wallace has commented that *Terminator 2: Judgment Day* inaugurated what he thought became a special new genre of big-budget film of the 1990s:

> Special Effects Porn. "Porn" because, if you substitute F/X for intercourse, the parallels between the two genres become so obvious they're eerie. Just like hard-core cheapies, [what] movies like *Terminator 2* and *Jurassic Park* . . . really are is half a dozen or so isolated, spectacular scenes—scenes comprising maybe twenty or thirty minutes of riveting, sensuous payoff—strung together via another sixty to ninety minutes of flat, dead, and often hilariously insipid narrative.[60]

And it is of no great importance that so-called "movie events" are quickly superseded by other forthcoming "movie events" soon after spectators have paid to watch them in theaters at least once. Hollywood cinema's business mode of operation and strategy shows an industry clearly intent upon producing films imbued with visual spectacularity and excess wherein anything goes . . . but only to an acceptable level, by virtue of limitations deliberately imposed by producers and filmmakers. This is achieved by adherence to formula, convention and generic specificity. Excess and extravagance are tolerated insofar that they remain "controllable" and have minimal risk of alienating audiences.

'The use of digital effects can provoke intense, almost visceral, complaints among film critics attached to the materiality of film,' some of whom often consider the appeal of many VFX-intensive films to lie more in sensual gratification, derived from visual artifice and display, than in coherent storytelling, consistent narrative, strong characterization and performance.[61] Such films are often criticized for simply flaunting VFX on screen in "exhibitionist" displays of expert craftsmanship and technical sophistication mainly to compensate for under-developed narrative, characterization and storytelling. Hence, for Barry Purves: 'Too many films are showing off the technology's capabilities, without necessarily finding a justification for them—the incessant morphing of the headgear in Roland Emmerich's *Stargate* comes immediately to mind.'[62] About *Stargate* (1994), film critic James Berardinelli also commented that

> Roland Emmerich appears infatuated with his film's look without caring if anything moderately substantial lies beneath the glitz. . . . Occasional, mostly-halfhearted attempts at character development are ignored whenever they don't impact directly on the plot. As has frequently become the case with movies of this genre, the visual effects and breathtaking cinematography far outstrip everything else offered by *Stargate*.[63]

The often negative critical response to *Stargate* indeed illustrates how some film critics have recurrently criticized VFX-intensive Hollywood movies for their marked tendency to project "flat" superficial characters and weak under-developed stories and plots. Another example is *Independence Day* (1996) about which James Berardinelli wrote: 'This is a spectacle, pure and simple [which] gets mired in syrupy, artificial character development. . . . Unfortunately, because the film makers mistakenly tried to inject a load of weak dramatic elements, *Independence Day* turns out to be overlong, overblown, and overdone.'[64] For his part, Roger Ebert wrote that 'for all of its huge budget, *Independence Day* is a timid movie when it comes to imagination. The aliens, when we finally see them, are a serious disappointment.'[65] In the case of *Starship Troopers* (1997), narrative and storytelling were thought to be somewhat lacking, and its dominant cast of American TV sitcom actors was not of much help either in conveying the film's interesting sub-commentary on American imperialism and fascism. Hence, James Berardinelli considered it as 'a prime time soap opera, complete with cheesy dialogue and unconvincing character development.'[66] And for Roger Ebert: 'Its action, characters and values are pitched at 11-year-old science-fiction fans. . . . The action sequences are heavily laden with special effects, but curiously joyless' while 'the Bugs aren't important except as props for the interminable action scenes.'[67] Considered to have taken the "excesses" of *Stargate, Independence Day* or *Starship Troopers* even further, *Armageddon* (1998) for its part was critically lambasted for being symptomatic of Hollywood cinema's over-indulgence in visual spectacularity and loss of attention to the non-techno-visual aspects of good filmmaking. This can be illustrated for instance by Roger Ebert's commentary that *Armageddon* is 'an assault on the eyes, the ears, the brain, common sense and the human desire to be entertained. . . . It is loud, ugly and fragmented. Action sequences are cut together at bewildering speed out of hundreds of short edits, so that we can't see for sure what's happening, or how, or why.'[68] *Armageddon*'s "gratuitous" VFX displays illustrated the Hollywood cinema industry's obsession with techno-visual perfection and sensual gratification through the privileging of visual spectacularity and excess over coherent storytelling, consistent narrative, strong characterization and performance.

As discussed above, many Hollywood directors and studios have been accused by film critics of over-indulgence and of forgetting how "good" stories were meant to be told, due to their films being nothing more than deliberately excessive visual pounding and "gratuitous" VFX displays. But filmmakers inevitably become quite defensive on this issue. About his VFX extravaganza *The Fifth Element* (1997) for instance, French film director Luc Besson pointed out that

> the most important thing about working in a movie with so many effects is to know that the movie is, above all else, a story. We put a great deal of work into the script, to see that the story is there—we were never interested in making a movie that is about some spectacular special effect, with the story built around it. The effects are there to do their job in relation to the story, not the other way around.[69]

Hence, from a film director's point of view, VFX have the inherent ability to expand the visual features of film, allowing these to incarnate storytelling aids and narrative orientation signposts, rather than simply being superficial signs of visual spectacularity, as illustrated by *Contact* in 1997. As commented by Robert Craig:

> Perhaps some of the most impressive shots [in the last part of the film] occur while Ellie [Jodie Foster] walks along the gangway into the pod to take her on her trip and she looks down into the computer-generated whirling rings that make up the device's energy source. Even her breath, which is visible in the morning's cold, exists courtesy of the computer. [T]he film's signature scene on the beach when she experiences contact with the alien in the form of her father also feature a variety of computerized elements that create both existing and hypothetical realities. . . .[70]

Contact hence represented 'a significant marker in film history, as another milestone in the utilization of animation and computerized special effects in telling a cinematic story.'[71] Australian filmmaker Alex Proyas (*The Crow, Dark City, I, Robot*) explains further that

> visual effects are really just a way of telling a story. Where I differ, a lot of people have pointed out to me . . . is that the way I use visual effects tends to be in a more narrative way. They do actually tell a story rather than being for purely spectacle or display purposes. A lot of filmmakers use it for the spectacle as opposed to the narrative.[72]

Countless low-budget as well as high-budget Hollywood movies have benefited immensely from the imagination of VFX teams in enhancing or productively supporting narrative, storytelling, re-presentation and characterization. VFX can make a strong contribution to the formal structure of a film and cannot always be considered as excessive "gratuitous" visual gimmickry. In proper dosage, they function to support the director's particular vision. To use *Contact* again as an example: its plot 'suggests a typical science fiction film trajectory, along with many possibilities for elaborate special effects, . . . in fact pointedly de-emphasizes large-scale special effects in favor of more subtle uses of computer-generated imagery.'[73] If VFX were simply considered as

> an intensification of fundamental filmmaking, then the gap between the two is slight, much smaller than in the stereotype of a special effects ghetto inhabited by technophiles, geeks, filmmakers hooked on spectacle and display, and odd bods who have forgotten that films tell stories.[74]

Arguably considered at first as tools of technical problem solving, VFX have progressively turned into a higher expressive form. They have done more than just contribute to the techno-visual glossification and sophistication of the Hollywood film. They have ended up tremendously expanding the range and nature of stories as well as the methods for telling them. VFX not only change the *form* of what is told and seen but also transform the *content* of what is told and seen. This is indicative of the fact that addressing VFX revolves not only around what they *are* but also around what they (can) *do* to the film viewing experience.

Functions and Objectives of VFX

VFX may be used like music to enhance the emotional impact of a scene, or they may contribute to actually telling the story directly. Often unidentifiable to the film viewer, VFX may be in the foreground or in the background. For example, in *Contact*, 'Zemeckis incorporates computer-generated images throughout the film in ways both obvious and subtle to help suspend audience disbelief, provide a sense of contemporary reality, and enhance the mise-en-scène and content.'[75] VFX also function generically in terms of how well they contribute to the "pleasures" familiarly associated with particular types of film. The generic classification of films and of their VFX—by filmmakers, producers, exhibitors and critics alike—is a way of signaling to audiences, and informing them, about the kind of sensual "pleasures" that can be expected from the film upon viewing. In disaster movies for instance, the disaster in question—be it a tornado, a volcano or an earthquake—is usually the film's central "antagonist," so to speak, in relation to which mayhem and destruction evolve. Outer space adventures often have to include drifting asteroids, laser beams, spaceship engine glow, computer consoles, information displays, hi-tech gadgets and gizmos. The action/adventure movie usually includes scenes of armed and unarmed combat, high-speed chases and explosions. Some VFX may come across on screen as being "innovative" in that what they re-present has not been seen previously in that particular form. In 1985 for example, *Young Sherlock Holmes* included a sequence involving the stained-glass figure on a church window of a sword-wielding knight that comes to life and stalks a priest. This sequence was a techno-visual breakthrough at the time since it contained the first moving digital character or "synthespian" ever seen in a Hollywood feature film.[76] Certain VFX may appear groundbreaking in specific films, but most come across as "variations" of generically defined and familiar types of VFX in the bulk of Hollywood movies. Being dominant characteristics in Hollywood cinema's generic, narrative and aesthetic proclivities, *predictability* and *iconography* inform and influence the conceptualization and creation of VFX.[77]

Most VFX in Hollywood cinema display the influence of antecedent VFX forms which they often re-adapt, re-contextualize, pay homage to, upgrade, copy or appropriate. The originality and appeal of VFX in Hollywood films are often situated in the degree of sophistication and competency displayed in the act of inter-textual referencing pivotal to the design and creation of VFX —which is why VFX rarely look "unfamiliar" on screen. For example, while Ray Harryhausen's stop-motion animated sword-fighting skeletons in *Jason and the Argonauts* (1963) could be related to sword-fighting scenes in Errol Flynn's earlier swashbuckling movies, Harryhausen's fighting skeletons are in turn re-produced in *Evil Dead III: Army of Darkness* (1992), except on a larger scale. The conceptualization, design and creation process of VFX in the Hollywood film inevitably begins with established formats and styles that have been standardized by means of their repeated use, recycling and circulation in earlier motion pictures as well as in other visual media (television, cartoons and so on). Hence, an aesthetic of inter-textual reference is an important component of a larger aesthetic of VFX design in that it allows the potential identification, in terms of proximity or affinity, with the relevant source(s) of reference. In other words, identifying the visual dynamics of an effect—in the very mechanics and aesthetics of its on-screen re-presentation—reveals its sources as well as its relationship with these sources. Concrete examples of this are abundant in Hollywood cinema: the indebtedness of *Jurassic Park* (1993) and *Godzilla* (1998) to the various animated "monsters" of pre-World War II American cinema, especially those created by Willis O'Brien [*The Lost World* (1925), *King Kong* (1933), *The Son of Kong* (1933)]; the influence of "dogfight" sequences in World War II movies [*The Dam Busters* (1955), *633 Squadron* (1964), *Battle of Britain* (1969)] on the choreography and look of space battle scenes in *Star Wars* and most space adventure movies since 1977; the upgrading of

Metropolis's (1927) projection of the human metropolis of the future by *Blade Runner* in 1982, itself becoming a reference for subsequent VFX practitioners and set designers in films as varied as *Back to the Future II* (1989), *Twelve Monkeys* (1995) and *The Fifth Element* (1997); *Alien*'s (1979) connection to *It Came from Outer Space* (1953) and *It! The Terror Beyond Space* (1958) and its subsequent influence on the design of outer-space alien monsters; the long-lasting trends set by *The Abyss* (1989) and *The Hunt for Red October* (1990) with regard to what underwater shots should "realistically" look like on screen. To be sure, VFX, like the principles of illusion vital to the magician's act, often depend on the element of surprise and tactics of "visual diversion"—through spectacularity—for the desired impact of the illusion to be complete and successful. Yet, through repetitive use certain types of VFX become iconographic "trend-setters" and create standards for how these should be displayed on screen and for what audiences expect to see. Anything radically different could be considered as either sub-standard or too "avant-garde."

Other than inter-textual referencing and iconography, the excessively mundane and yet impressive surface of the Hollywood film as it presents itself to viewers is also the outcome of observation, research, imagination, experimentation and skill. As such, the production of VFX involves the synthesis of technical skills, visual aesthetics and conceptual approaches so as to tell a story, to re-present the "realistic" and to generate "pleasures": excitement, wonder, awe, shock, recognition, and so on. The conceptualization of VFX to re-present the "realistic" is a combined outcome of what is *seen* and what is *imagined* by the VFX practitioner and the film director. VFX creation is shaped not only by the characteristics of the filmic medium, the techno-industrial constraints and economic parameters of the production process and by generic requirements, but also by what interests and preoccupies filmmakers and VFX practitioners alike. They depend on their surroundings, on what they see and hear, in conjunction with what they imagine, to create VFX that re-present and reproduce "realistically" what is often visually inexpressible—whether contributing to the narrative or not—in forms and ways considered most likely to make unambiguous sense to the cinema spectator.

After stripping away the hype of visual "magic" and all 'the fascination with *technovelty*, the "how-did-they-do-it" factor, the charisma of the gizmo,' the application of VFX in Hollywood filmmaking can be said to essentially fulfill a handful of functions and objectives.[78] To achieve visual manipulation, trickery and illusion, filmmakers resort to VFX in the Hollywood cinema industry for the purposes of:

- filmic "fantasy" construction—creating what does not actually exist in everyday reality;
- imitative reproduction, creative simulation and storytelling support—substituting the materially "real" with the artificially created

"realistic" as a time- and money-saving strategy, and facilitating narrative flow, building characterization, reinforcing action and spectacle;
- visual enhancement and correction—improving picture quality as well as concealing the actual filmmaking devices and techniques applied in achieving the above two functions.

Each of these will respectively be the topic of the next three sections.

Filmic "Fantasy" Construction

The construction of filmic "fantasy" was indeed the first major function of VFX applications in the early days of cinema—as suggested and illustrated by the turn-of-the-century films of Georges Méliès; for instance: *Escamotage d'une Dame au Théâtre Robert-Houdin* (1896), *La Lune à Un Mètre* (1898), *L'homme à la Tête de Caoutchouc* (1902), *Le Voyage dans la Lune* (1902), *Le Voyage de Gulliver à Lilliput et Chez les Géants* (1902), *Illusions Funambulesques* (1903), and *Le Puits Fantastique* (1903).[79] Through the construction and display of the *fantastic* in particular, VFX enable the synthesis on film of the "realistic" and the imaginary, of dreams and nightmares, of foresight and paranoia, of perception and hallucination.

VFX-intensive movies made during the studio system era in particular further reinforced this function of VFX in Hollywood: the construction of filmic "fantasy." Such movies depend, more than any other Hollywood vehicles, on VFX to express (sometimes "to give life to") what does not exist: a flight of fancy, a leap of the imagination—landscapes, foregrounds and backgrounds, things and beings that are often quite easily recognized as VFX when seen on screen. VFX have indeed been more prolifically used in the creation of imagined, ethereal, bizarre, extraordinary, prehistoric and futuristic worlds, characters and objects belonging to the realm of the fantastic.

> In films of the fantastic, the impression of unreality is convincing only if the public has the feeling of partaking, not of some plausible illustration of a process obeying a nonhuman logic, but of a series of disquieting or "impossible" events which nevertheless unfold before him in the guise of event-like appearances.[80]

Such genres as science fiction and horror in particular usually project the best alchemy between the wide-ranging potential of VFX and their respective generic requirements.

By the end of the 1990s the creation of fantastic worlds, characters and objects had become a major activity for the VFX domain in Hollywood. As Janet Wasko puts it: 'Technological developments also influence the types of film produced by Hollywood companies. Certainly, the number of science fiction or space epics has increased with the evolution of sophisticated special effects techniques.'[81] And Barry Langford further explicates that the rise of science fiction and fantasy

> offers an obvious showcase for spectacular state-of-the-art technologies of visual, sound and above all *special effects design*, the key attractions that provide a summer release with crucial market leverage. The genre is well suited to the construction of simplified, action-oriented narratives with accordingly enhanced worldwide audience appeal, potential for the facile generation of profitable sequels . . ., and ready adaptability into profitable tributary media such as computer games and rides at studio-owned amusement parks.[82] [emphasis added]

Alien stories, extra-terrestrial invasions, doomsday predictions, mythical tales and space operas, futuristic technology-intensive worlds, space explorations and adventures indeed belong to the realm of science fiction and filmic "fantasy," where the overwhelming presence of VFX, along with the extensive techno-industrial applications that go into re-producing them, is inevitable, less "questionable" and contextually more justified, compared to other popular generic vehicles of Hollywood cinema.

Without the construction of filmic "fantasy" through VFX application, many popular Hollywood A- and B-movies would never have been made:

- the "alien invasion" flick—*Invaders from Mars* (1953), *The War of the Worlds* (1953), *Earth vs. the Flying Saucers* (1956), *Independence Day* (1996), *Men in Black* (1997).
- the "dinosaur" movie—*The Lost World* (1925), *Dinosaurus!* (1960), *One Million Years BC* (1966), *Jurassic Park* (1993).
- the "horror" movie—*Dracula* (1931), *Frankenstein* (1931), *The Mummy* (1932), *Werewolf of London* (1935), *The Fly* (1958), *The Birds* (1963), *The Exorcist* (1973), *Poltergeist* (1982), *The Relic* (1997).
- the "magical and mythological creature" movie—*Alice in Wonderland* (1933), *Jack the Giant Killer* (1962), *Jason and the Argonauts* (1963), *Sinbad and the Eye of the Tiger* (1977), *Clash of the Titans* (1981), *Dragonslayer* (1981), *Willow* (1988), *Hook* (1991), *Dragonheart* (1996).
- the "monster" movie—*King Kong* (1933), *The Thing from Another World* (1951), *It Came from Beneath The Sea* (1955), *Jaws* (1975), *Gremlins* (1984), *Anaconda* (1997), *Deep Rising* (1998), *Godzilla* (1998), *Deep Blue Sea* (1999).[83]
- the "sci-fi" film—*Forbidden Planet* (1956), *Close Encounters of the Third Kind* (1977), *Innerspace* (1987), *The Abyss* (1989), *Total Recall* (1990), *Terminator 2: Judgment Day* (1991), *Contact* (1997), *The Matrix* (1999).
- the "space adventure" film—*Conquest of Space* (1955), *Star Wars* (1977), *Alien* (1979), *Star Trek* (1979), *Armageddon* (1998).
- the "superhero" movie—*Superman* (1978), *Batman* (1989), *The Rocketeer* (1991), *The Shadow* (1994), *Blade* (1998).

Commercial success has generally been the norm rather than the exception for these Hollywood films that rely mostly on VFX-driven "fantasy" construction, compared to other popular Hollywood genres ("gangster," "musical," or "melodrama").[84]

The popularity of "monster" and "dinosaur" movies with cinema audiences, for example, has more to do with the visual sophistication, impact and spectacularity of their VFX-driven creatures than with the performance of their human counterparts. In "monster" and "dinosaur" films such as *Jaws, King Kong, Jurassic Park* and *The Lost World,* VFX creatures of one kind or another are clearly the central protagonists. Their actions and behavior motivate what happens or is done to the real human actors, as illustrated by the Velociraptors and Tyrannosaurus Rex that are constantly chasing, and sometimes eating, humans in *Jurassic Park*. Another example is the great white shark that, in its case, affects the behavior of the entire community of the coastal town of Amity throughout all of the *Jaws* installments. And although having far less screen time, the shark is one of the most memorable characters of *Jaws* as a result of its VFX of 'astonishing complexity, needing both careful story-boarding and great cunning; to make a polyurethane shark leap into a boat and savage a man requires either crude cheating or extreme sophistication, and Spielberg does not cheat.'[85]

The pivotal function of VFX in the construction of filmic fantasy can be illustrated by the production of the Hollywood sci-fi blockbuster *Total Recall,* which involved the use of nothing fewer than 35 stage sets to produce hundreds of VFX shots for the depiction of the story's central Martian setting. As *Total Recall*'s director of photography Josh Vacano relates:

> The film is set in the future, but not very far into the future. . . . There is no atmosphere on Mars. The people have to be either in spacesuits or in the protected colonies. This is not a picture about people in space suits. So, we imagined a colony where the miners live digging special metals out of the planet. These people must have their own environment. . . . The second thing that makes Mars very special is that everything is red.[86]

Only massive VFX application could allow for the particular visual specificity required by *Total Recall* in projecting the mixed imagined, artistic and scientific vision of planet Mars colonized by humans, as derived from popular knowledge about what it might look like and from assumptions relative to how things on Earth look like. In the case of *Close Encounters of the Third Kind* in the 1970s, optical compositing and VFX shooting were the dominant components during its two years of production to achieve the desired innovative visualization of extra-terrestrials different from anything else done before ("alien encounters" by themselves being nothing new in Hollywood movies) to entice cinema spectators to watch the movie. A little more than a decade later, James Cameron would follow a similar principle with the sequence of first contact between humans and the non-terrestrial water alien in *The Abyss*. The planning and production of digital VFX for this sequence involved the first-time application of the Virtus Walk-through program.[87] The software contributed to the manipulation of digitized photographic data of background plate sets and to the creation of a digital model with which to integrate the film's famous computer-generated non-terrestrial water "pseudopod" as well as to perfecting visually its digital environment.[88] Lasting only about 75 seconds on screen but requiring eight months of work, the groundbreaking digitally animated "pseudopod" would significantly influence the conceptualization as well as the production of future Hollywood cinema extra-terrestrials on screen.

VFX also allow for the creative projection on screen of objects moving in ways that defy human visual perception of "reality," as illustrated by the "bullet-time" sequences in various sections of all *The Matrix* movies. Through highly innovative use of VFX, the "bullet-time" technique—of circling bodies and objects stilled in flight—simultaneously combines moving and stilled objects, real human characters, freeze frames, slow and fast motion, with the application of VFX meant to suggest the sense of action so hyper-fast that it needed to be slowed down in parts to be actually seen by the spectator. Before *The Matrix*, *Blade* in 1998 actually contained a "bullet-time" sequence with similar graphic themes, action design, spectacularity and motivation; see for example, the first encounter between Blade (played by Wesley Snipes) and Deacon Frost (played by Stephen Dorff). Within the subsequent decade, "bullet-time" VFX sequences would have become commonplace, noticeable in productions as varied as *Sherlock Holmes: A Game of Shadows* (2011), *Max Payne* (2008), *Crime Scene Investigation*, Smirnoff and The Gap television commercials.

Fantastic worlds, beings and objects—as constructed by VFX practitioners—can at times impact negatively upon the filmic experience of the spectator which, in order to

> fonctionne sans heurts, il lui faut un peu de réalité. Trop peu de réalité et l'expérience ne prend pas. Trop de réalité, et elle devient lourdingue.[89]

> work consistently, needs some reality. Too little reality and the filmic experience is not fulfilled. Too much reality, the experience becomes bloated.

This has been the case for *The Matrix* that repeatedly confronts the spectator with intertwined "real," virtual and fantastic worlds, beings and objects. In *Positif*, Yannick Dahan explains that *The Matrix*

> se compose . . . d'incessants va-et-vient entre le réel du futur et notre société virtuelle. En empruntant à la fois deux directions, l'anticipation

> pyrrhoniste et la réflexion métatextuelle sur le pouvoir des images, les frères Wachowski ont accouché d'un film hybride, chancelant, hésitant sans cesse entre la métaphore sérieuse et la jouissance d'une série B à la mise en scène fluide et innovatrice. Tout l'intérêt de *Matrix* réside dans cette indécision et dans la façon dont les réalisateurs s'amusent de notre réel, quand l'action se situe essentiellement dans la Matrice. La vision des personages, n'existe qu'à travers le prisme d'une multitude d'écrans, ordinateurs, télévisions, vitres, pare-brise . . ., un divers de réalités se superposant comme d'infinies couches altérant notre perception du réel.[90]

> is made up . . . of relentless comings and goings between the real of the future and our virtual society. By taking two directions simultaneously, the pyrrhonistic anticipation of and the meta-textual reflection on the power of images, the Wachowski brothers have created a hybrid and unsteady film which continuously wavers between serious metaphor and the pleasures of a B-movie endowed with a fluid and innovative narrative. The *Matrix*'s interest lies in this indecision and in the way the directors make fun of our reality, especially when the action occurs mainly in the Matrix. What the characters see, which exists only through the prism of countless screens, computers, television sets, window panes, windshield . . ., is a diversity of realities overlapping each other like infinite layers that modify our perception of the real.

The re-presentation of fantastic worlds, objects and beings in VFX-driven Hollywood movies in general often tends to follow a "story book," a "comic book" or, in the case of *The Matrix* in particular, a "video game" and "cyberspace" strategy of storytelling and visual display—something which the majority of its principal audiences tend to be already familiar with. As explained, for instance, by Andrew Sussex:

> In the Silicon Valleys of the world, *The Matrix* is the apotheosis of cyber-geek chic. It's a movie for gearheads—so cruelly and repeatedly tricked by Hollywood—that genuinely rewards computer savvy. Says Jamie Pallot, 39, an exec producer with Microsoft: "It does everything that *Johnny Mnemonic* so dismally failed to do. . . ."[91]

The degree of credibility accordingly attributed by spectators to the fantastic worlds, beings and objects of *The Matrix*—to its overall filmic experience—is ultimately shaped by cinéphilic and aesthetic interests. This usually occurs through filmic inter-textual referencing and in accordance with established generic rules and conventions of storytelling, narrative and characterization.

But however exaggerated the "fantasy" worlds of Hollywood films might seem to be, their construction is always bound to precise rules and conventions, as promoted by producers, filmmakers and VFX professionals alike, and as expected by Hollywood's vast audiences. Worlds, characters and objects of "fantasy"—whether futuristic, medieval, extra-terrestrial or imaginary—always have to look cinematically "realistic" and credible to the spectator. In other words, their creation has to be founded in an established visual culture of the (American) "everyday," that circulates through television, comic books and so on, and with which audiences will already be familiar upon viewing. 'Audiences may tend to enjoy the familiar, especially if narratives are pleasurable in part as a way to master an overwhelming array of stimuli. What Hollywood may do is offer the security of ritual, of routine, with even the deviations being well within the bounds of the conventional.'[92] Hence, however "bizarre" they might appear to be, the giant alien bugs of *Starship Troopers* for instance were based upon the VFX practitioners' observation of real bugs and insects that eventually received a particularly imaginative treatment from the VFX department. They are extremely absurd yet somehow familiarly "realistic." For its part, *The Company of Wolves* (1984) 'creates an effective blurring between what is real and unreal through its combination of studio set and live inhabitants [in a] world that only *seems* real by fusing the authentic (live wolves, owls and frogs) with the fake (giant toadstools, stylized tree trunks and plumes of studio mist).'[93]

Filmmakers and VFX practitioners are always conscious that the fantastic worlds re-presented in the Hollywood film have to be at the edge of fiction and on the border of verisimilitude. Digital VFX, cinematographer Dean Cundey claims,

> allow you to create this illusion of reality. You can take a really improbable event—say, the presence of a dinosaur—and, as long as you present it in a way that the audience accepts as real—the way the light hits the skin, the way it moves, the way the world reacts around it—they will accept the fact that the dinosaur really appears to be there. . . . One of the tasks of the visual effects person, and the cinematographer and director, is to create this illusion, using this sort of heightened reality, using reality but always expanded and stretched and twisted.[94]

Excessive artistic license and flights of fancy in the creation of VFX can alienate many sections of the viewing public, hence affecting the potential profitability of the film. Even in re-presenting and re-producing the fantastic, VFX *cannot* be endlessly open to interpretation. They have to be as unequivocal as possible with regard to what they re-present in order to avoid any distracting confusion and speculation over what they might be and so to avoid the alienation of the audience from the filmic text. This particular function of digital VFX—the construction of filmic "fantasy"—would correspond to Metz's 'visible *trucages*.'

Imitative Reproduction, Creative Simulation and Storytelling Support

Creative simulation in Hollywood cinema thrives upon the viewer's constant oscillation between the rational and irrational interpretation of the filmic worlds re-presented on screen. The process of creative simulation is one that straddles fiction and "reality." The oscillating and mutating "space" between these paradigms is where the current Hollywood movie viewing experience is located. This kind of in-between space on the screen can be characterized in terms of Jean Baudrillard's "modern un-reality," which 'no longer implies the imaginary, it engages more reference, more truth, more exactitude—it consists in having everything pass into the absolute evidence of the real.'[95]

VFX enable the imitative reproduction and creative simulation of what would be too difficult, too dangerous or too expensive to record on film in natural or "real" conditions: natural calamities, human-made catastrophes, explosions, crashes. In the words of VFX supervisor Thomas Smith,

> effects are not only a way of filming spaceships and laser blasts; there are many applications of this branch of film technology. Special effects are used in motion pictures when scenes are desired that would be impractical, expensive, dangerous, or even impossible to film in a normal manner.[96]

This particular function of VFX is especially pertinent to a common genre of Hollywood cinema, namely the disaster movie, as illustrated by: *Airport* (1970), *The Poseidon Adventure* (1972), *Earthquake* (1974), *The Towering Inferno* (1974), *The Hindenburg* (1975), *Daylight* (1996), *Twister* (1996), *Dante's Peak* (1997), *Titanic* (1997) and *Volcano* (1997). Considered by Roger Ebert as 'the best of the mid-1970s wave of disaster films,' *The Towering Inferno* constitutes a very good example of Hollywood VFX-intensive filmmaking at its finest during that decade.[97] Compared to the somewhat "phony" feel of *Earthquake*, for instance, wherein the cinema spectator could clearly see on screen where the high-rise set stopped and where the painted flats began—as such affecting the intended visual illusion and screen impact—*The Towering Inferno* is instead a masterpiece of stunt coordination and VFX production. The details of the fire are convincing; the explosions, the wreckage, the bombed-out fire escapes all look real in *The Towering Inferno*. The latter disaster movie illustrates well Scott Bukatman's commentary that 'special effects are an important part of cinema's experiential dimension: they can bring the visual, auditory and even tactile and kinaesthetic conditions of perception to the foreground of the viewer's consciousness.'[98] What makes disaster movies interesting is that

> we go to see the characters in the PROCESS of experiencing the disaster. As the disaster develops, so do the characters. The skyscraper in *The Towering Inferno* catches on fire in the first 15 minutes and burns for two hours. The good ship *Poseidon* is hit by a tidal wave 10 minutes after we walk into the theater; it turns over, and the characters spend the rest of the movie fighting fire and flood. The airplanes in the two *Airport* movies fly for 90 minutes after things go wrong. . . .[99]

Hollywood studios have routinely produced disaster films primarily because of their potential for generating quick profit—often on the basis of their visual spectacularity, intensive marketing and advertising, and saturation distribution, as illustrated by Philip Mora's equation: 'Big star names + Absurd accident or Outrageously improbable catastrophe + Big Budget + Soap Opera = Disaster Film.'[100] And disaster Hollywood movies generally do well at the box-office.

> These films represent commercial filmmaking at its peak of cunning. They have been constructed with the cinematic equivalent of Machiavellian precision. They exude a remarkable confidence in their almost total mastery of mass audience manipulation. The audience is placed in the stance of a car accident voyeur eating popcorn.[101]

Indeed, disaster films leave no room for ambiguity in exploiting the cinema spectator's scopophilic fascination with mayhem, destruction and death.

VFX application as imitative reproduction and creative simulation has become a normal aspect of filmmaking in the Hollywood cinema industry. As Charles Tahiro points out:

> All films, in order to persuade, must convince as compelling illusions of space. It is that capacity, structured in time, that distinguishes cinema from other media. Thus, the "great special effect," no matter how it is achieved, is so because of its capacity to persuade viewers momentarily that the event the effect represents "really" happened. That impact can be created only by measuring the manufactured effect against real space.[102]

Digital VFX in particular have enabled the imitative reproduction and creative simulation of much more convincing photo-realistic disasters, catastrophes and calamities (even if narrative, plot and characterization might not always be particularly "refined"), compared to the days of producing similar analog VFX by relying mostly on magnified fast-burning miniature sets, model ships in wave tanks, gallons of orange-dyed porridge (for simulating volcano lava), water on dry ice (to simulate mist), radio-controlled vehicles and flying objects. These "traditional" analog VFX are still sometimes resorted to in the digital era, except as a viable option rather than as necessity, with the additional advantage of a CG team being available to later correct, polish or enhance them further during post-production in order to obtain sophisticated visuals.

Digital VFX have also been particularly effective in terms of enhancing dramatic impact and supporting storytelling through imitative reproduction of the "realistic" while tremendously reducing film production costs.[103] Citing *In the Line of Fire* (1993), Sony Pictures ImageWorks' VFX supervisor Tim McGovern explains that for

> one of the crowd scenes, we shot 1500 people in five different groupings. We then took the five shots with hundreds of extras and weaved them into one shot with thousands of extras. As a result, when you look at the shots, it looks as if there are ten thousand people. . . . Without these techniques, it really would have been too costly. It would have dramatically affected the scope and authenticity of the film. If you figure that you want 10,000 extras and you have to pay each extra $50 a day, that's $500,000 for one day of shooting.[104]

Hence, one of the most important advantages of resorting to digital VFX production is that it often becomes unnecessary to shoot on real locations, or even to use real people, animals and props—hence allowing much production time to be gained and money to be saved without compromising the director's vision, film narrative and storytelling integrity.

Putting on screen a person about to fall while clinging to a high building, for example, constitutes a basic form of VFX, and this can be said to fulfill the function of imitative reproduction, creative simulation and storytelling support. Such VFX can be achieved by first shooting the two different elements. First, the actor is shot against a blue screen while providing the action that the director requires. Then, a second element is shot: the high building without the actor. These two film elements are then combined to create the final composite. The blue areas around the actor are removed, allowing the picture of the building to become the new background for the actor. When both elements have been carefully planned and executed, the audience will believe they are seeing the actor hanging from the edge of the building. As such, VFX function as momentous signs and markers of storytelling flow, of narrative progress, of setting and characterization. Either applied throughout an entire movie or in specific parts, such VFX have been designed to complement the precise demands of particular styles of storytelling, narrative and characterization, as opposed to merely generating visually spectacular images. In the Hollywood action film *True Lies* (1994) for example, burning fuel vapor trails which are expected to emanate from a flying missile were not shot during production but computer-generated and added to the live-action shot during post-production. This kind of digital VFX, while complex in its creation, subtly adds to the overall believability of the shot and of that action sequence.[105] In *Dead Ringers* (1988) VFX, achieved through motion-control photography in particular, were crucial to characterization and performance. They enabled actor Jeremy Irons to simultaneously play the dual roles of Beverly Mantle and of his twin brother Elliot Mantle in many scenes of the film. And in *The Crow* (1994), a number of uncompleted scenes yet to be shot had to be digitally created after the untimely death of actor Brandon Lee during principal photography. To enable this, Lee's face was digitally lifted from one scene and placed on a live "double" in other different backgrounds.[106]

VFX can be said to have a more sober supportive role when they do not visually overwhelm, minimize and undermine but instead fortify, highlight or subtly emphasize the principal components of the film: by facilitating narrative flow, building characterization, reinforcing action and spectacle, through imitative reproduction, creative simulation, to support storytelling and increase spectator engagement with the film. This particular function of VFX would correspond to Metz's 'perceptible but invisible *trucages*.'

Visual Enhancement and Correction

A less "visible" function of VFX has been the improvement of picture quality as well as the correction of visual "flaws" in the Hollywood film. This was originally achieved during the analog era through the time-consuming and expensive use of optical printing and photochemical color processing for instance. Following the appearance of digital non-linear editing systems, this particular function of VFX would be extended to practically all Hollywood movie productions in any generic category from the late 1980s onwards. Since the Hollywood film is always expected to be technically clean, visually uniform and of high picture quality upon theatrical release, digital VFX indeed favor adherence to these requirements by encouraging additional possibilities of correcting—adding, removing or altering—anything on film directly, as a more viable alternative to re-shooting scenes. In this sense, digital VFX constitute a kind of "filmic floss" for the efficient cleaning up of images in post-production, as opposed to their more obvious involvement in the creation of spectacular images. This function of VFX relies on a chain of film scanning, the digital storing of images, digital manipulation and the (re-) recording of manipulated digital images back to film—better known as the film-digital-film process.

> As images are stored to computer disk, they subsequently exist as digital files that consist of a specific matrix of pixels. There can be thousands or millions of pixels that make up one digitized film frame depending upon the resolution size. By applying certain pixel manipulation techniques to a digitized frame, the look of the image can be significantly changed.[107]

The film-digital-film process hence enabled each frame and each pixel to be addressed separately and individually through the isolation of specific colors to remove any unwanted item in the frame. The more common applications of digital VFX for visual enhancement and correction include wire removal and pixel manipulation. Among the first uses of this film-digital-film process was often the removal of items in a frame that the filmmaker did not have time to fix during the production process; digitally deleting, as well as adding, items in movies became a common occurrence during post-production in the Hollywood cinema industry from the late 1980s onwards.

Safety wires are commonly used during film production to safely suspend and hold stunt persons or actors at dangerous heights and in other situations. Shooting scenes involving wires has traditionally been a very "hit-and-miss" and time-consuming affair, requiring complicated planning for shots to be achieved in a specific way. Digital VFX application allowed for the shoot to proceed quickly with the knowledge that wires would be digitally removed during post-production. For the motorcycle jump in the Los Angeles canals in *Terminator 2: Judgment Day* for example, 'a large support crane and clearly obvious cables were used to safely support the stunt rider. Later, after digitizing the shot into a digital workstation, the suspension mechanisms were entirely painted out of the required frames.'[108] By using the pixels from a clean background and substituting them for the pixel areas that define the wires, the latter were essentially removed from view; additional techniques were then used to paint and blur certain sections in order to reconstruct the original tracking of the camera lens.[109] Not only wire removal but also removing objects from the film frame is a common requisite for many Hollywood films. In many cases, it would have taken too long to remove something from the frame that did not belong in the finished shot that would end up on the screen. Director of photography Steven Poster notes: 'On *Cemetery Club* [1993] we had a car shot and there was a piece of steel rigging in the shot and we were not in a position to re-photograph the shot. So I was able to scan the frames digitally and remove the steel.'[110] Other than visible rigging, wires and equipment used during production itself, everyday objects are routinely digitally painted out of the frame while a background is put in their place—the removal of electric poles and road signs in historical period films for instance. Through the film-digital-film process in particular, digital VFX would hence function as problem-solving tricks implemented mostly during post-production to eliminate, supplement or camouflage undesirable visible equipment, techniques, and other visual components, in order to obtain techno-visually flawless images. This particular function of VFX constitutes a deliberate intervention to alter visual elements, in any way and to any extent possible, from their original form, model or state, in a way that is "believable" on screen to the film viewer. The Hollywood film's "realism" can be obtained through varying degrees of enhancement, distortion, kinetization, obliteration and so on—whichever way is chosen to deliberately re-model specific visuals.

The effectiveness of VFX application for enhancement and corrective purposes is judged by virtue of the degree of adhesion to, and interaction of, filmic components constituting the frame, not on the basis of the individual effect in isolation from other elements in the frame. Effective visual enhancements and corrections attract the least attention possible to themselves and go mostly unnoticed, so that the intended on-screen visual is not disturbed. Their presence is most filmically productive in a state of pseudo-absence on screen. They are most effective when visually imperceptible—*optically* but not *perceptually* "there." This particular function of digital VFX would correspond to Christian Metz's 'imperceptible *trucages*' which he considers to be the 'only pure *trucages*. Only with these can we be certain that the spectator has been fooled since he has noticed nothing.'[111]

NOTES

1. Metz, 'Trucage' 670.
2. Metz, 'Trucage' 667 & 661–62.
3. Metz, 'Trucage' 662–63.
4. Pierson, *Special Effects* 1.
5. AMPAS's 'Academy Awards Database' (online).
6. The Academy Award for "Special Effects" from 1940 to 1964 is described as an "archaic category" in AMPAS's 'Academy Awards Database' (online).
7. For more details, see AMPAS's 'Academy Awards Database' (online).
8. See AMPAS's 'Visual Effects—Seeing is Believing' (online).
9. Blandford et al., *The Film Studies Dictionary* 221.
10. Bordwell & Thompson, *Film Art* 497; Monaco, *How to Read a Film* 136.
11. Smith, *Industrial Light & Magic* 268.
12. Smith, *Industrial Light & Magic* 270. See Appendix C for more details.
13. Dunlop et al., *The State of Visual Effects* (online).
14. See Appendix C for more details.
15. Zucker, 'New Media' (online).
16. Quoted in Ohanian & Phillips, *Digital Filmmaking* 187.
17. Pierson, *Special Effects* 11.
18. Pierson, *Special Effects* 137.
19. Purves, 'The Emperor's New Clothes' (online).
20. Harbord, 'Digital Film' 21.
21. Quoted in Vaz, *Industrial Light & Magic* vii.
22. See Vaz, *Industrial Light & Magic* 220–21.
23. See Appendix C for more details.
24. Morphing was originally developed and applied for the first time by software engineer and animator Doug Smythe at Industrial Light & Magic during VFX production for fantasy movie *Willow*, a few years earlier.
25. See Magid, 'Many Hands.'
26. Quoted in Magid, 'Many Hands' 64.
27. Wasko, *Hollywood in the Age* 30.
28. See Appendix C for more details.
29. Rodowick, *The Virtual Life* 6.
30. Keane, *CineTech* 59.
31. Cubitt, *The Cinema Effect* 11.
32. Benjamin, 'The Work of Art' 235.
33. Andrew, *Concepts* 27.
34. Benjamin, 'The Work of Art' 225.
35. Andrew, *Concepts* 19.
36. Mulvey, 'Visual Pleasure' 838-9.
37. Bazin, *What Is Cinema? (vol. I)* 24.
38. Andrew, *Concepts* 19.
39. Morris, 'Learning from Bruce Lee' 8.

40. Bazin, *What Is Cinema? (vol. I)* 158.
André Bazin made this comment in relation to the studio-based production of *Scott of the Antartic* (1948).
41. Quoted in Blanchard, 'Unseen Camera Aces' 268.
Bordwell et al. point out that during the late 1930s, 'the RKO Special Effects Unit, under Vernon Walker, had become famous for its realistic matte and optical printer work. In 1941, no writers acknowledged that many of *Citizen Kane*'s deep-focus effects had been created by Walker's unit.'
Bordwell et al., *The Classical Hollywood Cinema* 361.
42. Robins, *Into the Image* 13.
43. Robins, *Into the Image* 3.
44. Baudrillard, *Seduction* 30.
45. Ihde, *Philosophy of Technology* 111.
46. Andrew, *Concepts* 20.
47. Andrew, *Concepts* 16; Neale, *Genre and Hollywood* 32.
For further discussion of the concept of cinematic "verisimilitude," see: Aumont, *Aesthetics of Film*; Brewster, 'Film'; Genette, 'Vraisemblance'; Todorov, *The Poetics of Prose*.
48. Todorov, *Introduction to Poetics* 118–19.
49. O'Regan & Venkatasawmy, 'Only One Day at the Beach' 19.
50. Pye & Myles, *The Movie Brats* 226.
51. Conn, '*E.T. The Extra-Terrestrial*' 564–65.
52. Berardinelli, '*Close Encounters of the Third Kind*' (online).
53. Quoted in Ohanian & Phillips, *Digital Filmmaking* 199.
54. Cubitt, 'Phalke, Méliès' 120.
55. Butler, *Fantasy Cinema* 81–82.
56. Staiger, 'Introduction'—*The Studio System* 3.
57. Quoted in Ohanian & Phillips, *Digital Filmmaking* 240–41.
58. See Vaz, *Industrial Light & Magic* 116.
59. Staiger, 'Introduction'—*The Studio System* 3.
See also Mora, 'Disaster Films.'
60. Wallace, 'F/X Porn' (online).
61. Tryon, *Reinventing Cinema* 40.
62. Purves, 'The Emperor's New Clothes' (online).
63. Berardinelli, '*Stargate*' (online).
64. Berardinelli, '*Independence Day*' (online).
65. Ebert, '*Independence Day*' (online).
66. Berardinelli, '*Starship Troopers*' (online).
67. Ebert, '*Starship Troopers*' (online).
68. Ebert, '*Armageddon*' (online).
69. 'The Visual Effects'—*The Fifth Element* (online).
70. Craig, 'Establishing New Boundaries' 164.
71. Craig, 'Establishing New Boundaries' 159.
72. Quoted in O'Regan & Venkatasawmy, 'I Make Films' 30.
73. Craig, 'Establishing New Boundaries' 159.
74. O'Regan & Venkatasawmy, 'Only One Day at the Beach' 24.
75. Craig, 'Establishing New Boundaries' 159.
76. See Smith, *Industrial Light & Magic* 212.
77. For further explications of the concept of "iconography" in the cinema, see Alloway, 'On the Iconography'; see also Panofsky, *Meaning in the Visual Arts*; Panofsky, 'Style and Medium.'
78. Hayward, 'Industrial Light and Magic' 135.
79. See Gaudreault, 'Theatricality, Narrativity'; Jenn, *George Méliès cineaste*; Malthete-Méliès, *Méliès et la Naissance*.
80. Metz, 'Trucage' 667.
81. Wasko, *Hollywood in the Age* 38.
82. Langford, *Post-Classical Hollywood* 207.

83. *The Thing from Another World* featured on film the very first monster from space in 1951.
84. See Appendix B.
85. Pye & Myles, *The Movie Brats* 232–33.
The mechanical shark measured twenty-four feet, weighed one-and-a-half tons, and sank when it was first launched. By the end of production, due to frequent technical problems, the shark ended up putting a dent of $3.5 million into the allocated budget of *Jaws*.
86. Quoted in Lee, '*Total Recall*' 47.
87. See Wasko, *Hollywood in the Age* 22.
88. See Vaz, *Industrial Light & Magic* 117, 193 & 195.
89. Jones, 'Hollywood et la saga du numérique' 36.
90. Dahan, '*Matrix*' 138.
91. Essex, 'MatrixMania' 41.
92. Staiger, 'Introduction'—*The Studio System* 6.
93. Butler, *Fantasy Cinema* 79–80.
94. Quoted in Amos, 'Reality Bites' 28.
95. Baudrillard, *Seduction* 29.
96. Smith, *Industrial Light & Magic* 4.
97. Ebert, '*The Towering Inferno*' (online).
98. Bukatman, '*Blade Runner*' 24.
99. Ebert, '*The Hindenburg*' (online).
100. Mora, 'Disaster Films' 11.
101. Mora, 'Disaster Films' 12.
102. Tahiro, 'The *Twilight Zone*' 92.
103. For example, see Beacham, 'Movies for the Future.'
104. Quoted in Ohanian & Phillips, *Digital Filmmaking* 31.
105. This example is from Ohanian & Phillips, *Digital Filmmaking* 208.
106. Again, the example is from Ohanian & Phillips, *Digital Filmmaking* 209.
107. Ohanian & Phillips, *Digital Filmmaking* 205.
108. Ohanian & Phillips, *Digital Filmmaking* 199.
109. Ohanian & Phillips, *Digital Filmmaking* 199.
110. Quoted in Ohanian & Phillips, *Digital Filmmaking* 207.
111. Metz, 'Trucage' 672.

FOUR

Science, Technology and Hollywood Cinema

Although marked by various imperial wars, socio-political agitations and maneuvers of global territorial expansion and colonization, the last two decades of the 19th century witnessed influential discoveries and inventions that enabled the emergence of "new" media of artistic expression, information and communication, such as the cinema, and of their associated technologies of mass re-production and distribution—all endowed with the potential for future extensive commercial exploitation.[1] Of equal significance during that period are the beginnings of corporate-industrial structured science which would (i) establish the model of practically all techno-scientific research and development to follow during the 20th century, and (ii) enable the fulfillment of extensive commercial exploitation for discoveries and inventions by engendering cheap, efficient techno-industrial applications.[2] Corporate-industrial science and technology R&D would indeed be instrumental to the emergence and development of the cinema, of analog and digital filmmaking and of VFX, as well as to the late 20th century digital convergence between the media, communication, information and entertainment industries. Despite large-scale wars and dramatic socio-political revolutions, by the mid-20th century, there was even more intense corporate-industrial research and development following from many of the techno-scientific discoveries, inventions, prototypes, theories and concepts that had surfaced towards the end of the previous century.[3] Due to the extensive capital and human resources at its disposal, the Hollywood cinema industry's ability to initiate and nurture filmmaking-related research, development and innovation—essentially meant to support strategies of market exploitation and expansion—has never been considerably disrupted nor compromised, not even in times of major crisis or after, by comparison with other cinema institutions and industries internationally.

Derived from the Greek *teknnologia* meaning 'systematic treatment,' and from *tekné* meaning 'art' and 'skill,' it is from the 17th century onward that the term "technology" acquired its modern connotation as the application of the sciences to industry and commerce, inclusive of the methods, theories, and practices governing such application. A micro-level definition of the term suggests that:

- a technology must have some concrete component, some material element, to count as a technology;
- a technology must enter into some set of praxes—"uses"—which humans may make of these components;
- there must be a relation between the technologies and the human beings who use, design, make, or modify the technologies in question.[4]

Hence, a conceptualization of scientifically derived "technology" implies 'a means and a human activity [that] can therefore be called the instrumental and anthropological definition of technology.'[5]

In the present context, the understanding of "technology" amalgamates all of the above and is additionally associated with the systematic study of human skills, techniques, instruments and tools associated with filmmaking and VFX production used in the Hollywood cinema industry context.

> When we are dealing with science and technology, it is hard to imagine for long that we are dealing with a text that is writing itself, a discourse that is speaking all by itself, a play of signifiers without signifieds. It is hard to reduce the entire cosmos to a grand narrative, the physics of subatomic particles to a text, subway systems to rhetorical devices, all social structures to discourse.[6]

Science is not a thing at all, like gravity, electricity or atoms: these are only topics subject to enquiry in a scientific manner and do not constitute science by themselves. Science is a *process*, the process of enquiring about and of studying these topics. It is a process for reaching conclusions about what is probably true and what is not; in other words, the objective of science is to produce reliable information that can be applied in practice in the (re-)production and application of technology, amongst other things. Most systems of knowledge do not attempt to simply accept whatever "reality" might be but set out to rationalize what "reality" is believed to be. Scientific experimentation does not try to establish what is believed to be true but is designed to falsify theories; if they cannot be proven to be false, theories can only become stronger.

"Modern science" does not automatically mean 21st or even 20th century science. It implies, more productively, the historical roots of the so-called "modern" era that in turn means the Renaissance and following.[7] The Renaissance inaugurated significant engineering practices and mechanization in Europe, culminating with the Industrial Revolution, mostly as the outcome of science becoming more pivotal to a whole range of "applied" disciplines. A pivotal aspect of scientific investigation from the Renaissance onwards would be the growing centrality of instrumentation and technology-intensive measurement in the experimentation process.[8]

> To be experimental in the modern sense entails a number of factors, including setting up a situation in which certain variables can be controlled; in which a measurement occurs thus implying a mathematical or quantitative judgement about something; but above all . . . , experiment entails technologies or instruments against which and in relation to which the phenomenon is compared.[9]

This would hence enable, amongst other things, technologically embodied science's investigation of the phenomena of magnetism and its technologies that would ultimately lead to the electrical technologies of the 20th century.

By mid-20th century, 'the now established nomenclature which relates the "pure" or "theoretical" sciences such as physics, to the "applied" sciences such as engineering were already in place,' with science becoming 'fully a *technoscience* now thought to be the "motor" which drove and developed what is called Modern Technology, presumably distinct from any ancient or traditional technologies.'[10] Hence, in relation to the Hollywood cinema industry in particular, "science" and "technology" are largely conceptualized here according to a techno-culture formulated by actual techno-scientific objects and by their respective definitions (in terms of origin, relations, impact, physical and industrial characteristics).[11]

Techno-scientific innovation and so-called "new" technologies in any era have created and expanded the range of creative and aesthetic possibilities for Hollywood cinema. 'Innovations in film technologies have influenced the artistic conditions within which films have evolved,' as illustrated by Hollywood's conversion to sound movies that proved to be a catalyst in the evolution of the musical and the gangster film, for example.[12] Techno-scientific innovations have also influenced how meaning and knowledge is produced, exchanged and received about it through the development of its very own "mythology of evolution." A "mythology of evolution" is often driven by the techno-scientific innovations and "new" technologies central to the artistic production, aesthetic and economic pursuits of an industry. AMPAS's "Sci-Tech Council" and its "Scientific and Technical Awards," for instance, often shape the formation of the Hollywood cinema industry's "mythology of evolution" with central and peripheral icons, preferred medium of expression, stories and storytellers, sites and tools of production, and so on.[13] The centrality of techno-scientific innovation and the emergence of "new" technologies in the materialization and development of Hollywood cinema, and of their subsequent "mythology of evolution," cannot be ignored nor taken for granted. Historical study has

> served to emphasize how developments in various technologies have both directly contributed to specific cultural forms or otherwise provided a range of influential precedents for cultural developments. In no real sense can there be said to have been any model 'organic' past where the expressive impulse or productive craft was untrammelled by various enabling and prescriptive uses of tools or technologies.[14]

If what is mainly intended through the art of filmmaking is the re-presentation, re-production and simulation of "real," "probable" and fantastic worlds and the projection of illusion to generate sensual pleasures for the film viewer, then addressing the impact and influence of techno-scientific innovation on visual re-presentation and expression in Hollywood cinema should clarify the relationships between techno-scientific filmic objects and the human senses. That is to ascribe some degree of objectivity to the "realistic" and fantastic worlds artificially constructed by the Hollywood cinema industry.

Science and Technology: From "Analog" to "Digital"

Computer graphics and digital images of all sorts have indeed invaded human thought processes, becoming so naturalized that human perception, cognition and interpretation cannot occur without the inevitable reference to the technologically mediated digital visual sign.[15] Digital filmmaking, video and cinema had become so entrenched in popular consciousness and in popular culture by the end of the 1990s that what "digital" might mean was often popularly conceptualized derivatively from, or by association with, computers or, simply, with the domain of information technology in general. As such, a common distinction between "analog" and "digital" is often reduced to the fact that "analog" is continuous while "digital" is discrete. By virtue of their use here—predominantly to qualify technologies of filmic production in the Hollywood cinema industry—the implications of "digital" and "analog" are extensions of this. While for Janet Harbord, 'digital film is at once an object, an affectual experience, an idea/memory, a system of code, and transferable data,' for David Rodowick, because the digital arts are

> without substance and therefore not easily identified as objects, no medium-specific ontology can fix them in place. The digital arts render all expressions as identical since they are all ultimately reducible to the same computational notation. The basis of all representation is virtuality: mathematical abstractions that render all signs as equivalent re-

> gardless of their output medium. Digital media are neither visual, nor textual, nor musical—they are simulations.[16]

"Digital" and "analog" *designate* media, the techno-industrial vehicles that connect distant and separated coordinates on the grid of human communication. They are neither technology nor science in themselves. "Digital" and "analog" in isolation are both quite ambiguous in meaning, gaining credence only in relation to techno-industrial processes or vessels that allow messages, information or communication of any kind to be conveyed across space and time. It becomes necessary to explore, albeit briefly, the potential meanings of these two terms, as informed by techno-scientific progress, before arriving at definitions of "analog" and "digital" that apply productively to the Hollywood cinema industry.

"Analog" has the Latin root *analogus* and the Greek root *analogos*, and means "proportionate." "Analogy," in relation to logic and mathematics, is a form of reasoning that allows for a parallel to be established between two or more things (objects, concepts and so on) on the basis of identifying similitude(s) between them in other respects.[17] The sharing of elements, the presence of similarity or commonality, is what connects things and allows them to entertain a bond of continuity. This constitutes the premise of an analog medium.

The term "analog" is most commonly used in relation to devices or methods that utilize non-discrete variations in amplitude, frequency, location, and so on, in order to transmit or symbolize mathematical data, signals, sounds or other types of information; instead of being in steps, these signals vary continuously. For example, the mechanical clock is an analog medium because, by and large, it displays the continuous passing of time. A particular time is an intersection of the continuous movement (although at varying speeds) of each hand of the mechanical clock, but it would still be impossible to pinpoint exact time.[18] Analog technologies "mimic" information so that, for example, vocal sounds are technologically represented as electrical signals—with amplitude and frequency being proportional to the volume and pitch of the vocal sounds in question. Timothy Binkley points out that, until recently, media conformed 'to what is called an *analog* paradigm characterized by an imprinting process. Analog media store information through some kind of *transcription* which transfers the configuration of one physical material into an analogous arrangement in another.'[19] As an analog medium, the traditional photograph stores information via the interaction between light and celluloid chemicals. This interaction establishes a first level of continuity and constitutes the transcription that transfers the configuration of any three-dimensional object in space into its analogic re-arrangement upon the flat surface of a two-dimensional plate. The latter cannot be fragmented into discrete or separate parts: everything is bound together on the surface of the paper by chemical emulsion, which constitutes a second level of continuity and which is what intrinsically makes photography an analog medium. The vinyl-based music record is yet another analog medium because the music that it contains is basically sound information recorded into one continuous groove, as opposed to the digital Compact Disc where the smaller tracks constitute a sequence of distinct bits—the music being recorded as discrete numerical combinations of 0s and 1s.

Just like still photography, motion picture celluloid requires various technological transformations—electrical and chemical processing, light projection and so on—until the recorded re-presentation can be visually perceived by the human eye. As an analog medium, traditional motion picture celluloid imbues 'objects with resilient marks perceivable either directly through the senses or indirectly through a display process that carries out an additional transcription.'[20] The recording of motion on celluloid has to be projected in a linear logical sequence otherwise it would make no sense. It is not the single still frame but a series of still frames moving in continuous regular order that transcribes motion into visually perceptible forms. As an extension of celluloid-based moving pictures, videotape-based motion picture systems are similarly analogic, functioning predominantly on the basis of continuous variable quantities, such as electric signals. 'When an image is transcribed into an electric signal . . . , a particular amount of light measured in luxes is matched to a different amount of current measured in volts. This kind of co-ordinated matching is the hallmark of an analog medium.'[21] Continuity is hence absolutely vital to the process of moving pictures, especially in relation to how the human eye perceives analogically.

"Digital" is a derivation of "digit" which has the Latin root *digitus*, meaning toe or finger; "digit" can also be applied to any of the first ten Arabic numerals. The original meaning of "digit" is already an indication of the discrete nature of the digital: each finger or each number is a very distinct, separate component which does not rely on continuity nor on its relation to other components, to be perceived as such—as opposed to an analogic hand gesture which relies on the continuity of motion of all components of the hand to be actually perceived as such. The digital clock shows only a set of numbers which pertain to very specific moments in time in opposition to the mechanical clock: digital time is never passing but "arrested" or frozen, so to speak, at precise points. And a number is ultimately an abstraction that does not enjoy any form of concrete physical existence. There is nothing concrete nor any malleable form within the electronic digital realm since everything in it is pure information or data of one kind or another. Hence, the digitally created image is originally rooted in some form of the "real," but is ultimately re-presented as intangible and yet logically coherent information.

The simplest understanding of a digital medium can be expressed in terms of what it is not: the digital medium is basically discrete as opposed to the continuity that characterizes the analog medium. The digital medium does not need continuous variable input: its input is discrete (derived from the Latin root *discretes*, meaning "separated" or "set apart"), made up of numbers, characters and letters combined and programmed in some form of language and processed internally to a device in binary notation—0s and 1s. The digital medium operates according to a method of signal representation by a set of discrete numerical values, as opposed to analog media that operate according to signals that continuously vary. Because it is all primarily constituted by strings of discrete separate numbers, digital information can be naturally translated from one digital medium into another: numerical values all belong to the same abstract dimension and can be replaced unproblematically by other numerical values.

An analog signal can usually be converted to digital by making use of an analog-to-digital converter chip that collects samples of the signal at a fixed time interval (such as sampling frequency).[22] A binary number is assigned to these samples that make up a digital stream which is then recorded. Upon playback, a digital-to-analog converter chip decodes the binary data and re-constructs the original (now modified) analog signal.[23] Because an analog signal is received and amplified at each repeater station, any noise accompanying the signal will hence also be amplified. A digital signal, on the other hand, is detected and re-generated (not amplified): any accompanying noise will be lost, unless it corresponds to a value that the re-generator interprets as a digital signal. Therefore, since digital signals are virtually impervious to noise, distortion and other quality problems, the problem of generation loss does not occur with digital-to-digital conversions: each duplicate being an exact copy without perceptible differences of the original master, multi-generational dubs can be made without signal degradation.

The analog image can be fragmented into variations of shade, gradations in linear or aerial perspective, with continuous unity prevailing. But the digital image, on the other hand, can be broken down to the single pixel, which can be extracted or replaced since it has no specific meaning or identity of its own—except as a number—unless it is attached to other strings of numbers or groups of pixels. To follow William Mitchell's explanation:

> images are encoded digitally by uniformly subdividing the picture plane into a finite Cartesian grid of cells (known as *pixels*) and specifying the intensity or color of each cell by means of an integer number drawn from some limited range. The resulting two-dimensional array of integers (the *raster* grid) can be stored in computer memory, transmitted electronically, and interpreted by various devices to produce displays and printed images. In such images, unlike photographs, fine details and smooth curves are approximated to the grid, and continuous tonal gradients are broken up into discrete steps.[24]

The analog image possesses an indefinitely large amount of information, although the automatic loss of generation from continuous reproduction will lead to increasing graininess, fuzziness and degradation, resulting in a noticeable difference between original and copy. The digital image, on the other hand, in being contrived within a grid can only but express limited information. The continuous reproduction of the digital image will suffer no loss of generation and no degradation precisely because of the numerical precision and specificity provided by the raster grid—hence the lack of noticeable difference between original and copy.

The technology that allows the digital manipulation of photographs (eventually developed to accommodate other visual media) was originally pioneered at MIT by William Schreiber and Donald Troxel.[25] The project started in 1969 with a ten-year contract from Associated Press to solve their problems with differences in wire-photo standards between the United States and the rest of the world. Schreiber and Troxel invented a laser photo facsimile system and then an "Electronic Darkroom" for Associated Press. 'Photographs and slides were read by a laser scanner in minute detail, translated into digital form, and stored in a computer, where the images could be reshaped and edited at convenience.'[26] In this way, any form of retouching or manipulation applied to it would be barely detectable on close inspection. For instance, physical dodging or airbrushing in a chemical photograph is quite easily noticeable as a foreign element "forcefully" introduced into organized continuity. In the case of celluloid film, blue outlines can sometimes be observed around characters and objects when the chroma key compositing technique is used or when matte backgrounds "leak" into foreground elements. Visual retouching or manipulation by the intermediary of the digital medium, on the other hand, is practically undetectable. One of the determining factors for picture quality is the extent to which the grain is or is not noticeable. According to William Schreiber:

> If you have a picture represented by a discrete set of numbers, and you change some of the numbers, you may not be able to tell that that was not a natural image. . . . Since the computer-manipulatable pixels can be smaller than the grain of film, there is no limit to how persuasive the detail can be in a retouched photo.[27]

The smallness of the pixel implies tremendous potential and advantages over the grain of motion picture celluloid in terms of both processing and manipulation. In the domain of VFX production, analogic manipulation and image compositing tend to increase the visibility of film grain—as an outcome of generation loss through repeated re-production—and, by extension, the visibility of the actual manipulation of the medium, which is usually unwanted. Each analogic alteration of the image implies the repetitive involvement or processing of the whole image since moving pictures cannot be separated into distinct parts but are bound together in continuity. But the pixel—the smallest element of the digital image—is much smaller than is visible to the human eye. The microstructure of grid-cells is the bottom line as far as alteration is concerned. Unless we go down to the atomic level, there is no more information beyond the pixel. William Brown explains that

> if we consider images in the same way that a computer does, the information that comprises the digital image is rendered simply as a series of pixels, each of which represents a certain color in a certain position. Whether each pixel is, when we see it on-screen, "part of" a human, a sky or a wall is, to the computer, irrelevant. . . . For the computer, therefore, a wall, a human, sky, an animal, a planet—each is simply pixels, with no pixel having greater or lesser prominence or meaning than any other.[28]

In the digital realm, the image is processed and stored as numerical information so that a partial alteration is translated only as an alteration of certain sequences of numbers with other sequences being unaffected since they are organized discretely. The process of visual alteration, translation and re-production is perceptually unnoticeable and suffers no generation loss, with the end-product copy being as good as the original source. As noted in the late 1980s by MIT's Media Laboratory founding director Nicholas Negroponte: 'Digital is a noise-free medium, and it can error-correct. I can see no reason for anyone to work in the analog domain anymore—sound, film, video. *All* transmission will be digital.'[29]

Techno-Scientific Innovation and Hollywood Cinema

Considered to have been instrumental in the development of monocular perspective painting, the invention of the camera obscura, for example, confirmed how even 'the most traditionally artistic forms of cultural production have been profoundly influenced by various mechanical and scientific innovations.'[30] A decisive moment in the evolution of what would later become the cinema

> undoubtedly came with the discovery of the first scientific and already, in a sense, mechanical system of reproduction, namely, perspective: the camera obscura of Da Vinci foreshadowed the camera of Niepce. The artist was now in a position to create the illusion of three-dimensional space within which things appeared to exist as our eyes in reality see them.[31]

Towards the end of the 19th century, the cinema materialized from the sharing of a common background of progress—in scientific discovery, techno-industrial innovation, visualization technique and artistic practice—with painting, photography and the visual arts in general. The

> guiding myth, then, inspiring the invention of cinema, is the accomplishment of that which dominated in a more or less vague fashion all

> the techniques of the mechanical reproduction of reality in the nineteenth century, from photography to the phonograph, namely an integral realism, a recreation of the world in its own image, an image unburdened by the freedom of the artist or the irreversibility of time.[32]

Before becoming a techno-scientifically informed art however, the cinema was primarily conceived as a technology meant to enable the visualization and recording of motion by virtue of the

> principles of vision and the manufacture of film, camera, and projector—these matters of optics, chemistry, and machinery are inherent in the motion picture as a device. The art of the motion pictures, depending on the instrument, had to wait for the invention of the device. The machine, however, was not invented to make the art possible. It was originated merely as device—a device to record and depict motion.[33]

The subsequent growth and evolution of the cinema as an art form was the outcome of scientific discovery in combination with a number of progressively emerging "new" technologies. Each of these found its essence in how it was made to contribute to the artistic expression and aesthetic pursuits of particular individuals in the process of making motion pictures.[34] This can be illustrated by the short fantasy film *The Motorist*—produced and shot by Robert W. Paul in 1905—which pioneered several VFX techniques that would later be developed more extensively. It featured a couple in a magical motorcar that drives up buildings, flies into outer space and travels through the solar system before coming back to Earth. From a background in building equipment as a mechanical engineer, Robert W. Paul first became involved with the cinema when he devised and built in 1896 a film projector that he would subsequently make and sell in large numbers (mainly because of existing restrictions on the number of Lumière projectors that could be sold). Robert W. Paul had constructed by 1897 Europe's first film studio which included a dolly track for camera movements, a hanging bridge and trap doors. According to André Bazin,

> up till about 1938 the black-and-white cinema made continuous progress. At first it was a technical progress—artificial lighting, panchromatic emulsions, travelling shots, sound—and in consequence an enriching of the means of expression—close-up, montage, parallel montage, ellipsis, re-framing, and so on. Side by side with this rapid evolution of language and in strict interdependence on it, filmmakers discovered original themes to which the new art gave substance. [This] was directly attributable to technical progress—it was the novelty of expression which paid the price for new themes.[35]

With regard to VFX in particular, *The Lost World* in 1925 was a techno-scientific breakthrough by virtue of pioneering stop-motion animation techniques and essentially ushering in

> a whole new genre of fantasy cinema, the "dinosaur" movie. The many plausible monsters were the work of the famous early master of stop-motion animation, Willis O'Brien, who was the first person to perfect the technique of photographing miniature models of monsters a frame at a time, and combining the results with live action.[36]

The application of emerging "new" technologies influenced the creation of motion pictures ranging from the excitingly innovative, like *The Lost World*, to the plain ordinary. All were works of art nevertheless and contributed productively to the techno-industrial practices and aesthetic development of the infant art of the cinema. Willis O'Brien's innovative VFX work also confirmed in the early days of cinema the status of the VFX practitioner as "artist-engineer-scientist" just as Leonardo Da Vinci and Galileo were considered as "artist-engineer-scientists" during the Renaissance period of history. Another parallel can also be drawn between early 20th century VFX practitioners of Hollywood cinema, like Willis O'Brien, and 18th century '*quadraturistas*, or illusionistic painters, [who] operated practically as a guild through Western Europe, along with the final phase of the Baroque fantasy—automata and "toys."'[37]

> Thus by 1938 or 1939 the talking film, particularly in France and in the United States, had reached a level of classical perfection as a result, on the one hand, of the maturing of different kinds of drama developed in part over the past ten years and in part inherited from the silent film, and, on the other, of the stabilization of technical progress. . . . In short, with panchromatic stock in common use, with an understanding of the potentials of the microphone, and with the crane as standard studio equipment, one can really say that *since 1930 all the technical requirements for the art of cinema have been available.*[38]

For Hollywood cinema, once the "right" technological structure was firmly in place, consequent techno-scientific R&D pertaining to motion picture-related equipment would become much more corporately driven—commercially motivated by the need to improve on-screen spectacle in order to attract more spectators to theaters and boost box-office profits. 'Adopting new technologies did not alter the structure of the industry the way talkies did; rather, they enabled Hollywood to operate more efficiently and enhanced the techniques of conventional storytelling.'[39] Corporate techno-scientific filmmaking R&D was rapidly normalized in the Hollywood cinema industry.[40] Before, during and after the studio system, there was no interruption in the industry's continuous reliance on intensive techno-scientific filmmaking R&D in devising strategies to stir up audience interest in Hollywood movies, most of them involving experiments with screen size and depth illusion.[41]

Hollywood cinema's standards of imaging and the related techno-visual R&D—film stock and processing, lenses, theatrical screening hardware, digital VFX and so on—have been characterized by a perpetual quest for flawless re-presentation and re-production onto the two-dimensional media of flat celluloid strips, theater and later television screens. Hollywood films, as well as the cinema theaters in which they are usually screened, illustrate the constant obsession with enabling the perfect audio-visual experience by the film viewer. Techno-visual innovations in film production and exhibition have indeed often been marketed with the justification of enabling the Hollywood movie experience to become increasingly more "realistic" as well as "larger than life." This can be illustrated, for instance, by the continuous competition in film format and screening R&D since the beginnings of the industry to achieve the perfect motion picture format of production and exhibition. The embryonic movie studios that would later constitute the Hollywood cinema industry mostly favored the 1.33:1 aspect ratio during the silent era until the advent of sound film around the mid-1920s. The Academy of Motion Picture Arts and Science had proclaimed by 1932 the Academy Ratio as the standard for motion picture production and exhibition in the Hollywood cinema industry: 35mm film projected at 24 frames per second with an aspect ratio of 1.37:1—which would be used in the Hollywood cinema industry and in cinema theaters, until the advent of widescreen with Cinerama and with Twentieth-Century Fox's CinemaScope process in the early 1950s.[42] CinemaScope 'compressed the image onto a conventional 35mm filmstrip through an anamorphic lens fitted to a single camera. A corresponding lens on the single projector then unsqeezed the image, expanding it over a wide screen. Apart from its single camera and projection system . . . , CinemaScope required a less severe curve (1.5m at its deepest) and its effects could be appreciated from any seat in the house.'[43] By 1956, CinemaScope—eventually to become Panavision—with an aspect ratio of 2.35:1, had become a standard film format in motion picture production and exhibition. With film gauge standards progressing from 35mm to 70mm, and with aspect ratios evolving from the Academy 1.37:1 to the CinemaScope 2.35:1, cinema theater screens similarly evolved to accommodate changes in film formats from 35mm (4.6m X 8.6m) to 70mm (9m X 19.7m).

Other than the development of film format and exhibition technologies, the introduction of Photoshop into filmmaking is an additional illustration of the importance attached to techno-scientific R&D, specifically in digital visual media, by the Hollywood cinema industry. Developed during the late 1980s by John Knoll at Industrial Light & Magic and his brother Thomas from the University of Michigan, Photoshop is a draw-and-paint software specifically designed for use on Macintosh computers.

> Photoshop edits digitized pictures, allowing computer artists to perform such image-processing functions as rotating, resizing, and color-correcting images and painting out scratches, dust, or other flaws. The

> software is versatile enough to be used for texture-map editing in computer graphics, creating matte paintings, and designing concepts in the art department.[44]

It was first extensively applied during the making of James Cameron's *The Abyss* in 1989 to help manipulate digitized photographic data of background plate sets in which to integrate the film's computer-generated non-terrestrial "pseudopod." After *The Abyss*, Photoshop would become a major digital tool in the domain of VFX production to develop visual concepts; digitally create matte paintings; and digitally paint shots to completion. It was subsequently published by Adobe and commercially released in 1990. Photoshop was emblematic of the increasing synergy of techno-visual R&D between the art of cinema and computer science. Originally intended for the desktop publishing industry, the cinematic application of Photoshop in the production of digital VFX illustrates Philip Hayward's notion of a 'system of enablements' that is inevitably generated by the combined use of sophisticated technologies:

> In the case of individually customized, adapted or combined systems, their operations often deviate considerably from the functions and potential uses originally conceived for them. In this way, advanced cultural technologies can be seen to set up a system of enablements within two polarities: 'preferred' or conventional uses (designated or indicated by the technology and its accompanying packaging, instructions or contexts of use); or more original applications (arising from both the manner and method of their employment of technologies and also the design and contextual application of the texts so produced).[45]

Intensive corporate techno-scientific R&D in 3D computer imaging and animation in application to filmmaking—begun at Lucasfilm's Computer Division in the early1980s—led to the release of the first entirely computer-generated feature-length 3D animation motion picture in 1995. The associated innovations would lead to the increasing use of animated 3D objects and characters in live action Hollywood filmmaking.

A more recent illustration of the importance of techno-scientific R&D in digital visual media is the emergence of computer-generated actors or "synthespians" within the context of late 20th century digital VFX production.[46] Not exactly "humanoid" nor "simulacrum" of anything in particular, *Terminator 2: Judgment Day*'s T-1000 "liquid mercury" android and Jar Jar Binks in *Star Wars: Episode I—The Phantom Menace*, for instance, are fundamentally concepts of visual "fantasy," eventually acquiring a filmic "existence" on-screen from being virtual quasi-objects made up of polygons, algorithms and binary information within the digital realm.

> Le cinéma, comme tout art, est avant tout une fabrique de corps et la synthèse reprend en l'état ce questionnement permanent sur la nature de l'homme. Qu'est-ce qu'un corps humain? Comment, à partir du plus

> immatériel possible (0 et 1) figurer le matériel par excellence, la référence commune à toute personne? La synthèse propose deux réponses . . . , la ressemblance et la dissemblance. Bien avant les figurants virtuels de *Titanic,* la synthèse s'efforce de "copier" à l'identique le corps de l'homme (comme c'est le cas, du moins conceptuellement, dans *Futureworld* (1976), *Looker* (1981), *RoboCop II* (1990)).[47]

> The cinema, like all art, is first about producing the human body and digital imaging takes on the permanent questioning of the nature of man. What is a human body? From the most immaterial (0 and 1), how to produce the utterly material, that is the common reference to any body? Digital imaging proposes two answers . . . , similarity and dissimilarity. Long before the virtual crowds of *Titanic,* digital imaging has been striving to identically "copy" the human body (as is the case, at least conceptually, in *Futureworld, Looker, RoboCop II*).

Synthespians also illustrate the kind of problem faced by digital cinema that similarly affected the evolution of painting after the invention of the camera obscura centuries ago. With computer-generated actors, the first psychological phase of perfecting duplication by the intermediary of techno-scientific advancement was well on its way with regard to the visual re-production of human characters. But what would be the next level of evolution (aesthetic and otherwise) for digital cinema once the digital synthespian had been techno-visually perfected in the re-presentation of its human model?

While the development and perfection of analog cinema-related technologies required almost a century, it took less than twenty-five years for digital cinema-related technologies to be developed and perfected in Hollywood. Digitization 'developed the platforms of delivery and exchange that have facilitated an increase in the circulation and accessibility of objects. The tension played out through digital technology replicates the larger tension in Hollywood film culture, of dispersion and gathering.'[48] By the end of the 20th century, digital filmmaking and the digital exhibition of Hollywood films in cinema theaters had become reality (albeit on a limited scale), while, for broadcast television, the transition from analog formats (NTSC, Pal and so on) to the digital HDTV standard had been initiated.

The rapid evolution within a couple of decades of the digital medium confirms Moore's Law which is based on Gordon Moore's prediction that the number of transistors on a computer chip would increase from 2,300 (as on the 1971 Intel 4004 processor) to 5.5 million (as on the 1995 PentiumPro processor for instance). The story goes that in 1965

> Gordon Moore was preparing a speech and made a memorable observation. When he started to graph data about the growth in memory chip performance, he realized there was a striking trend. Each new chip contained roughly twice as much capacity as its predecessor, and

> each chip was released within 18–24 months of the previous chip. If this trend continued, he reasoned, *computing power would rise exponentially over relatively brief periods of time.*[49]

Moore's Law about the progress of computing power could equally apply to the accelerated development and use of increasingly complex digital technologies in filmmaking and VFX production.[50]

The disorderly poly-temporal and poly-spatial development of computers equally applies to the evolution of digital VFX that are, after all, the outcome of a combination of temporally and spatially overlapping industrial production domains and techno-scientific spaces of enquiry such as physics, computer science and filmmaking R&D, amongst others. Digital VFX are poly-temporal and poly-spatial. They are ultimately the result of at least three overlapping spirals: progress in scientific knowledge, in computer science and in cinema technology. This "spiral" imagery (borrowed from Bruno Latour) implies the co-existence of different temporal and spatial dimensions, as illustrated for example by the continuing significance of the "old" technologies of analog photography, television and cinema alongside that of their "new" digital counterparts.[51] Similarly, VFX are still produced through the interaction of both "old" and "new" analog and digital technologies. For example, digital VFX production for miniature shots in general practically follows similar rules and principles to their analog counterparts in the early days of cinema—just as 20th century digital photography essentially follows the same principles that had emerged during the 19th century.

The digitization of filmmaking in Hollywood did not emerge without points of origin. It was the outcome of many years of experimentation, discoveries and prototypes in the intersecting domains of VFX production, motion picture production and computer technologies. Cinema technology in general has never stopped evolving since the public demonstration of Thomas Edison's and William Dickson's Kinetograph and Kinetoscope in 1893, and of the Lumière brothers' Cinématographe in 1895; and in the case of computer technology, experimentation, research and development have intensified since World War II in particular.[52] But, except for a handful of exceptional instances of experimentation, it is only from the late 1970s that the film/digital synergy started to gain impetus, becoming increasingly important within the Hollywood cinema industry—by the intermediary of specific VFX requirements in certain VFX-intensive films in particular.

Since digital naturally "feeds" on moving images, digitization was an inevitable phase in the techno-industrial evolution of Hollywood cinema. By the early 1980s, visual outcomes resulting from the application of digital technologies in certain types of VFX-intensive productions prompted filmmakers, producers and major studios alike to consider seriously the long-term potential and implications of digitizing filmmaking tools for the whole industry. Many became interested in the potential advantages of the film/digital synergy for filmmaking by embracing the digital realm. Digitization in the Hollywood cinema industry became a sufficient concern to influence R&D investment priorities. This improved the convergence of cinema's technologies with those of the digital medium and re-directed the evolution of filmmaking technologies in terms of closer connection with innovations in the field of computer technology.

The Hollywood cinema industry's digital transition has never achieved an absolute sense of closure since analog filmmaking technologies have not been entirely phased out by their digital counterparts. For Ohanian and Phillips, in the mid-1990s, the term "digital filmmaking" referred to 'a methodology that combines certain traditional filmmaking techniques with new capabilities that have come about through the integration of computers, digital image manipulation, disk recording, and networking.'[53] Analogic difference still exists and persists in and alongside digital continuity. Thus, techno-industrial transition to the digital medium has occurred in and alongside analogic continuity within the Hollywood cinema industry.

Ultimately, change and continuity can only co-exist with regard to techno-scientific development in the Hollywood cinema industry. One techno-industrial standard does not necessarily or logically lead to, or depart from, another simply on the basis of sharing similar foundations and coordinates of departure. And there are many spatio-temporal gaps and unaccounted for parts of the overall narratives of Hollywood's "mythology of evolution" as well as overlaps between research, inventions, release, standardization and use of particular technologies, devices and products because there are rarely pre-defined standards or rules to follow.

Techno-scientific innovation can be said to have progressed—according to the Hollywood cinema industry's "mythology of evolution"—from an "analog" (mechanical, chemical or electrical) state, through an intermediate electronic phase, to a digital state. Economically motivated patterns of techno-industrial change to the art of filmmaking within Hollywood have usually been triggered by the overlapping interactivity of domains in continuous evolution—corporate R&D in physics, computer science, filmmaking and VFX production. The digitization of filmmaking technologies has indeed propelled the art, practice and consumption of Hollywood cinema into economic, artistic and aesthetic directions that would have been unimaginable when there was no alternative to analog filmmaking.

Techno-Industrial Transformations: "Revolution" or "Transition"?

For some, modern technology displays a clear pattern of development and progress, as illustrated by Langdon Winner for instance. 'If there is a distinctive path that modern technological change has followed, it is that *technology goes where it has never been*. Technological development proceeds steadily from what it has already transformed and used up toward that which is still untouched.'[54] But there seems to be limited patterned regularity, spatio-temporal order and "neatness" in the occurrence of events' relative techno-scientific innovation, transformation and progress. According to Brian Winston's version of the history of technoscience for instance, innovations and inventions have extensive poly-temporal and poly-spatial "family-trees" and phases of development.[55] The computer as we know it is a good example of disorderly poly-temporal and poly-spatial techno-scientific evolution. In 1642, Blaise Pascal made the first mechanical calculator in France, which constituted an opening towards artificial or non-human reasoning. In 1834, Charles Babbage's "difference engine" in England would be an even stronger move in the direction of the 20th century computer. In 1849, based on Ramon Lull's 13th century logic diagrams, George Boole invented what would become known as "Boolean algebra," the foundation of binary logic in 20th century computing. And in 1889, Herman Hollerith in the United States created a tabulating machine that used punch cards for computation. Another important breakthrough in the development of computer technology was Lee de Forest's thermionic triode valve in 1906 that applied Boolean algebra in enabling the comparison of two electronic inputs to produce a logical output.[56] And the last crucial event in the genealogy of the computer would be Alan Turing's 1937 theoretical computing machine in England. Hence, before its actual materialization, the computer and most of its fundamental principles and characteristics had existed conceptually, albeit often separately, over many centuries.[57]

> The idea of the computer as a stored program device was first made in 1944, although glimmers of it as a possibility can be traced back over the previous century. Five years later, the first giant machines worked: 10 years later, there were a couple of hundred computers in the world: 20 years later, some 30,000. In 1976, the first microcomputer appears, although throughout the 40-year period of the giants, various 'baby' machines had been stillborn. Five more years pass before the PC achieves a significant measure of diffusion. [A] similar time frame is described for the microprocessor heart of the minicomputer. A suggestion for a solid-state device is made in 1925 and various partial prototypes are demonstrated from the 1930s to the 1950s. The idea for a solid-state circuit is articulated in 1952. Contrary to popular opinion, the move from primitive transistor to CPU takes place virtually independently of the computer industry; and it takes two decades. Then a

> further six years elapse before anybody puts a CPU into a small computer.[58]

The early 1950s also witnessed the parallel emergence of the first interactive computer graphics, as information actually displayed on a computer screen. This was an outcome of cybernetics pioneer Joseph Licklider's programming work—a subsidiary innovation that occurred while the technological medium itself was still in the making. By 1958—the year in which the first fully transistorized supercomputer was built—while computer graphics were then very recent, the microprocessor and the CPU were already decades old, and Boolean algebra and algorithms were much older, but they all cohabitated poly-temporally and interacted productively through the development of the computer as a techno-scientific object. Here we see change in and alongside sameness.[59]

Through techno-industrial mediation, experience, knowledge, information and meaning have been turned into manufactured, evaluated, deconstructed, adapted, regulated and infinitely reproducible quantifiable audio-visual entities in the age of digital reproduction. Mark Taylor and Esa Saarinen's distinction between "meaning" and "information" deserves mention here:

> It is important not to confuse information and meaning. Though not precisely opposites, they are inversely proportional: as information increases, meaning decreases. One of the distinctive features of the information age is the proliferation of data whose meaning remains obscure. The more we accumulate the less we have.[60]

The term "information age" came into use in the 1990s to characterize the proliferation of visual data in particular as a source of meaning and knowledge—enabled by electronic and digital technologies of communication and reproduction—that supersedes the written or printed word.[61] Eliminating the need for translation across linguistic boundaries, the digital medium conveys information recorded as 0s and 1s, hence constituting a truly universal techno-industrial medium. In being designed to be information-based and "image-centric," nearly all technologies pivotal to mass communication traffic had gone digital by the end of the 1990s.

The contribution of the digital medium to universalizing meaning and knowledge as information-based and "image-centric" subsequently influenced the increasing use in popular discourse and elsewhere of such terms—as an extension of "information age"—as: "computer age," "cybernetic age," "new age of information," or "new era of digital revolution."[62] These terms are validated by their users on the basis of the proliferation and fast circulation of information, the abundance of its accompanying technologies and their vast potential to rapidly achieve anything previously considered complicated. Stewart Brand, for instance, writes that

> [w]ith digitization all of the media become translatable into each other—computer bits migrate merrily—and they escape from their traditional means of transmission. A movie, phone call, letter, or magazine article may be sent digitally via phone line, coaxial cable, fibre-optic cable, microwave, satellite, the broadcast air, or a physical storage medium such as tape or disk. If that's not a revolution enough, with digitization the content becomes totally plastic—any message, sound, or image may be edited from anything into anything else.[63]

According to Russell Neuman (head of the Audience Research Facility at MIT's Media Laboratory in 1988), '[t]here's a natural instinct to see either a revolution or a conspiracy in every new technology that comes down the pike.'[64] But the semantic dissection of the above terms brings about certain conceptual paradoxes. The "computer age," "digital revolution," or "new age of information" do not particularly describe "new" phenomena as they were so enthusiastically deemed to be. Brian Winston, for instance, believes that 'we are now entering the fifth decade of the so-called computer age'—hence shattering popular misconceptions about the "computer age" or "new age of information" as "revolutionary" phenomena.[65]

Societies have mostly operated on the basis of shared available information and of its reproducibility. As such, information and the reproducibility of art—at various stages of techno-industrial development since Gütenberg's invention of the printing press in 1450—have been in a state of rapid proliferation and circulation in close alignment with whatever communication technologies and media might be available during any chosen historical period. According to Walter Benjamin, 'Historically, [the mechanical reproduction of the work of art] advanced intermittently and in leaps at long intervals, but with accelerated intensity.'[66] In the age of digital reproduction, Benjamin's commentary is relevant to the techno-industrial evolution of VFX production in the Hollywood cinema industry. The last two decades of the 20th century constituted its phase of 'accelerated intensity.'

By the 1990s, it became fairly common for production personnel and filmmakers alike, such as James Cameron, to employ such terms as "digital renaissance" and "digital age" while witnessing and embracing the techno-industrial digitization of their craft.[67] Coming from James Cameron—as illustrated by his constant use of innovative, state-of-the-art VFX in his movies (*Titanic*, the *Terminator* films, *The Abyss*, *Aliens*)—such terms sound extremely plausible and exciting indeed. But concepts of digital 'age' and 'renaissance'—well promoted and validated by the intensified presence of flash VFX and slick images on cinema and television screens—created a strong *impression* that the Hollywood cinema industry and film audiences alike might well be in the midst of a "revolution," of an extraordinary occurrence blindingly radiating a newly found sense of techno-visual freedom and originality.

Talking about a digital "revolution" or "renaissance" to describe the techno-industrial transformation of Hollywood cinema can be problematic. Such terms overshadow the conceptualization of digitization as a continuously evolving process that diachronically combines and integrates technologies and scientific discoveries belonging to separate spatio-temporal dimensions. There may not in fact have been any digital "renaissance" or "revolution" with regard to Hollywood cinema if technologies of the screen are deemed to be *always already* in a state of evolution and are characterized by the constant cohabitation and interaction of so-called "new" with "old" technologies.

It is often too convenient to associate the term "new" with any recently or currently emerging technologies. This is often done on the double popular assumption that "new" technologies only emerge to replace technologies that have apparently become "old," and that "new" technologies are necessarily better, in one way or another, than those they are supposedly replacing. While the term "new technology" signifies the recent development of specific technologies and their application to techno-industrial contexts, it has an additional associative meaning. The term also connotes these new technologies as innately "progressive," as beneficial to industry, communication and the general advancement of societies increasingly dependent on a standardized techno-scientific industrial base.[68]

> Progress . . . lies deeply embedded in the culture of science. Knowledge is thought to be progressive, accumulative, and quantitatively on a trajectory of either evolutionary or revolutionary improvement. And insofar as technologies are today associated with science and its culture, the same belief in progressivism is often held.[69]

This kind of perspective is usually favored by those for whom the enthusiastic embracing of "new technologies" has meant the capacity to produce more, faster and cheaper: ultimately to achieve substantial economic returns on investment.

Technologies and media have always been in such a state of frenzied evolution that it would be difficult to decide, with utmost accuracy, when anything becomes "new" while something else becomes "old." As pointed out by Janet Wasko for instance, 'The myriad of technological changes that have prompted discussions of a "new age of information" has been introduced into societies which remain fundamentally the same. In other words, there is as much continuity as there is change in any so-called "new age."'[70] For Janet Harbord, '[t]he "new" in new media is then the realization of a particular mix of components, which in turn shape and are shaped by the particular traditions of film cultures. . . . What we are experiencing is less than a technological revolution and more of an incremental cultural shift and remix.'[71] And for Bruno Latour, 'Technologies are not new, they are not modern in the banal sense of the world, since they have always constituted our world.'[72] It is hence unproductive to locate and characterize precisely a technology, a medium or an era as "new," since whatever previously existed does not automatically disappear into oblivion but often co-exists with the "new."

It is difficult to talk about the emergence of "new" technologies, "new" filmmaking, of a "new" aesthetic or of a "new" cinema in relation to Hollywood cinema's techno-industrial digitization. There was no overnight replacement of the "old" by the "new" as such. "New" commonly functions as a convenient, all-purpose substitute for the act of specifying and qualifying media functionally or techno-scientifically for instance. It often becomes a convenient replacement for such terms as "analog," "electronic," "digital," and so on, which do not contribute to making media "attractive" to the extent that "new" does by the intermediary of its inherent but un-stated "promise" to be better, faster and more effective than "old media."

The Hollywood industry cannot be said to have undergone a "revolution" either, if the term were used according to its commonly understood sense of "radical change." According to Raymond Williams:

> *Revolution* and *revolutionary* and *revolutionize* have of course also come to be used, outside of political contexts, to indicate fundamental changes, or fundamentally new developments, in a very wide range of activities. . . . Once the factory system and the new technology of the late eighteenth century and early nineteenth century had been called, by analogy with the French Revolution, the Industrial *Revolution*, one basis for description of new institutions and new technologies as *revolutionary* had been laid.[73]

Celebrating "new technologies" as "revolutionary" is a prerequisite for upholding the capitalist imperative in the pursuit of capitalist ideals. Modern industrial capitalism involves processes of techno-industrial centralization and standardization, which presuppose mass production and mass consumption. They also idealize the universalizing and homogenizing of corporate technoscience that generates "revolutionary" "new technologies."[74] Media, communication, information and entertainment industries in particular are pivotal to the creation, circulation and maintenance of the apparent universality and homogeneity of "new technologies" and so-called techno-industrial "revolutions." Janet Staiger explains that

> capitalism markets its products in ways that work against pure repetition of the product. This is especially true for certain items such as those in the luxury field of expenditures. Hollywood (in addition to most other types of industries) has discovered that product differentiation is valuable in re-creating demand. Thus, it is completely within the capitalist system to cultivate innovation in products, particularly if the

> novelty can be advertised. The value placed on change encourages the industry to seek creative ideas and differences.[75]

The means of production of "new technologies" are inevitably the propriety of a handful of corporations and conglomerates. Hence, the latter are ultimately the biggest beneficiaries when the "advantages" of so-called "revolutionary" "new technologies" are marketed and promoted, even if the use of and access to "new technologies" still remain the privilege of a few at the top of the ladder. In other words, the constant promotion of emerging technologies as "new" or "revolutionary" follows an economic capitalist rationale more than anything else. In the middle of a network of conglomerate synergies and corporate partnerships, the Hollywood cinema industry operates in ways that justify this economic rationale in the qualification of emerging technologies as "new." In the case of digital VFX, the emergence of so-called "new" technologies for their production did not suddenly phase out all the analog forms but, rather, cohabited with them productively—the main purpose ultimately being economic: to increase profits by trying to attract larger audiences to cinema theaters with more visually appealing filmic spectacle.

For his part, Brian Winston establishes three basic transformations that affect techno-scientific innovation and progress in general:

> - the first of these *transformations* moves the technology from the phase of *scientific competence* into the phase of *technological performance*. The first transformation . . . thus moves from science to technology, its effect being to activate the technologist;
> - the second transformation (the transformation occasioned by *supervening social necessity*) pushes the work of the technologist from prototypes into what is popularly conceived as "invention";
> - the third (a transformation which will be called *the 'law' of the suppression of radical potential*) moves from the invention of devices to their diffusion.[76]

To apply the above model of techno-scientific innovation and progress to the early uses of computer-generated images in the context of film, Winston's first stage of transformation can be identified in the development of basic CG systems during the 1960s—as illustrated by Ivan Sutherland's *Sketchpad* (a method of real-time interactive computer graphics) and by Charles Csuri's *Hummingbird*, one of the first films to incorporate computer-generated representational figures—more details are provided about these in the next chapter. Winston's second stage of transformation can be exemplified by *Futureworld* in 1976, and by *Tron* and *Star Trek II: The Wrath of Khan* at the beginning of the 1980s. *Futureworld* included the scanned and animated head (by computer graphics company Motion Pictures Product Group) of lead actor Peter Fonda as well as a computer-generated animated version of a left hand. In 1982, *Tron*'s sixteen minutes of computer-generated images and *Star Trek II: The Wrath of Khan*'s computer-generated "Genesis" sequence both also illustrate how techno-scientific competence in early CG application was developed very specifically in relation to VFX production; the only objective then being to reproduce precise digital VFX as opposed to other filmmaking applications (which would begin to occur a few years later).[77] These three films involved direct applications to filmmaking of innovations in computer science and technology, as well as indicated a move from 'scientific competence' to 'technological performance.'

Winston's third stage of transformation occurred in two phases in relation to the Hollywood cinema industry's techno-industrial evolution. First, the entirely computer-generated 1.8 minutes long animated short film *The Adventures of André and Wally B.* in 1984, produced by Lucasfilm's Computer Division (later known as Pixar Animation Studios), established the *potential* of diffusion for the various digital CG technologies and devices especially created during its making. This would indeed happen next, as confirmed by the intensive involvement of Pixar in the production of computer graphics, animation and digital VFX for many big-budget Hollywood films from the mid-1980s. The release of the first entirely computer-generated 3D animation feature-length movie *Toy Story* in 1995 ultimately confirmed the *possibilities* of diffusing digital CG technologies and devices for the production of more computer-generated feature-length movies. *A Bug's Life*, *Antz* and *Toy Story 2* would become further evidence of this before the end of the 1990s.[78]

Rather than employing the term "revolution"—there has not exactly been one—it would be more appropriate to describe the techno-industrial transformations that have occurred in the Hollywood cinema industry, as a result of techno-scientific innovation, as logical "transition" and "upgrading."

NOTES

1. See Appendix A.
2. To better understand the impact and ramifications of corporate structured science in relation to techno-scientific innovation in the 20th century, see de Solla Price, *Big Science*.
3. 'World War II began to introduce genuine hi-tech processes and histories of engineering still rate that period as the highest period of technological innovation in history. In terms of recent engineering history, World War II produced many of the inventions and developments which were to drive the Euro-American technological engine for several decades.'

Ihde, *Philosophy of Technology* 30.

4. This list is based on Ihde, *Philosophy of Technology* 47.
5. Heidegger, 'The Question' 4.
6. Latour, *We Have Never Been Modern* 64.
7. See Ihde, *Philosophy of Technology* 5.

8. Alfred North Whitehead writes that 'the reason we are on a higher imaginative level is not because we have a finer imagination, but because we have better instruments. In science the most important thing that has happened in the last forty years is the advance in instrumental design.'
Whitehead, *Science* 107.
See also: Ackermann, *Data*; Galison, *How Experiments End*; Latour, *Science in Action*.
9. Ihde, *Philosophy of Technology* 6.
10. Ihde, *Philosophy of Technology* 5 & 17.
The term "technoscience" gained popularity predominantly from Bruno Latour's *Science in Action*, but may have originally been coined by Gaston Bachelard in the 1930s.
11. For more critically diverse conceptualizations of "science," "technology" and of their manifold relationships, see: Ballard, *Man and Technology*; Barrett, *The Illusion of Technique*; Borgmann, *Technology and the Character*; Ellul, *The Technological Society*; Feibleman, *Technology and Reality*; Habermas, *Toward a Rational Society*; Hacking, *Representing and Intervening*; Heelan, *Space Perception*; Ihde, *Instrumental Realism*; Ihde, *Technics and Praxis*; Kuhn, *The Structure of Scientific Revolutions*; Marcuse, *One Dimensional Man*; Rapp, *An Analytical Philosophy*; Winner, *Autonomous Technology*; Winner, *The Whale*.
12. Kapsis, 'Hollywood Genres' 13.
13. In existence since the 1930s and conferred in a separate event from the Academy Awards ceremony since 1975, "Scientific & Technical" Awards are decided by AMPAS's Science & Technology Council and include: the "Academy Award of Merit," the "Scientific and Engineering Award," the "Technical Achievement Award," the "Gordon E. Sawyer Award," the "John A. Bonner Medal" and "Special Awards."
For more details, see AMPAS's 'Scientific and Technical Awards Database' (online).
14. Hayward, 'Introduction'—*Culture, Technology and Creativity in the Late Twentieth Century* 5.
15. The novelty and speed of delivery as good enough motivations for the excessive amount of visual information, and the technology going with it, to become vitally desirable and important are dismissed by Brian Winston as constituting primarily an artificially persuasive media-driven rationale, with little or no input from the consumers. This is because 'there is no scrap of evidence that our capacity to absorb information has markedly changed because of this monstrous regiment of data, print and images, nor that the regular provision of 'facts' beyond our limited capacity to absorb them materially alters our ability to run our lives and societies.'
Winston, *Misunderstanding Media* 367.
16. Harbord, *The Evolution of Film* 9; Rodowick, *The Virtual Life* 10.
17. See Hesse, *Models and Analogies*.
18. Much depends here on whether the clock in question has a discrete or a continuous sweep-second hand.
19. Binkley, 'Reconfiguring Culture' 93.
See also Binkley's analysis of Leonardo Da Vinci's *Mona Lisa* to explicate and visually illustrate his concept of "transcription" and "analog paradigm," in Binkley, 'Reconfiguring Culture.'
20. Binkley, 'Reconfiguring Culture' 95.
21. Binkley, 'Reconfiguring Culture' 96.
22. See Appendix C.
23. See Binkley, 'Reconfiguring Culture' 98.
24. Mitchell, *The Reconfigured Eye* 5.
See Appendix C for explanation of terms.
25. See Brand, *The Media Lab* 220.
26. Brand, *The Media Lab* 220.
27. Quoted in Brand, *The Media Lab* 221.
28. Brown, 'Man without a Movie Camera' 78.
29. Quoted in Brand, *The Media Lab* 19.
In the 1980s MIT's Media Laboratory was, according to Stewart Brand,

> taking a leading role in a complex array of communication technologies which are increasingly interlocked and all encompassing. Communications media are so fundamental to a society that when their structure changes, everything is affected, hence the sheer pervasiveness of all that gives meaning to the Media Lab cliché about "inventing the future."

Brand, *The Media Lab* xiii.

30. Hayward, 'Introduction'—*Culture, Technology and Creativity in the Late Twentieth Century* 5.

31. Bazin, *What Is Cinema? (vol. I)* 11.

According to A. R. Fulton,

> some of the principles of motion picture machinery were understood long before the device was perfected, and crude variations of it were devised. Apart from such early devices as Leonardo da Vinci's camera obscura, its origin is the magic lantern, invented by the Dutch scientist Christian Huygens about 1655. Almost two centuries later, in 1832, Simon Ritter von Stampfer, of Vienna, made a device he called the Stroboscope, whereby drawings on the rim of a disc viewed through slits in a second disc simulated motion. Then, in 1853, another Viennese, Franz von Uchatius, used a magic lantern to project the Stroboscope's pictures onto a wall. One of the most popular early versions of the motion picture machine was the Zoetrope, or wheel of life, [d]evised in 1833 by an Englishman, William George Horner . . .—it consisted of a shallow cylinder about one foot in diameter with vertical slots in the edge and, on the inside, a series of pictures which, seen through the slots, seemed to move when the wheel was turned. Another kind of wheel machine, patented in 1861 by Coleman Sellers, a Philadelphia machinist, was an arrangement whereby photographs were mounted on paddles. Sellers called this paddle-wheel machine the Kinematoscope. Such were the early gropings toward the motion pictures. They were, however, gropings primarily in the direction of motion picture projection. The motion picture camera had to wait for the invention not only of photography [with Louis Daguerre's invention in 1837] but also of photographic film.

Fulton, 'The Machine' 21.

32. Bazin, *What Is Cinema? (vol. I)* 21.

33. Fulton, 'The Machine' 21.

34. See Appendix A.

35. Bazin, *What Is Cinema?* (*vol. I*) 73.

36. Nicholls, *Fantastic Cinema* 15; see also Archer, *Willis O'Brien*.

37. Klein, *The Vatican to Vegas* 5; see also Sjostrom, *Quadratura*.

38. Bazin, *What Is Cinema? (vol. I)* 30. Emphasis added.

39. Balio, 'Introduction'—*Grand Design* 10.

40. See, for instance: Allen, 'The Industrial Context' and Allen, 'From *Bwana Devil*.'

41. Balio, *United Artists* 49.

42. Thomas Edison had already developed by 1889 a type of projector (that he called a *kinetograph*) which projected moving pictures on a 35mm film strip that had four perforations on each side, at 24 frames per second with an aspect ratio of 1.33:1. Following Edison's prototype, the Lumière brothers had launched by 1895 their *cinématographe* which consisted of a single camera using 35mm film for both photographing and projecting moving pictures at 16 frames per second, with an aspect ratio of 1.33:1 (by 1903, the Lumière brothers had already patented the *Autochrome Lumière* color photography process which was launched in 1907). Over the next decades, different types of film gauge (ranging from 9mm to 75mm) with varying aspect ratios would be used by dozens of emerging motion picture production companies in the burgeoning Hollywood cinema industry and elsewhere.

43. Wollen, 'The Bigger the Better' 13.

Created in 1953 and copyrighted to Twentieth Century-Fox, CinemaScope is based on the "Anamorphoscope" film process originally patented in 1926 by Professor Henri Chrétien. The CinemaScope picture is photographed on standard 35mm film with an anamorphic lens. When projected in the cinema theater through another anamorphic lens, the picture subsequently has an aspect ratio of 2.35:1 and a screen size two-and-a-half times the size of the conventional screen aspect ratio of 1.37:1.
See Barr, 'CinemaScope: Before and After'; Belton, 'CinemaScope: the Economics'; Chisholm, 'Widescreen Technologies.'

44. Vaz, *Industrial Light & Magic* 117.

45. Hayward, 'Introduction'—*Culture, Technology and Creativity in the Late Twentieth Century* 6.

46. While producing the experimental short film *Sextone for President* in 1988, Jeff Kleiser and Diana Walczak coined the term "synthespian." See Appendix C for more details.

47. Hamus, 'Retour vers le passé' 25.

48. Harbord, *The Evolution of Film* 55.

49. 'Intel's History' (online). Emphasis added.

50. See Appendix A.

51. See Latour, *We Have Never Been Modern.*

52. See Appendix A.

53. Ohanian & Phillips, *Digital Filmmaking* 3.

54. Winner, *The Whale* 174.

55. See Winston, *Misunderstanding Media.*

56. The thermionic triode valve would also influence the cathode ray tube's development that led to television. See Appendix A for more details.

57. What crucially distinguishes the computer from the numerous earlier calculating devices that preceded it is its ability to store data and instructions. Brian Winston's pertinent explication in that regard should be noted:

> A symbol-manipulator, as opposed to a number cruncher, must have an in-built set of instructions that can be varied by the operation of the machine itself as a result of its actual processes of manipulation; and it must therefore have an extensive data-store. In practice, meeting these requirements most likely means that the symbol-manipulator must be entirely electronic. This is a computer; everything else, even if entirely electronic and massively fast, is a calculator.

Winston, *Misunderstanding Media* 115.

58. Winston, *Misunderstanding Media* 183.

59. This can be further illustrated by Stewart Brand's commentary in 1988 that 'nearly all computers to date are "serial"—they do one thing at a time, as quickly as possible, and move on to the next. This was the "architecture" laid down by mathematician John Von Neumann in the late 40s, the foundation on which the vast and otherwise various superstructure of computer technology has been built.'
Brand, *The Media Lab* 181.

60. Taylor & Saarinen, *Imagologies* 185.

61. Although for Mark Taylor and Esa Saarinen, 'the word is never simply a word but is always also an image. The audio-visual trace of the word involves an inescapable materiality that can be thought only if it is disfigured.'
Taylor & Saarinen, *Imagologies* 45.

62. See: Castells, *End of Millenium*; Everette, *Reshaping the Media*; Jones, *CyberSociety*; Poster, *The Second Media Age*; Wasko, *Democratic Communications*; Youngblood, *Expanded Cinema.*

63. Brand, *The Media Lab* 18.

64. Quoted in Brand, *The Media Lab* 15.

65. Winston, *Misunderstanding Media* 103.

66. Benjamin, 'The Work of Art' 220.

67. See Ohanian & Phillips, *Digital Filmmaking* xxiii.
68. See Hayward, 'Introduction'—*Culture, Technology and Creativity in the Late Twentieth Century* 1.
69. Ihde, *Philosophy of Technology* 62.
70. Wasko, *Hollywood in the Age* 2.
71. Harbord, 'Digital Film' 24.
72. Latour, *We Have Never Been Modern* 126. With regard to his use of "modern," Bruno Latour points out that the 'adjective "modern" designates a new régime, an acceleration, a rupture, a revolution in time. When the word "modern," "modernization," or "modernity" appears, we are defining, by contrast, an archaic and stable past. . . . "Modern" is thus doubly asymmetrical: it designates a break in the regular passage of time, and it designates a combat in which there are victors and vanquished.'
Latour, *We Have Never Been Modern* 10.
73. Williams, *Keywords* 229–230.
For a more complex problematization of the concept of "revolution" in the context of scientific innovation and progress, see Kuhn, *The Structure of Scientific Revolutions.*
74. The explication of Mark Taylor and Esa Saarinen about the danger of technological standardization is worth mentioning:

> On the technological level, standardization creates vulnerability. Consider, for example, an analogy drawn from the domain of physical organisms. One of the most effective defenses against diseases and infection is biodiversity. Differences among organisms make certain species vulnerable to some diseases but resistant to others. If all organisms had the same genetic constitution, it would be possible for a single virus to wipe out an entire species. The standardization of electronic technology poses a similar threat. It is no accident that the term "virus" has been transferred from biology to computers.

Taylor & Saarinen, *Imagologies* 55.
75. Staiger, 'Introduction'—*The Studio System* 4–5.
76. See Winston, *Misunderstanding Media* 17.
It should be noted that Brian Winston later developed a seven-phase model of technological transition; see Winston, *Technologies of Seeing.*
77. The practical filmic application of the then new RenderMan 3D software during the production of *Star Trek II: The Wrath of Khan*'s "Genesis" sequence allowed its transition from prototype to *invention*. For more details about the RenderMan software, see Apodaca & Gritz, *Advanced Renderman.*
78. But Winston's third stage of transformation is not simply about 'diffusion,' as marked by the release of all-digital productions. By coupling 'diffusion' to what he calls 'the 'law' of the suppression of radical potential,' Winston is in effect arguing that the mainstreaming of technology is often linked to maneuvers by existing players to limit the extent to which new technologies will in fact change the existing industry. In this context, the end of the 1990s was marked not by the digital resolution of all filmmaking problems and the 'diffusion' of digital cinema, but by the possible fracturing of the universality of existing 35mm film-related technologies into a series of competing proprietary systems devoid of common standards and therefore of interoperability. Hollywood majors subsequently formed the Digital Cinema Initiative to formulate and mandate common standards for all aspects of digital cinema. Although the Digital Cinema Initiative finally released its standards in 2005, many national territories (including large parts of Asia and Europe) chose not to follow them. This caused a technological fragmentation of the global market that the Hollywood majors wanted to avoid in the first place. That is why digital distribution and exhibition has been progressing very slowly despite the technological capacity being available for at

least a decade. Even in the domestic U.S. market, digital screens still constitute only about 25 to 30 percent of the market, although the recent success of digital 3D theatrical exhibition has been a significant driver of faster uptake.

FIVE

The Hollywood Cinema Industry's Coming of Digital Age

Since its invention and diffusion at the end of the 19th century, the cinema has essentially progressed by being continuously subjected to various forms of techno-industrial transformation, as exemplified by the digitization of VFX production and of filmmaking generally towards the end of the 20th century. What set of circumstances enabled major techno-industrial transformations in the VFX domain to occur, and to be consequently extended to film production? Who are the individuals and organizations that influenced such major techno-industrial changes and who benefited? The Hollywood cinema's progressive techno-industrial transition to digital was not an outcome of "fate" or "coincidence": the filmmaking practices, ambitions, achievements, and partnerships of a handful of people changed the "rules of the game" for everybody else in Hollywood. To follow Raymond Williams's advice:

> We have to break from the common procedure of isolating an object and then discovering its components. On the contrary, we have to discover the nature of a practice and then its conditions. . . . But as we discover the nature of a particular practice, and the nature of the relation between an individual product and a collective mode, we find that we are analyzing, as two forms of the same process, both its active composition and its conditions of composition, and in either direction this is a complex of extending active relationships.[1]

The parallel narrative about the rise to power of the so-called "movie brats" and their associates from the 1970s on is a confirmation of the centrality of personal friendships, of professional alliances and collaborations, of commercial and industrial synergies in the techno-visual transformations that would rapidly shake the foundations of filmmaking practice in Hollywood.

Although computer-generated images and digital VFX progressively became common occurrences in the Hollywood cinema industry, their emergence was not the outcome of simply digitizing existing analog tools and technologies. This did happen, although much later during the process of technological digitization in the VFX production industry. Their evolution only partially occurred within the industry while their origins are located outside the domain of filmmaking altogether. To consider the historical origins and ramifications of the techno-industrial evolution of VFX production from analog to digital implies travelling back in (officially-recorded) history to identify the people involved in the origins of computer imaging and graphics and in the early stages of its development, as well as their connections with Hollywood filmmaking. In other words, it is about investigating how the digital medium (and its associated technologies of enablement) was borrowed from military and scientific research domains, transformed into a *bona fide* tool for artistic expression and production, eventually applied to VFX production and ultimately integrated into commercial filmmaking in Hollywood.

The Techno-Industrial Evolution of Digital VFX

While the "movie brats" were just beginning to slowly make their way in Hollywood by networking and collaborating with each other during the 1960s and early 1970s, complex electronic media and computer imaging technologies began to become available outside the U.S. military domain. This allowed for various forms of experimentation with, and application of, industrial and scientific computer imaging hardware and software that would gradually evolve into the kind of computer-generated graphics and cinematic VFX that would start appearing from the late 1970s. Before its more extensive use in the Hollywood cinema industry, the application of computer imaging to film was mostly limited to the experimental work of graphic artists, and to industrial design and scientific modeling.

> Two somewhat different types of research and development appear to constitute and characterize work on computer imaging in the initial period. In the first place, there is the applied research of computer scientists and engineers; in the second, there is the experimental work undertaken in the area by artists. Things were never quite as simple as this however, for almost without exception, the artists involved worked in close collaboration with computer scientists and program researchers. At the same time, many of the engineers who had come into contact with artists became producers of 'computer art' (as opposed to computer graphics) in their own right. Other engineers however carried on research into graphics' potentialities that were of a

> much more functional and applied kind and showed little or no interest nor involvement with art.[2]

As was also the case later for what became the Internet, the U.S. military originally created and nurtured interest in the use of computer imaging before it spread into the domains of industrial and scientific research.[3] Hence, military R&D in, and application of, interactive computer graphics, mainly to develop combat flight simulators and air defense tracking systems, in the early 1960s suggested the vast potential of computer imaging for commercial aviation and other industries. They brought together significant funds, resources and computer scientists to further research and develop the desired hardware and software. Line animation films were being computer-generated by 1963 at Bell Laboratories to visually express and simulate scientific and technical ideas using primitive vector display or "wire frame" techniques. Meant to simulate the motion of a communication satellite so as to ensure that the same side of the satellite always faces Earth, Edvard Zajac's computer-generated film "wass" (Wavefront Analysis of Spatial Sampling—a term associated with R&D in aircraft landing and approach) for Bell Laboratories in 1963 is possibly the earliest instance of convergence between computer and motion picture technologies. In the same year at MIT, postgraduate student Ivan Sutherland was developing the precursor of contemporary "touch-screen" technology: *Sketchpad*. It was a method of real-time interactive computer graphics involving

> drawing directly on to a cathode ray tube display screen with a 'light-pen' and then modifying the geometrical image possibilities so obtained with a keyboard. *Sketchpad* is considered by many today as a crucial breakthrough, from which have sprung most of the other technical developments to date, up to and including the various 'paintbox' systems which are so prevalent today (such as the *Quantel Paintbox* and its moving image adjunct *Harry*).[4]

By 1964, experimental filmmaker and computer artist Stan Vanderbeek together with computer imaging research engineer Kenneth Knowlton were collaborating on creating short films using a graphics programming software called *Beflix* (one of the earliest languages designed for computer imaging purposes) under development at Bell Laboratories.[5] And in 1967 Charles Csuri's *Hummingbird*, made in collaboration with James Shaffer at Ohio State University, would be one of the first films to incorporate computer-generated representational figures.[6] The first theatrical film to feature computer graphics prominently was *Futureworld* in 1976. Produced by independent motion picture production company American International Pictures as a sequel to MGM's sci-fi thriller *Westworld*, *Futureworld* incorporated the head of lead actor Peter Fonda that had been scanned and animated by computer graphics company Motion Pictures Product Group. The film also included a computer-generated animated version of a left hand which computer scientist Ed Catmull—a former postgraduate student of interactive computer graphics pioneer Ivan Sutherland—had created in 1973. Only by the early 1980s would theatrical feature films produced by Hollywood majors actually start including extensive computer graphics and computer-generated animation sequences, as illustrated by Paramount's *Star Trek II: The Wrath of Khan* and Disney's *Tron*.

Amongst the pioneering computer artist experimenters of the 1960s—such as Stan Vanderbeek, Lillian Schwartz, Charles Csuri and John Stehura—John Whitney Sr. would become the better known, especially within the Hollywood cinema industry context. Considered by many as the father of computer-generated art, Whitney Sr.'s contributions to the advancement of computer imaging application to film would eventually be officially recognized by the cinema industry when he became a "John A. Bonner Medal of Commendation" recipient (an AMPAS Scientific & Technical Award) for cinematic pioneering at the 58th Academy Awards ceremony in 1985.[7]

Whitney Sr. started building computerized drawing machines by recycling obsolete computing hardware discarded as surplus by the U.S. military after World War II. 'This eventually led to the development of a fully automated system which involved high precision integrated coordination and control of the entire production process (including drawing, motions, lighting and exposures)—a mechanical analog computer, specifically designed to produce complex abstract film animations.'[8] During most of the 1950s, Whitney Sr. used animation techniques that he had himself developed to create sequences for television programs and commercials. He also directed engineering demonstration films for the military and collaborated with renowned graphic designer Saul Bass to produce the animated title sequence for Hitchcock's *Vertigo* in 1958. Two years later, he founded Motion Graphics Incorporated to produce commercials and motion picture and television title sequences using the mechanical analog computer he had created in the 1950s. Whitney Sr.'s experimentations with the filmic medium allowed for early crucial steps to be made in the later convergence of computer imaging, motion picture technologies and cinematic VFX: the development of motion-control and slit-scan photography techniques.[9] These techniques would have an important influence on the production and aesthetics of spectacular visuals in commercial feature-length movies from the late 1960s, starting with the "stargate-corridor" sequence seen towards the end of Kubrick's landmark film *2001: A Space Odyssey* (1968). The production of VFX for this sequence was mostly inspired by Whitney Sr.'s early efforts in developing abstract VFX for his experimental film animation work and involved the use of a slit-scan device that was similar to the one he originally invented.

Because of its application of computer imaging for the first time to a theatrical feature film, the "stargate-corridor" sequence of *2001: A Space Odyssey* was groundbreaking for the late 1960s, thereby acquiring pioneer status in the evolution of computer imaging and VFX for motion pictures. It suggested the immense potential of computer imaging for future filmmaking, for visual aesthetic development in the cinema and for VFX production. But the techno-visual achievements of *2001: A Space Odyssey* would turn out to be a "one-off" event in the sense that they were not immediately followed up in subsequent commercial filmmaking. It would take another seven to eight years for its VFX achievements to be further developed, refined and implemented, especially in the Hollywood cinema industry. And this state-of-affairs occurred for a number of reasons: the techniques involved in producing the "stargate-corridor" sequence were very time-consuming and considered far too experimental and expensive for cost-effective implementation in commercial filmmaking at the time. In any case, the computer technology available was then too limited in terms of processing power to enable similar techno-visual outcomes within shorter time frames. Furthermore, Kubrick's techno-visually "obsessive" style of filmmaking was far ahead of its time and could hardly be emulated by other filmmakers or easily endorsed by most Hollywood producers at the time. The production of *2001: A Space Odyssey*'s VFX constituted a solid "training ground" for two individuals who would later become pivotal in developing the use of computers in VFX production for Hollywood movies: Douglas Trumbull and John Dykstra.

An inventor, cameraman and VFX master, Douglas Trumbull originally studied architecture before making films. A small graphic arts and animation studio where he worked, called Graphic Films, made a demonstration film on spaceflight that caught the attention of director Kubrick who subsequently hired Trumbull as special photography VFX supervisor for the making of *2001: A Space Odyssey*. The film's "stargate-corridor" sequence was Trumbull's outstanding achievement: the result, as we have seen, of a revolutionary camera design involving motion-control and slit-scan photography techniques originally pioneered by Whitney Sr. Motion-control is a technique that essentially enables the camera shutter to remain open while the subject of the camera is moving, creating as a result a blurring that occurs naturally in normal cinematography but was unknown prior to this in single-frame photography. The slit-scan technique involves a moveable slide, into which a slit has been cut. The slide is inserted between the camera and the subject to be photographed to achieve blur or deformity, and the creation of a psychedelic flow of colors and spectacular animations.[10]

During the early 1970s, other than producing VFX for various film and television projects by using a number of VFX techniques developed for *2001: A Space Odyssey*, Trumbull would also direct *Silent Running* (1972). In the mid-70s, he turned down Lucas's offer to supervise VFX production for *Star Wars* due to other commitments. By the late 1970s, he was contributing VFX to Spielberg's *Close Encounters of the Third Kind* (1977) and to *Star Trek: The Motion Picture* (1979). In the early 1980s, he supervised VFX production for *Blade Runner*. Trumbull would be nominated in the Academy Awards VFX category for *Close Encounters of the Third Kind, Star Trek: The Motion Picture* and *Blade Runner*. In 1983, he directed his second major motion picture *Brainstorm* showcasing his newly developed Showscan system that he considered the optimum way of producing an almost three-dimensional big-screen image. The film was shot at 60 frames per second (compared to the standard 24 frames per second for 35mm film), and what Trumbull was ultimately trying to achieve was to overwhelm the cinema spectator with 150 percent more visual information than conventional films by making use of 70mm film stock, a larger screen set closer to the audience and a powerful state-of-the-art sound system—arguably a cousin of the widescreen IMAX system. But Trumbull and his financial backers never persuaded moviegoers at the time to accept his proposed new cinema technology of "super realism."[11]

John Dykstra began his career at VFX production company Trumbull Film Effects under the tutelage of Trumbull, building models and doing VFX photography for such sci-fi films as *The Andromeda Strain* (1971) and *Silent Running*.[12] While recruiting for VFX production for *Star Wars*, Lucas approached Trumbull who pointed him towards Dykstra. In June 1976, Lucas contacted and hired Dykstra as special photography VFX supervisor. Dykstra would become instrumental in the development of the Dykstraflex motion-controlled camera system responsible for *Star Wars*'s groundbreaking VFX shots and in the early foundations of Industrial Light & Magic as a VFX company. Dennis Nicholson explains this:

> Lucas's script called for fast and complex battle sequences in outer space, and this meant that an extremely versatile camera system would be needed. So Dykstra created the "Dykstraflex." A VistaVision camera was converted to utilize an eight-perforation, horizontal 35mm film format. This gave the larger negative area required for optimum clarity. Several servomotors drove the camera mounted on a crane, along a straight 42 feet length of track, simultaneously raising or lowering, panning and tilting, around a static miniature. The focus was adjusted by a built-in motor-driven follow focus mechanism. All this was then controlled by a pre-programmed computer bank, a system so accurate that it could at any time retrace its movements, to the frame, over a previously plotted and filmed set-up. After each shot was completed, the film, together with the corresponding computer data, was forwarded to the control department where all the relevant information was catalogued for possible future reference.[13]

The all-computer controlled Dykstraflex camera system—which enables seven axes of camera motion (boom, lens focus, motor drive, pan, roll, shutter control, swing, tilt, track, traverse, as well as their repetition in multiple takes)—came about as a result of the availability of relatively low cost integrated-circuit RAM, second-hand VistaVision cameras (which deliver higher image resolution than normal cameras) and a PDP-2 computer (hooked up to the camera shutter and a dolly mover to control the camera's motions). As Dykstra recalls: 'We took archaic cameras built before we were even born, and we created hybrids of them by bolting different parts together. Nobody else was inventing cameras to make films in 1975. We were there when a genre was being born and reborn.'[14] The Dykstraflex camera system was clearly influenced by motion-control and slit-scan photography techniques previously developed by John Whitney Sr. and later applied by Trumbull during the production of *2001: A Space Odyssey* and of *Silent Running* (on which Dykstra had also worked).

In the aftermath of the tremendous commercial success of *Star Wars*—which won an Academy Award for Best VFX and an AMPAS Scientific and Engineering Award for technical achievement in VFX photography—Dykstra moved on from Lucas's Industrial Light & Magic to create Apogee Inc. his very own VFX production company. Supported in Apogee Inc. by his key creative team from *Star Wars*, Dykstra supervised VFX for the TV series *Battlestar Galactica* (1978–1979) which he also produced. Apogee would develop innovative motion-control and blue-screen technologies under Dykstra's guidance. Apogee and Dysktra subsequently produced VFX for *Star Trek: The Motion Picture*; some of these VFX (such as the "warp drive" sequences) would be recycled in later *Star Trek* movies. *Firefox*'s VFX in 1982 constituted the next major achievement for Dykstra: he had successfully taken on the difficult challenge of bringing together actual backgrounds, matte work on white backgrounds and miniature VFX by creatively applying a reverse blue-screen technique.[15] In spite of winning Academy Awards, the film was a modest hit at the box-office. Following VFX production for *Firefox*, Dykstra worked on the film's laserdisc-based arcade game spin-off which became a huge hit. For the rest of the 1980s, he focused on designing video games, as well as directing theme park attractions and commercials. In 1992, Dykstra directed *Sewer Shark*, a full-motion video game which turned out to be a critical and commercial flop. He then returned to active filmmaking during the 1990s to supervise VFX production for such films as *Batman Forever* (1995), *Batman and Robin* (1995) and *Stuart Little* (1999).

Other than their very own groundbreaking VFX work and achievements, Trumbull and Dykstra would further contribute to the techno-industrial evolution of VFX production in the Hollywood cinema industry by recruiting and supervising a number of individuals who would later become innovators in their own right, especially in the domain of digital VFX during the 1980s and 1990s. One of them is Dennis Muren, who was hired by Dykstra in 1976 to work on VFX production for *Star*

Wars and especially for building and photographing miniatures of the iconic Millennium Falcon under Dykstra's supervision.[16] After *Star Wars*, Muren would be involved in VFX production, under the supervision of Dykstra's earlier boss Trumbull, for Spielberg's *Close Encounters of the Third Kind*. He would then join Dykstra's company Apogee to work on VFX for the TV series *Battlestar Galactica* after which he moved back to Lucas's Industrial Light & Magic to work on VFX production for *Star Wars: The Empire Strikes Back*. Muren would subsequently remain at Industrial Light & Magic to work as VFX supervisor on such films as *Dragonslayer, E.T. The Extra-Terrestrial, Indiana Jones and the Temple of Doom, Star Wars: Return of the Jedi, Innerspace* and *Willow*, amongst others. Convinced that computer graphics were the future for VFX production, Muren took a six-month sabbatical to study the technology fully. In 1989, he would set up "The Mac Squad" at Industrial Light & Magic, as an adjunct to the optical VFX department, using Macintosh computers and Photoshop to bring so-called "traditional" analog film production departments—such as optical, animation and rotoscoping—into the digital realm.[17] During the 1990s, as well as being a VFX supervisor, Muren would also be involved in the development of innovative digital VFX and CG character and creature design for such films as *The Abyss, Terminator 2: Judgment Day, Casper* and *Jurassic Park*. Muren hence spearheaded Industrial Light & Magic's transition from the use of physical models and miniatures to CGI modeling applications for VFX production. And he contributed to ushering in a new era of digital VFX with the computer-generated dinosaurs of *Jurassic Park* in particular. This was the techno-visual breakthrough that convinced Lucas that the technology had advanced far enough to make the next *Star Wars* prequels trilogy. After his groundbreaking CG and digital VFX work for *Jurassic Park*, as well as being a consultant for digital 3D animation provider Pixar Animation Studios, Muren would continue to work for Industrial Light & Magic as Senior VFX Supervisor and was involved in VFX production for *Star Wars: Episode I—The Phantom Menace* in the late 1990s. Since his beginnings with the *Star Wars*'s VFX production crew, Muren has been the recipient of nine Academy Awards (including Special Achievement awards) and has led the pack in developing cutting-edge VFX techniques, especially through his pioneering work at Industrial Light & Magic in CGI and digital VFX-related technologies of production.

Richard Edlund was also hired by Dykstra to assist in VFX photography during the production of *Star Wars*. He was a key figure in equipping the then new Industrial Light & Magic facilities with the right cameras for filming the required VFX. After *Star Wars*, Edlund also followed Dykstra to Apogee to work on VFX production for the TV series *Battlestar Galactica* but he would be invited back to Industrial Light & Magic by Lucas to work on the optical compositing of miniatures during VFX production for *Star Wars: The Empire Strikes Back*. At Industrial Light & Magic, Edlund would subsequently be involved in VFX production for *Raiders of the Lost Ark*, *Poltergeist* and *Star Wars: Return of the Jedi* after which he would set up, in 1983, his own VFX company Boss Film Studios, whose first VFX production credit would be *Ghostbusters* (1984). Until it ceased operation in 1997, Boss Film Studios achieved ten Academy Award nominations in the VFX category for such movies as *Poltergeist II: The Other Side* (1986), *Die Hard* (1988), *Ghost* (1990), *Alien 3* (1992), *Batman Returns* (1992), *Cliff-hanger* (1993), *Species* (1995), *Multiplicity* (1996), and *Air Force One* (1997), amongst others. After the closure of Boss Film Studios, Edlund has continued working in Hollywood as a freelance VFX supervisor.[18]

Ken Ralston was hired by Dykstra to work as assistant cameraman in the miniature and optical VFX production unit of *Star Wars*. Staying with Industrial Light & Magic after its completion, he would subsequently be involved in VFX production for *Star Wars: The Empire Strikes Back* and *Dragonslayer*. He later worked as VFX supervisor on *Star Wars: Return of the Jedi*, *Star Trek II: The Wrath of Khan*, *Star Trek III: The Search for Spock*, *Cocoon*, the *Back to the Future* trilogy, *Who Framed Roger Rabbit?*, *Forrest Gump* and *Death Becomes Her*, amongst others. In 1995, Ralston left Industrial Light & Magic to join VFX provider Sony Pictures ImageWorks where he was appointed Senior VFX Supervisor for *Jumanji* (1995), *Phenomenon* (1996), *Contact* (1997), *What Lies Beneath* (2000), and *The Polar Express* (2004), inter alia. Throughout his career, Ralston has contributed considerably to the development, perfection and the digitization of many VFX techniques used in Hollywood.[19]

A far less celebrated digital pioneer of computer-generated VFX and CG animation in the Hollywood cinema industry is Steve Lisberger. A young animator in the late 1970s, producing commercials and various animations for television, his only feature film credit by 1980 was the independent 2D animation *Animalympics*. Increasingly interested in the new worlds of electronics, computers and videogames, Lisberger moved away from traditional 2D animation and began writing *Tron*, a story in which much of the action would occur in a virtual reality world within computer programs—a concept that nobody in Hollywood had even thought of at the time. After developing the script on his own and despite limited filmmaking credentials, Lisberger eventually managed to obtain a production budget from Disney executives—even though his concept was mostly incomprehensible to them—who had arguably caught a glimpse of what could be Disney's next level of industrial development after decades of "traditional" 2D animation. While Disney animators refused to work on *Tron* because they perceived computer-generated animation as a threat to their craft, Lisberger managed to bring together *Blade Runner*'s Visual Futurist Syd Mead and French comic book artist Moebius to design sets, costumes, vehicles and props for *Tron*'s fully-realized CG world populated by live-action human characters. The innovative 16 minutes of entirely computer-generated visuals, background mattes and digital VFX in the action sequences of *Tron*—the longest such sequences ever seen in a Hollywood movie at the time—were achieved through a combination of live-action footage being composited frame by frame with CG elements and involved four main American computer graphics companies: Digital Effects Inc. MAGI-Synthavision, Robert Abel & Associates, and Stargate Films Inc. From an investment of $17 million by Disney, against all predictions, *Tron* generated only $33 million from the box-office upon its theatrical release in 1982—even though the arcade video game derived from it became a huge hit. This significant financial disappointment dampened Disney's enthusiasm for digital VFX and their related technologies. Disney was discouraged, at least in the immediate future, to further foray into the digital realm and preferred to remain with its proven recipe of family-oriented traditional 2D animation films. And Lisberger never got a second chance in Hollywood after *Tron* (although he would be credited as "producer" in 2010 for *Tron: Legacy*, the sequel to *Tron* more than twenty-five years later). But *Tron* was nevertheless an important techno-visual landmark, in terms of displaying the possibilities and future potential of computer-generated imaging for the Hollywood cinema industry, as well as influencing the evolution of gaming.

The Production and Impact of Star Wars

Even though the origins of digital VFX production and its filmic application can be identified in various pioneering artistic experimentations with industrial and scientific computer imaging systems during the late 1960s—as illustrated by embryonic traces in *2001: A Space Odyssey* for instance—*Star Wars* in 1977 constituted the Hollywood cinema industry's major "springboard" into the digital realm. It is as a result of the massive box-office success of this particular movie that Lucasfilm Inc. and Industrial Light & Magic would subsequently influence and spearhead the rest of the VFX production industry's transition from its analogic state into the digital realm—expanding in the process the range of visual re-presentations, artistic expressions, and aesthetic aspirations of the Hollywood film in general. As expressed by Barry Langford, for instance,

> the opening shot of *Star Wars*—the vast bulk of the Imperial star cruiser grinding (and growling, in Dolby surround sound) endlessly, crushingly overhead—has become the foundational moment of a new audio-visual regime. *Star Wars* is widely held to have relegitimated the spectacular, after the relative restraint of (some) Hollywood Renaissance films.[20]

It is hence important to acknowledge the "snowball" impact of the very first *Star Wars* in 1977. All the *Star Wars* films ever made might perhaps simply be considered as VFX-intensive sci-fi movies with big budgets or well-marketed and packaged, slick, B-grade "space operas." But what is of much more interest about the *Star Wars* movies are the transformations they have triggered in relation to filmmaking in the Hollywood cinema industry, in terms of: expanding the techno-visual and aesthetic range of the motion picture; introducing new tools and techniques to the benefit of both VFX and film production; and re-orienting filmmaking practices and strategies for filmmakers and VFX practitioners alike. After *Star Wars*, Hollywood VFX-intensive movies began to be made differently as well as to look (and sound) "better" than before with the progressive industrial implementation and standardization of digital technology, inevitably transforming in the process the spectator's viewing experience of the Hollywood film. Following the massive commercial success of *Star Wars* in 1977, its technology R&D-intensive approach to filmmaking and to VFX production would be replicated in the making of other movies. To start with, this was but for a handful of films, but it progressively extended to many more until it became normal practice across the Hollywood cinema industry by the end of the 1990s.

As for *Star Wars: Episode I—The Phantom Menace*, it became the first feature film to be theatrically exhibited in digital—albeit as a test in a few movie theaters—upon its release in 1999. After more than a century of being an analog medium, not only had film production gone digital but the very exhibition of film had also done so. Such an event in itself signposted a crucial turning point not only for the Hollywood cinema industry but also for cinema in general: the "end" of analog cinema and the "beginnings" of digital cinema. In the words of Douglas Gomery: 'Twentieth-century cinema ended on 19 May 1999 with the premiere of George Lucas's much awaited *Star Wars: Episode I—The Phantom Menace*. . . .'[21] If *Star Wars* in 1977 inaugurated the beginning of a new era of "better things to come," *Star Wars: Episode I—The Phantom Menace* in 1999 closed that era, in being the apex of "better things to come," while simultaneously laying the foundations for yet another new era of "better things to come" for the industry.

Without the phenomenal commercial success of *American Graffiti* in 1973, Lucas would have never been able to get *Star Wars* produced by any Hollywood studio because everybody had the same apprehensions: a relatively unknown young director's B-grade style sci-fi film set in outer space and with a "good versus evil" storyline would never be sufficiently interesting for most cinema viewers in order to generate profits. As Lucas explains: 'We made the deal on *Star Wars* on the first of May, and *Graffiti* came out in August. But the film was building before release. And it was really in Hollywood that it was beginning to build.'[22] Pye and Myles continue:

> All Twentieth Century-Fox promised in the May deal was the money to start developing a script. Like all Hollywood deals, this one moved step by cautious step. It did not guarantee the film would ever be

written, let alone made. But by the second and third steps in the contract, *American Graffiti* was in release.[23]

While *American Graffiti* was circulating theatrically nationwide, Lucas did research to refine his ideas for *Star Wars*: 'I researched kids' movies and how they work and how myths work; and I looked very carefully at the elements of films within that fairy tale genre which made them successful.'[24] Lucas was essentially doing market research for a film that he fundamentally treated as a product to be manufactured rather than created because he had merchandizing—t-shirts, records, models, kits, figurines, dolls and toys—in mind from the beginning.[25] And he decided early on that *Star Wars* was going to be a visually driven high adventure "gee-whiz movie" full of VFX "eye candy" for children.[26]

Hollywood industry analysts did not fully predict *Star Wars*'s vast popularity and profitability—which was in fact still in doubt until after it had opened theatrically. Its success was

neither mystic nor supernatural. . . . From the start, [Lucas] was determined to control the selling of the film and its by-products. "Normally you just sign a standard contract with a studio," he says, "but we wanted merchandizing, sequels, all those things. I didn't ask for another $1 million, just the merchandizing rights. And Fox thought that was fair trade." When the film appeared, the numbers became otherworldly: $100,000 worth of t-shirts sold in a month, $260,000 worth of intergalactic bubblegum, a $3 million advertising budget for ready-sweetened *Star Wars* breakfast cereals.[27]

Lucasfilm Inc. had a merchandizing department—under Lucas Licensing—that was as big as that of Twentieth Century-Fox after *Star Wars*. George Lucas would consequently make more money from the merchandizing of the *Star Wars* franchise—for which he remained the sole owner of the rights—than from theatrical box-office revenues over the following decades.

Although very much influenced by B-grade sci-fi films of the 1940s and 1950s, in terms of the desired look for *Stars Wars*, Lucas was more inclined towards the visual achievements of Kubrick's *2001: A Space Odyssey* which he held in high regard. But the allocated budget from Twentieth Century-Fox, while the largest amount Lucas had ever worked with, was clearly too limited to do what Kubrick had achieved previously in the late 1960s, causing Lucas and producer Kurtz to constantly fight with studio executives who would not let them go over the allocated budget. Quoting Gary Kurtz, Pye and Myles tell us that

[c]ompared with *2001*—Lucas calls Kubrick's film "the ultimate science fiction movie"—the special effects in *Star Wars* were cheap. Where Kubrick could allow his space stations to circle elegantly for a minute, Lucas always has to cut swiftly between individual effects. But that became part of the film's design. Where Kubrick's camera was static,

> Lucas and Kurtz encouraged their special effects team to develop ways to present a dogfight in space with the same realism as any documentary about World War II. As is usual in animation, they prepared storyboards, precise drawings of how each frame was to look; but, unlike most animation, they based their drawings on meticulous study of real war footage. They looked for the elements that made an audience believe what they were seeing.[28]

To achieve the desired look of *Star Wars* with its limited budget, it became obvious during pre-production that a lot of miniatures and other related VFX shots would have to be incorporated into the film, and that it would be cheaper to produce the required VFX "in-house" as opposed to sub-contracting them to external optical VFX specialists, a common practice in the Hollywood cinema industry then. Hence, VFX supervisor Dykstra decided against normal Hollywood production methods by combining under one roof all the facilities, equipment and personnel that would be needed for the film's VFX production.[29] He then proceeded to set up various VFX production crews by hiring people according to their abilities in highly-specialized domains such as matte painting, optical printing, pyrotechnics, miniatures, model-making, and stop-motion animation. *Stars Wars* would end up involving 900 people to work on its 365 VFX scenes, compared to the 106 people hired to work on *2001: A Space Odyssey*'s 35 VFX scenes (which had a budget of more than $10 million).[30] The facilities, personnel and equipment brought together as a result of more than $1 million being spent in eight months to produce *Star Wars*'s VFX shots would end up becoming the foundations of Industrial Light & Magic, Lucas's very own VFX company under the umbrella of Lucasfilm Inc.

Financially boosted by the tremendous commercial success of *Star Wars*, George Lucas would subsequently expand the operations of Lucasfilm Inc. and Industrial Light & Magic: by diversifying the range of services they could provide; by constantly upholding high techno-visual standards of imaging for all film projects undertaken; and by setting up in-house filmmaking R&D teams. By the end of the 1970s, Lucas was beginning to realize that the future of VFX and film production would most likely be associated with computers and with the digital medium. The fortune he was making from *Star Wars* merchandizing as well as from VFX production work undertaken by Industrial Light & Magic for various Hollywood movies would enable him to recruit personnel and finance the setting up of a computer division within Lucasfilm Inc. New York Institute of Technology's Computer Graphics Lab—with Ed Catmull as director—had attracted Lucas's attention in the mid-1970s. When the Computer Graphics Lab began struggling financially by the end of the 1970s despite its technological achievements, Lucas offered Catmull the position of vice president at the new Computer Research and Development Division of Lucasfilm Inc. created in 1979.

Catmull did not start his professional career in the movie industry: he originally studied mathematics, physics and computer science. He worked as a computer programmer for a short period after college graduation before going to graduate school in 1970 where he was one of Ivan Sutherland's students. Catmull considered the new field of computer graphics in general and Sutherland's computer drawing program *Sketchpad* as pivotal to the evolution of animation which he loved. He became part of the computer-generated animation revolution from the beginning. Under Sutherland's supervision, Catmull made some significant discoveries in computer graphics, such as texture mapping and bicubic patches, and he also devised innovative algorithms to refine subdivision surfaces.[31] His earliest contribution to the motion picture industry was a computer-generated animated version of his left hand, made in 1973, that was later incorporated in the 1976 sci-fi thriller *Futureworld*. In 1974, Catmull took the position of director at the new Computer Graphics Lab at NYIT where he put together a talented research group mostly engaged in the development of innovative 2D animation tools that could assist computer animators. The group's many inventions included: the *Paint* program, commercial animation program *Tween* (to automate the process of producing in-between frames), and animation program *SoftCel*. Catmull and his research group eventually shifted from 2D animation to focus on 3D computer graphics and animation in application to motion picture production in particular.

After joining Lucasfilm Inc. Catmull contributed to developing digital image compositing technologies utilized in the combination of multiple images in a convincing way. Lucasfilm's Computer Division would be responsible for the computer-generated images and animation sequences of *Star Trek II: The Wrath of Khan* in 1982, for the landmark computer-generated 3D animation short film *The Adventures of André and Wally B.* in 1984, and for the digital VFX of *Young Sherlock Holmes* in 1985. In the same year, Catmull and his team would begin to operate under the name of Pixar—an image-processing computer system developed at Lucasfilm's Computer Division—choosing to focus solely on 3D computer animation and its related R&D. In 1986 Apple Inc. co-founder Steve Jobs bought Pixar from Lucas, assigning Catmull the position of Chief Technical Officer.[32] Over the next ten years at Pixar (eventually re-named Pixar Animation Studios), Catmull would be a key developer of the photo-realistic *RenderMan* system, amongst other software, that was used in the first entirely computer-generated 3D animation feature film *Toy Story* in 1995. Catmull has been a multi-Academy Award winner for contributions to the Hollywood cinema industry in terms of his pioneering innovations in computer graphics, computer-generated 3D animation software, and digital image compositing in the domain of VFX production.

During his tenure with Lucasfilm's Computer Research and Development Division, Catmull was also responsible for hiring John Lasseter who would become a key figure in the production of entirely computer-generated 3D animation theatrical feature films. Lasseter developed his interest in animation from watching Chuck Jones cartoons on television at an early age. In 1975, he enrolled in California Institute of the Arts' then new animation course where he received training from some of Disney's master animators such as Eric Larson, Frank Burgess and Ollie Johnston. After graduation, Lasseter worked as an animator in the Walt Disney Feature Animation division. In the very early 1980s, he came across videotapes from the then fairly new SIGGRAPH (Special Interest Group on GRAPHics and Interactive Techniques) conference that contained some of the very early stages of computer-generated animation on film. Shortly after that, Lasseter saw the first "lightcycle" animation sequences that had been produced by Lisberger for Disney's upcoming *Tron*, featuring some innovative computer imaging. He then really began to see the huge potential of computer graphics technology for film animation and to realize how computers could be used in making films with 3D backgrounds wherein traditionally animated characters could interact—as such bringing in a new visually striking depth that had not been previously thought of. As a follow up to this realization, he began developing an animated feature project where the background would be computer animated. But Lasseter would be fired in the process by Disney which, at the time, did not share his enthusiasm for computer-generated animation. He subsequently met with Catmull, the head of Lucasfilm's Computer Division. Lasseter made a deal with Catmull and his colleagues to work as an interface designer on a project that would become the Computer Division's first entirely computer animated short film, namely *The Adventures of André and Wally B.* It turned out to be far more revolutionary than Lasseter himself had imagined prior to joining Lucasfilm Inc. He originally intended to use computers to create only backgrounds, but in the final short film that appeared in 1984 everything was in fact computer-generated including the animated 3D characters. After that, it would take Lasseter about 10 years of continuous computer animation R&D and the production of many short CGI films with the same group of people to eventually deliver *Toy Story* in 1995. Lasseter has been the executive producer for all of Pixar's animated feature films and associated computer-generated animation shorts, and also personally directed *Toy Story, A Bug's Life* and *Toy Story 2*. Lasseter has received two Academy Awards for his computer-animated short films as well as a Special Achievement award for *Toy Story*. He eventually became Chief Creative Officer at Pixar and at Walt Disney Animation Studios as well as Walt Disney Imagineering's Principal Creative Advisor.

The reputation of Lucasfilm's Industrial Light & Magic as a reliable provider of high-end VFX was rapidly established not just through the *Star Wars* movies but also from its involvement in the production of many big-budget and VFX-intensive Hollywood movies of the time.

Spielberg's films in particular, as well as those of certain movie brats and of some of their "protégés," were often the privileged first beneficiaries of the groundbreaking techno-visual innovations of Industrial Light & Magic, especially with regard to computer-generated VFX. The expertise and efficiency of Industrial Light & Magic would hence benefit a "select few" in the entourage of Lucas and Spielberg, such as Robert Zemeckis and Ron Howard, whose films also contributed to the digital VFX boom in Hollywood. By the mid-1980s, the profitability of the *Star Wars* franchise as well as the box-office success of Spielberg's movies enabled Lucasfilm and Industrial Light & Magic to expand extensively, to the point of being able to cover the whole spectrum of filmmaking services—analog and digital VFX production, computer imaging, CG animation, digital sound engineering, and digital filmmaking R&D and application, amongst many others. More importantly, Lucasfilm, Industrial Light & Magic and Pixar had begun to demonstrate that the economic, artistic and techno-industrial future of the Hollywood cinema industry would be inextricably tied to computer imaging and graphics, to digital VFX and to digital filmmaking.

Like *Star Wars* in 1977, the impact of *Star Wars: Episode I—The Phantom Menace* on popular culture twenty-two years later was second to none. In spite of all the hype and anticipation, *The Phantom Menace* was not 'the greatest film ever made, as some fans would have you believe' nor did it 'signal the death knell of artistic motion pictures, as high-brow critic-prophets cry out.'[33] As a "space opera" or "space western," with borrowings from Errol Flynn swashbucklers, from movies of chivalry and medieval knights, and from a few Italian "sword-and-sandal" epics—mostly of the 1930s to 1950s period—*Star Wars: Episode I—The Phantom Menace* was essentially very similar to all the previous installments. George Lucas did not weaken nor considerably reinforce the *Star Wars* mythology, he neither dramatically altered nor strictly copied: what he did was to provide the established mythology with a good polish and additional directions in which it could evolve perfectly into something else if required. There were no surprises and not much to disappoint either. Like all the other installments, *Star Wars: Episode I—The Phantom Menace* included a similar mix of swashbuckling heroics, easily accessible mysticism (the Force), easily identifiable "goodies" and "baddies," and various "loose strings" in the plot to enable follow up in sequels. Other than being far more techno-visually sophisticated, the form of *Star Wars: Episode I—The Phantom Menace* was conceptually very similar to the original. As illustrated by director of photography David Tattersall:

> While planning the lighting style for *Phantom*, . . . it soon became obvious that the only relevant source of information would be the original *Star Wars* trilogy. To fly off on a tangent with another style would have really been wrong. George and I always came back to the idea that the

> feeling of *Phantom*—especially in the scenes set on Tatooine—should be very much the same as what we've seen before.[34]

With a formula, structure and mythology that people already seem to enjoy very much, *Star Wars* is clearly not the kind of franchise wherein risks can be taken.

The groundbreaking VFX achievements of *Jurassic Park* in 1993 had influenced his decision to develop a new *Star Wars* trilogy, but Lucas only began production on *Star Wars: Episode I—The Phantom Menace* in June 1997 when he felt that VFX had advanced to the level of what he had envisioned for the film. During the production of *Star Wars* in the 1970s, the new Dykstraflex motion-controlled camera system had to be developed to shoot the numerous fast and complex battle sequences in outer space and to resolve a number of VFX cinematography problems. Similarly,

> a few technological hurdles needed to be overcome before principal photography could commence on *The Phantom Menace*. One was the classic post problem of matching live-action photography with CG effects. This is normally done under the supervision of an effects unit, most often employing motion-control cameras. However, because the film would feature several all-CG characters, virtual sets and other effects, nearly every shot in the film was an effects shot. Seeking to avoid the sloth-like production pace usually dictated by motion-control systems, Lucasfilm turned to Arri Media—the London-based Arri rental house supplying cameras for *Episode I*—to devise a solution. In 1990, when Arri unveiled the 535, then camera assistant Marc Shipman-Mueller came up with the notion of linking the camera to his laptop computer via the provided serial port. Arri supported this idea, which resulted in the Laptop Camera Controller, and later inspired the Data Capture System (DCS) used on *The Phantom Menace*.[35]

Although pivotal in adding a degree of realism to the drama, location shooting for *Star Wars: Episode I—The Phantom Menace* was minimal: at the Reggia Palace in Caserta, Italy, for the Naboo Palace interiors, and in the desert near Tozeur, Tunisia, for the Tatooine scenes. But most of the work was actually carried out at Industrial Light & Magic in California and at Leavesden Film Studios in England.[36] As director of photography David Tattersall tells us: 'One has to realize that *The Phantom Menace* was shot in just 65 days, while the postproduction work took another 20 months or so. . . . Many sequences were created entirely at Industrial Light & Magic, so it's a film that was defined not only in preproduction and on the set, but in post.'[37] Because so much of the film was in effect finished in post-production, the shooting crew would often end up working on a mostly virtual production involving large amounts of green/blue-screen material.

Ranging from complex computer-generated characters that interact seamlessly with human actors to large-scale self-contained sequences, the

VFX of *Star Wars: Episode I—The Phantom Menace* were often the outcome of "traditional" analog techniques being combined with some of the newest digital techniques. The creation of certain shots required new computer software to be written by John Knoll and his VFX team at ILM. With 90 percent of the film featuring digital VFX shots (fully computer-generated scenes, synthetic environments, digital terrain generation, computer-generated lead characters and thousands of digital extras), *Star Wars: Episode I—The Phantom Menace* was indeed the culmination of many years of R& D in digital VFX.

> Most of the scenes were digitally created (the final Gungan battle) or enhanced (by extending the standing sets, built only 6 ft. or 7 ft. high, into palaces and Senate chambers). "A typical summer movie has maybe 2,000 shots, with, say, 250 effects shots," says Knoll. *Titanic* had about 500. "This one is backward. Of the 2,200 shots, only about 250 shots are not effects shots." There is just one sequence totally untouched by the digitalizers.[38]

And such large amounts of VFX in the film did not go unnoticed by critics, as illustrated, for instance, by James Berardinelli's commentary that

> *The Phantom Menace* is a testimony to how far special effects have come. Does Lucas overdo it? Yes. Every scene is crammed with as many aliens, otherworldly creatures, and CGI synthetics as space will allow. Sometimes, it's breathtaking, but there are occasions when Lucas seems to be saying, "See! Look what I can do!"[39]

But in spite of various qualms about characterization, performance and plot, most critics were in agreement about the sheer spectacularity of *Star Wars: Episode I—The Phantom Menace*. Hence, Janet Maslin wrote in *The New York Times* that 'stripped of hype and breathless expectations, Lucas' first installment offers a happy surprise: it's up to snuff. It sustains the gee-whiz spirit of the series and offers a swashbuckling extragalactic getaway, creating illusions that are even more plausible than the kitchen-raiding raptors of *Jurassic Park*.'[40]

Star Wars: Episode I—The Phantom Menace would turn out to be the very last *Star Wars* movie shot on 35mm celluloid film. The next two installments would be shot using Sony CineAlta high-definition video cameras (at 24 frames-per-second equivalent). *Star Wars: Episode II—Attack of the Clones* in 2002 would hence become the first major Hollywood motion picture completely filmed in digital video. And from July 18, 1999, (and for the next four weeks), digital screenings of *Star Wars: Episode I—The Phantom Menace* were held at selected cinema venues in the U.S. The theaters of Loew's Meadows 6 (New Jersey) and of the AMC Burbank Media Center 14 (California) were specially equipped with Texas Instrument's Emmy award-winning Digital Light Processor (DLP) installed in a prototype projector. The Pacific Winnetka Theater (California) and Loew's Cineplex Odeon (New Jersey) for their part used Hughes-JVC's Image Light Amplifier (ILA) 12K Projectors. All these special presentations were jointly organized by Lucasfilm Inc., THX, CineComm, Texas Instruments, and Hughes-JVC essentially to showcase and promote the potential for a drastic change with regards to the future of film distribution. In 1999, *Star Wars: Episode I—The Phantom Menace* was, in many ways, simultaneously closing a long chapter in the history of cinema and beginning a new one.

The Normalization of Digital VFX and Digital Filmmaking

With Lucasfilm, Industrial Light & Magic and Pixar setting the pace in terms of digital VFX production and post-production, and with the rising need for sophisticated VFX by the Hollywood cinema industry—VFX-driven movies were after all generating huge box-office profits—numerous VFX production companies inevitably mushroomed all over California to meet increasing demands. Digital Productions was founded in 1982 by John Whitney Jr. and Gary Demos, both ex-staff of computer graphics provider Tripe-I. Boss Film Studios was created in 1983 by ex-ILM VFX supervisor Richard Edlund. Tippett Studio was created in 1984 by ex-ILM VFX artist Phil Tippett. VisionArt was founded in 1985 by David Rose and Todd Hess. Blue Sky Studios was founded in 1987 by a group of ex-MAGI/Synthavision graphic artists and technicians who had met while working on *Tron*'s VFX sequences. Rhythm & Hues Studios was also founded in 1987 by six former employees of computer graphics and VFX provider Robert Abel and Associates. Matte World Digital was co-founded in 1988 by producer Krystyna Demkowicz, VFX supervisor Craig Barron and matte painter Michael Pangrazio, both ex-ILM staff since 1979. Stargate Digital was founded in 1989 by cinematographer, VFX supervisor and producer Sam Nicholson. And more digital VFX production companies would continue emerging during the 1990s: Cinesite Digital Studios, created in 1991 as a Kodak subsidiary (later to become Laser Pacific in 2003); Digital Domain, started in 1993 by ex-ILM General Manager Scott Ross; Sony Pictures ImageWorks began operations in 1993 under Sony Pictures Entertainment; Flash Film Works, created in 1993 by ex-Introvision International VFX supervisor William Mesa; Vision Crew Unlimited (VCU), founded in 1994 by VFX artists Evan Jacobs, Jon Warren and Douglas Miller; Warner Digital Studios, started operating in 1994 as a subsidiary of Warner; Mass. Illusion started in 1995 by VFX supervisor Joel Hynek as a subsidiary of Cinergi Pictures Entertainment.

The indispensability of digital technologies might not have seemed so obvious to the dominant Hollywood cinema industry players from the start. Resorting to digital technologies was often essentially perceived as problem solving because it usually took a technical difficulty, impasse or crisis situation during film production for digital solutions to be sought and applied. The boom in digital technologies and digital film service providers during the late 1980s to early 1990s was a dominant influence on the progressive convergence of digital and analog media and on the digitization of analog VFX production technologies, of film post-production and of practically all aspects of production in Hollywood before the end of the 1990s decade.[41]

> In mid-1993, filmmakers rapidly adopted digital, nonlinear editing methods. One fascinating statistic is that by April 1994, there were 70 feature films in various phases of post-production in Hollywood. Of these 70 films, 30 were being edited on film. An astounding 40 films, or 57 percent were being edited digitally! These films ran the gamut from low budget films to budgets over $100 million. In just a few years, the acceptance of digital editing in the world center of filmmaking has seen an amazing progression.[42]

From a handful of additional new resources becoming available through the digital medium during the early 1980s, the filmmaking process in Hollywood swiftly progressed from specific and limited applications of digital technologies to the multi-lateral infiltration of the entire filmmaking process by the digital medium before the end of the 1990s.

> What has changed, and will undoubtedly continue to change, is that there will be different production options to achieve a desired end. . . . Most digital options have been concentrated in post-production, but increasingly the digital threshold is demanding a re-thinking of the whole production process, including the relation between production and post-production.[43]

The less obvious applications of digital technologies—part of the film and yet invisible to the spectator—which have benefited immensely from the computerization and digitization of analog filmmaking include, amongst others: digital stop-motion cameras for more accurate and smooth animation of models and miniatures; digital image processing, which allows the manipulation and correction of color, contrast, saturation, sharpness and shape of images; digital motion picture retouching systems, which allow the removal of rigs, wires, damaged or unwanted artifacts from the film frame; digital colorization which enables the adding of color to particular elements or whole films.[44] During pre-production, computerized storyboarding has become useful by allowing decisive pre-visualizations of scenes or sequences before the actual live-action shooting process gets under way. The pre-visualization stage during pre-production usually consists of the following categories: set design and construction; lighting considerations; costume and make-up design; casting roles; location scouting and property searches; story-boarding and script visualization with some images and audio; planning cinematography, camera moves and shots; feasibility studies to determine the possibility of achieving specific effects.[45] For each of these categories, some form of software or computer program became available to assist the various personnel involved in making crucial decisions during pre-production. The use of software programs allows time and money to be saved by nullifying the "trial-and-error" factor, especially with regard to expensive and complex shots involving complicated camera movements, explosions, car chases, aerial and underwater scenes, inter alia. While perfectly complementing moving pictures in general, new digital tools and their associated technologies (fiber optics, data storage, processing and so on) have been easily accepted by industry end-users because they are fundamentally artist-friendly. They were also relatively cheap, flexible and more efficient compared with the analog tools and technologies that had traditionally equipped movie studios.

The progressive integration of digital media into filmmaking proved to be a worthwhile venture indeed, first through the relatively fast and cost-effective production of sophisticated VFX and animation, later through the ability to accelerate post-production until the entire filmmaking process had gone digital. Investment in computer imaging, digital VFX and digital filmmaking R&D and the progressive upgrading of production and post-production technologies from analog to digital was at first mainly considered as a strategic move to boost motion picture exhibition revenues while reducing production costs in the long run. Hence, by 1988, it was not surprising for Nicholas Negroponte to comment that the 'movie industry doesn't want to make prints and ship them around. Film is costly to make, costly to deliver, and it deteriorates. They want to get into distributing digital signal by satellite and broadcasting into movie theaters.'[46] This certainly justified the existence in the late 1980s of the "Movies of the Future" research group in the Media Laboratory at MIT that, at the time, was receiving about $1 million from Warner, Columbia and Paramount. They had all begun to realize that the future of the Hollywood cinema industry would be in computer imaging and digitization which promised cost-cutting, time-saving, higher efficiency and quality, amongst other things, with regard to motion picture production, exhibition and distribution.[47]

Standardizing and normalizing the digitization of filmmaking technologies was not implemented overnight, so to speak, nor without some form of consensus or agreement reached by the various studios; nor did the technologies concerned emerge from nowhere. The digitization of filmmaking in Hollywood did not occur without experimentation, tests, monitored trials and board meetings. But setting industrial standards, however, especially in such a closely-knit professional community as Hollywood, is rarely extensively publicized simply because of the vast amount of money, work and resources involved. Stewart Brand asserts that 'nobody votes for technology. . . . But standards are agreements. They are a political process that takes place far from the political arena of the public the standards will affect.'[48] And standardizing and normalizing digital filmmaking in Hollywood might possibly have occurred faster had it not been for the proliferation of many different technologies simultaneously emerging and competing to control the same market—as illustrated by two different DVD formats being respectively introduced in 1995 by the partnerships of Sony-Philips and of Toshiba-Time-Warner; both parties would eventually reach an agreement in the same year to standardize one particular format for worldwide commercialization. According to Robert Samuelson, the

> most significant competition doesn't involve identical products sparring over price. It involves rival technologies struggling for superiority. Cable TV competes against satellite TV. Wireless communication competes with landlines. The Linux operating system is beginning to challenge Windows. For most technologies, standards are vital. Without them, mass markets are impossible. Sometimes standards arise by voluntary agreements among firms; sometimes they result from the triumph of one or a few firms. The check on this dominance—if there is a check—is the threat of a new technology.[49]

In the case of the Hollywood cinema industry, 'the triumph of one or a few firms' could be equated with the continuous dominance of digital VFX application and of digital filmmaking R&D by Lucasfilm, Industrial Light & Magic and Pixar—a dominance powered by their numerous breakthroughs in the development and use of digital technologies, to the point of setting techno-visual standards for others to follow in the industry.

Hollywood decision-makers obviously invested their money and hopes in the right direction with regard to the future of the studios as well as that of commercial filmmaking in general. As well as easier recouping of investment and increased profit margins enabled through the cost-effective production of more profitable higher quality motion pictures, digitization caused a much faster convergence of practically all communication media.

> Along with the tendency for once separate cultural media to become increasingly integrated into (technological) 'meta-media,' . . . there is the often awkward emergence of new creative engagements with contemporary technologies in contexts outside those usually seen to constitute a privileged cultural terrain.[50]

The transition to digital, in conjunction with conglomeratic tendencies across the Hollywood cinema industry and the American entertainment and communication industries at large, encouraged major improvements in the economic fundamentals of film production and distribution. Digitization hence broke the perceived barriers of low-budget production values. Working in digital allowed low-cost productions to achieve high quality, stylized images with levels of craftsmanship and precision usually associated with big studio feature films. It also considerably expanded their access to a wider range of distribution networks and exhibition outlets. The existing process of "price tiering" in the movie business became even more efficient and profitable with the increasing centralization of the digital medium in the distribution of movies. As explained by Tino Balio, when a new motion picture is released

> the goal of the distributor is to wring top dollar from each market, or "window," as it is called in the trade. Going through the distribution pipeline, the picture is exploited in one market at a time, with the exception of home video, which has a window that remains open almost indefinitely. . . . Movies are first released to theaters at top prices to "high value" consumers, which is to say, those who are most anxious to see them and are willing to pay seven dollars and more a ticket; movies are then released to "lower value" consumers at prices that decline with time. . . . Distributing pictures in this manner allows a distributor to tap every segment of the market in an orderly way and at a price commensurate with its demand.[51]

Prior to going digital, the Hollywood cinema industry was already in the habit of releasing movies through all ancillary media available, in response to the programming requirements of the VCR and videodisc ventures of the early 1980s in particular—although the cost of converting, transferring and re-producing celluloid film master content to other distribution media tended to be fairly high. With the standardization of the digital medium in motion picture production, circulation and distribution, one effect was an increase in exhibition outlets in addition to lower costs of conversion, transfer and re-production.

During the 1990s, electronic commerce as carried out on the Internet represented more than $100 billion of global business.[52] The Hollywood cinema industry could hardly ignore such a lucrative market. Had it not gone digital, it would have been unable to benefit extensively from global technological convergence and would have been stuck in the lower profit margins of more "traditional" forms of film exploitation. The state of the commercial film distribution market became such that a major company devoid of an in-house multi-media production unit would mean being less competitive in having to spend more on sub-contracting to other multi-media facilities. Hence, every major motion picture company rapidly organized its own multi-tasking multi-media unit soon after the required technologies emerged. It was the efficiency of multi-media units that ensured a film-generated revenue in many more ways than just theatrical release, merchandizing, video rental and retail.[53]

> Instead of the manufacturer simply imposing standardized patterns of consumption on seemingly passive consumers, consumers play a more active role in determining what is produced. In other words, there is a two-way relation between supplier and consumer. In this altered econ-

> omy, the consumer produces the producer as much as the producer produces the consumer. To respond to changing patterns of consumption quickly and effectively, a less centralized manufacturing structure is necessary. As power becomes decentralized, hierarchical relations give way to lateral associations until standardization and homogenization are replaced by diversification and localization. Products proliferate to meet ever-changing patterns of consumption.[54]

For example, spin-off multimedia products of sci-fi film *Stargate* during the 1990s included *Secrets of Stargate* (an interactive CD-ROM containing movie footage, biographies, stills, and so on) and *Stargate: The Game* (compatible with such platforms as Game Boy, Sega Megadrive and Game Gear). And the collector's edition of sci-fi blockbuster *Men in Black*'s DVD in 1997 launched the innovative additional option of allowing the viewer to alter several specific scenes in the movie by choosing alternative shots from rushes and extra takes provided in the DVD's "special features" menu. Ten years later, seven of the top-ten grossing Hollywood films in 2007 had video game releases; as Robert Brookey explains: 'Video game spin-offs are often tied to the most successful films in the market and are an important tactic in the larger marketing strategy of establishing a film franchise. Indeed where the most profitable films are concerned, video game releases are becoming the rule rather than the exception'—as a result, video games have become more like movies and more movies become video games.[55]

Normalizing the digital medium as the standard technological platform reinforced the prominence of synergy within the conglomerated Hollywood industry by expanding even further the commercial exploitation of film across a number of different yet related media. The concept of "digital cinema" would hence incorporate 'not only film culture as it is practised in theaters but also as it is re-imagined by the new devices that place emphasis on instantaneous one-click access to what appears to be an unlimited amount of content.'[56] The additional advantage of lower production costs for film spin-offs, destined for retail in supermarkets, specialty shops and their virtual counterparts on the Internet, conclusively established the viability and profitability of integrating digital media into filmmaking, and of setting up multi-media units to manage this integration. 'Now that viewing environments, audiences, technology, and genres are so multiple, the cinema is again in its mixed-medium mode. The American screen has extended its reach so that it now can be said to be [in the words of Thomas Elsaesser] "colonizing outer space, inner space, and virtual space."'[57] With regard to the future of motion picture distribution and exhibition and to make the best out of the Internet's future potential, the online "theatrical" release of feature films was being explored from the late 1990s onward. As Robert Sickels comments, 'The industry has to reinvent itself in the wake of the onslaught of new technologies, as the tried and true process of giving a film a grand opening to garner the reviews that will bolster its subsequent wide distribution seems positively antique in this day and age of instantaneous communication and social networking.'[58]

There is as much continuity as discontinuity when it comes to techno-visual innovation, in the wake of Hollywood's entry into the digital realm. What the consequences might have been, had the process of techno-industrial digitization been more radical and totalizing, are endlessly speculative. What matters is that digital transition permitted an extremely productive cohabitation of analog and digital tools and technologies, thus enabling a major expansion of the range of what could be visually achieved on film, and how it could be commercially exploited in multiple ways.

Consequential to transformations that have occurred in techno-industrial practices and personnel, there has been a noticeably extensive improvement of the standards of imaging in Hollywood cinema. VFX within two decades became increasingly refined on screen, hence emphasizing the highly sophisticated look of the Hollywood film in general. The need for spectacularity in Hollywood films—other than to satisfy requirements of techno-visual sophistication and objectives of profit maximization—has contributed to the normalization of VFX-oriented film-making in the Hollywood industry. Its constant preoccupation with all "new" cinema-related technologies was ultimately fulfilled through the transition to digital: more aesthetic possibilities, more output and more profit potential with less financial liability.

NOTES

1. Williams, *Problems in Materialism* 47–48.
2. Darley, 'From Abstraction' 41.
3. Andy Darley comments that 'the vast sums of money given over to the military for "defense" purposes—and later as part of the "space race"—was (arguably) the most important factor in ensuring rapid developments in the field of electronics, and particularly in the area of information and computer technology.'

Darley, 'From Abstraction' 41.

For a better idea of the influence and contribution of military research to cultural industries, see: Virilio, *War and Cinema*; White, *The Fruits of War*.

4. Darley, 'From Abstraction' 42.
5. See Darley, 'From Abstraction' 42.
6. Charles Csuri would later co-found one of the world's first computer animation production companies: Cranston/Csuri Productions (CCP).
7. 'The Medal of Commendation is awarded by the Academy Board of Governors upon the recommendation of the Scientific and Technical Awards Committee. It is given in appreciation for outstanding service and dedication in upholding the high standards of the Academy.'

See 'Academy Awards Database' (online).

8. Darley, 'From Abstraction' 43.
9. See Whitney Sr., 'Motion Control.'
10. See Appendix C for more details.

11. Subsequently, Trumbull's career shifted from traditional Hollywood film projects to the development of technology for use in the exhibition industry and in theme-park rides. He also focused more on the evolution of VFX, especially on virtual digital sets and electronic cinematography. An innovator in the domain of optical and computer-based VFX in the cinema, he would eventually be the recipient of a Scientific and Technical Award at the 65th Academy Award ceremony in 1992 for the concept of the camera system design of the CP-65 Showscan Camera System for 65mm motion picture photography.
See 'Academy Awards Database' (online).

12. After studying industrial design and still photography, Dykstra had begun experimenting in the early 1970s with the use of computers to facilitate filming at the National Science Foundation at the University of California–Berkeley where he would design, build, and operate a computer-controlled camera system—later to be acknowledged as one of the foundations of camera motion-control technology. Much of his earlier experimental work would end up being applied during VFX production for *Star Wars*.

13. Nicholson, 'Special Effects in *Star Wars*' 119 & 192.

14. Quoted in Baxter, *George Lucas* 34.

15. See Appendix C.

16. Dennis Muren developed an interest in filmmaking at an early age from watching countless monster and science fiction movies, and from reading VFX-related publications such as *Famous Monsters of Filmland* (1958–1983)—all of which fuelled his imagination and sparked his interest in VFX. While studying business, Muren co-directed the sci-fi film *Equinox* (aka *The Beast*) in 1970. Tonylyn Productions liked the picture enough to distribute it in theaters. Despite mixed reviews, the movie generated enough money from ticket sales for Muren to recover his original investment; it even became a minor cult classic in later years. After completing his college studies, Muren worked full-time as a freelance VFX artist in the early 1970s until he was hired for *Star Wars*.

17. Vaz, *Industrial Light & Magic* 119. See Appendix C for more details.

18. A multi–Academy Award–winning VFX cinematographer, Richard Edlund has also served as a governor for AMPAS for twelve years, as chairman of AMPAS's VFX Branch since its inception, and as chairman of AMPAS's Scientific and Technical Awards Committee for eight years. He has also actively served on the boards of the American Society of Cinematographers and the Visual Effects Society. With a background in photography training and camera repair in the U.S. Navy, and in film studies at the University of Southern California in the 1960s, Edlund started his cinema career at Joe Westheimer's optical effects studio and, for a brief stint, at Robert Abel's special effects studio in the 1970s, until he was recruited for *Star Wars*.

19. Multi–Academy Award winner Ken Ralston made garage movies using claymation and a Kodak 8mm camera in his childhood. He started his career at VFX provider Cascade Pictures in the early 1970s, working mostly on TV commercials before being hired for *Star Wars*.

20. Langford, *Post-Classical Hollywood* 248.

21. Gomery, 'Economic and Institutional' 31.

22. Quoted in Pye & Myles, *The Movie Brats* 130.

23. Pye & Myles, *The Movie Brats* 130–31.

24. Quoted in Pye & Myles, *The Movie Brats* 133.

25. See Earnest, '*Star Wars*: A Case Study.'

26. See Pye & Myles, *The Movie Brats* 132.

27. Pye & Myles, *The Movie Brats* 132.

28. Pye & Myles, *The Movie Brats* 135.

29. For a comprehensive study of *Star Wars*, see Brooker, *Star Wars*.

30. About VFX production for *2001: A Space Odyssey* and for *Star Wars*, see: Agel, *The Making of Kubrick's 2001*; Bizony, *2001: Filming the Future*; Nicholson, 'Special Ef-

fects in *Star Wars*'; Schwam & Cocks, *The Making of 2001*; Smith, *Industrial Light & Magic*; Vaz, *From* Star Wars *to* Indiana Jones.

31. See Appendix C for more details.

32. By 1991, Pixar Animation Studios would start a close collaboration with Disney on the production of animation feature films, and in 2006 Disney would purchase Pixar from Steve Jobs through an all-stock transaction worth $7.4 billion.

33. Berardinelli, '*Star Wars: Episode I—The Phantom Menace*' (online).

34. David Tattersall is quoted in *The Complete Coverage* (online).

35. *The Complete Coverage* (online).

36. A 286-acre filmmaking facility located outside London, Leavesden Film Studios was originally an old Rolls Royce aviation factory before it was converted for film work.

37. David Tattersall is quoted in *The Complete Coverage* (online).

38. Corliss and Booth, 'Cinema: Ready' (online).
The scene where toxic gas is released on the Jedi is the only sequence in *Star Wars: Episode I—The Phantom Menace* without any digital alteration.

39. Berardinelli, '*Star Wars: Episode I—The Phantom Menace*' (online).

40. Maslin, '*The Phantom Menace*' (online).

41. See Fisher, 'Dawning of the Digital Age.'

42. Ohanian & Phillips, *Digital Filmmaking* 173.

43. McQuire, 'Technicalities' 25.

44. For instance, Huston's *The Maltese Falcon* (1941) and Curtiz's *Casablanca* (1942) both underwent the digital colorization process after their rights were bought by Ted Turner.

45. See Ohanian & Phillips, *Digital Filmmaking* 31.

46. Quoted in Brand, *The Media Lab* 81.

47. Brand, *The Media Lab* 12.

48. Brand, *The Media Lab* 76.

49. Samuelson, 'Puzzles' 58.

50. Hayward, 'Introduction' 11.

51. Balio, 'Introduction to Part II' 263.

52. The top three websites during the 1990s were run by America On-Line, Yahoo, and Netscape. Defining itself as a media company, Yahoo, for instance, had built up a $5.3 billion market cap in only three years—doing so with just a handful of traditional media people among its 470 employees. See Roth, 'New-Media Nightmare' 82.

53. See Curtin, 'On Edge: Culture Industries.'

54. Taylor & Saarinen, *Imagologies* 58.

55. Brookey, *Hollywood Gamers* 5.

56. Tryon, *Reinventing Cinema* 175.

57. Miller, 'Introducing Screening' 25; Elsaesser, 'Film Studies' 44.

58. Sickels, *American Film* 7.

Afterword

While digital VFX had been criticized in the past for being intrusive and distracting in drawing attention to themselves, after the 1990s they simply became a standard component of motion picture production in Hollywood and an expected aspect of the Hollywood film viewing experience. And the Hollywood movie is now just one media platform in a multimedia environment, with the theatrical release of a film ceasing 'to be an end in itself: it became simply a means to a much larger end, the exploitation of the copyrighted intellectual property it represented across an expanding cascade of profit-making opportunities, the overwhelming bulk of them centred not in theaters but in the home.'[1] Hence, the proliferation of Hollywood film-related spin-off television serials, video games, cartoons, online content, novels, comics, and most peculiar cinematic objects that are the digitally re-mastered motion picture and, more recently, the 3D converted motion picture.

The year 1997 saw the release of the digitally re-mastered *Star Wars* movies containing additional unseen footage, characters, creatures, objects and VFX.[2] Having attained a particularly high degree of sophistication by the mid-1990s, the digital technologies available enabled this particular trilogy of films to demonstrate convincingly how just *one* motion picture could be "re-made" many times over decades and (re-)circulated as a "unique" and "authentic" filmic text in its own right. After the digital re-mastering of the *Star Wars* trilogy, some other motion pictures originally made during the 1970s and 1980s would also be re-mastered (or "restored") with new digital VFX, imaging and sound replacing their original analog counterparts. Examples are: *The Exorcist, THX 1138, E.T The Extra-Terrestrial* and *Jaws*. And the 2010s would witness the emerging practice of wide theatrical releases of "3D makeovers," such as *The Lion King 3D* in 2011, *Titanic 3D* and *Star Wars: Episode I—The Phantom Menace 3D* in 2012—all obtained from converting the 1990s original films to 3D. Could this imply that the original analog Hollywood films of the 20th century will eventually disappear entirely to be replaced by their digital re-masters with added digital VFX, imaging and sound and by their digital 3D versions—which arguably make them different films altogether? Digital re-mastering and 3D conversion have caused collective cinematic memory to be continuously re-designed, re-created, and re-produced by the Hollywood cinema industry, as such modifying long established concepts of cinéphilia, archiving, memory and history. As such, digitization has not affected 'a radical break or aesthetic disjunction from earlier forms. Rather, the digital contributes to new hybrids of cinematic culture, which return to earlier cultural traditions in the forging of the "new".'[3]

Since the beginnings of Hollywood cinema, the creation of a heightened apprehension of experiential possibilities for the spectator has always relied on trickery and illusion. Remarkable transformations undergone by screen-related technologies in enabling the re-presentation, reproduction, recording as well as projection of visual trickery and illusion have irreversibly altered the relationships between cinema spectators and filmic texts. This is illustrated by one of the most significant outcomes of VFX and filmmaking going digital: the emergence of computer-generated actors or "synthespians" in Hollywood cinema. The creation of realistic synthetic actors has been the objective of many in the domain of computer graphics from the late 1970s onwards—as exemplified by two short films in particular that were first screened at SIGGRAPH conferences during the 1980s: *Rendez-Vous à Montreal* (which essentially re-created Marilyn Monroe and Humphrey Bogart as interacting synthetic characters involved in a romance) and *Tony de Peltrie* (the first time on film whereby a synthetic actor conveys complex emotions).[4] Since then, synthetic characters have become increasingly common on the screen, particularly on television: a long departed Gene Kelly dancing in commercials for cars and household appliances; Natalie Cole dueting live with her deceased father Nat King Cole, as a holographic projection, at the Grammy Awards; Céline Dion performing live next to the holographic projection of Elvis Presley at the Academy Awards; and the deceased Tupac Shakur performing live on stage at the Coachella Valley Music and Arts Festival as a full hologram.[5] And gaming and cyberspace have also been populated by many more "realistic" computer-generated synthetic characters or "avatars."

With regard to Hollywood feature films, at first limited to seconds in terms of screen appearance—such as T-1000 in *Terminator 2: Judgment Day* and some of the "passengers" in the ship sinking sequence of *Titanic*—from the end of the 1990s onwards, talking "synthespians" appeared on the theatrical screen for an increasingly longer time, as illustrated by Jar Jar Binks in *Star Wars: Episode I—The Phantom Menace*, Gollum in *The Lord of the Rings* trilogy and Dobby in *Harry Potter and the Chamber of Secrets*.

> While audiences have, in certain circumstances, rebelled against the digitization of certain characters, such as Jar Jar Binks in *The Phantom Menace*, the tremendous popularity of digitized characters in Peter Jackson's *Lord of the Rings* films, among many others, illustrates the degree to which audiences can embrace digital effects when wrapped in an enticing narrative.[6]

More significantly however, by 2001, the first entirely computer-generated feature-length movie to attempt realistic human characters was released in cinema theaters, namely *Final Fantasy: The Spirits Within*. Creating 'a world that is neither live action nor animation, but some parallel cyberuniverse,' *Final Fantasy: The Spirits Within* led film critic Robert Ebert to comment:

> Is there a future for this kind of expensive filmmaking ($140 million, I've heard)? I hope so, because I want to see more movies like this, and see how much further they can push the technology. Maybe someday I'll actually be fooled by a computer-generated actor (but I doubt it). The point anyway is not to replace actors and the real world, but to transcend them—to penetrate into a new creative space based primarily on images and ideas.[7]

Derived from a popular series of video games originally developed in Japan, *Final Fantasy: The Spirits Within* was not produced by any of the big Hollywood players. It was nevertheless distributed by one of the Hollywood majors and involved capital investment, screen stars (voicing characters) and digital VFX production personnel commonly associated with the Hollywood cinema industry.

By the end of the first decade of the 21st century, various computer graphics and animation companies were able to create fully computer-generated synthetic but reasonably "realistic" human characters—in feature films, cyberspace, video games, TV commercials and music videos. While some movies and video games still include relatively clunky animated CG human characters, it is most likely that, before 2020, the Hollywood cinema industry will be producing entirely computer-generated animated films with characters and sets that are practically indistinguishable from those of so-called "live-action" films, with *Avatar* providing a glimpse into the possible futures of digital cinema and VFX.

One of the cinema's many possible futures is perhaps embedded in the contemporarily more pronounced synergistic convergence of moving pictures, gaming and online distribution—as illustrated by the online release and sale in 2011 of the *Star Wars: The Old Republic* interactive game that enables the player/viewer to be 'the hero of your own *Star Wars* saga in a story-driven massively-multiplayer online game. . . . Choose to be a Jedi, a Sith, or from a variety of other classic *Star Wars* roles, and make decisions which define your personal story and determine your path down the light or dark side of the Force.'[8] As such, has the cinema, as promised for more than a hundred years, finally been delivered in its optimal form to the spectator? But . . .

. . . is this still cinema? Est-ce encore du cinéma? For Charles Tesson:

> S'aventurer aux frontières du cinéma revient à reformuler *la* question d'André Bazin: qu'est-ce que le cinéma, à la lueur de ce qu'il a été (l'enregistrement, la projection) et de ce qu'il est devenu, par exemple

> avec le numérique? Bazin avait répondu en prenant position: "*Pour un cinéma impur.*" Sauf que l'impureté a changé de nature (dans le cinéma) au risque de changer sa nature (le numérique), même si on oublie le grand principe ("*Pas de gain de réalité sans perte*") qui définit le cinéma selon Bazin.[9]

> Exploring the boundaries of cinema boils down to re-formulating *the* question according to André Bazin: what is cinema, in the light of what it has been (recording, exhibition) and of what it has become, as a result of digitization for example? Bazin responded by taking a specific stand: "For a mixed [*impur*] cinema." Except that what a "mixed [*impur*] cinema" implies has changed (within the cinema) at the risk of the transformation of its very nature (digital), even if we forget the underlying principle ("No gain in reality without loss") that defines the cinema according to Bazin.

Although having its roots in celluloid-based cinema, digital cinema has been progressively effacing the traces of its analogic origins. By the 1990s, analog cinema had become part of domains and media that are *a priori* un-cinematic—such as video games and virtual reality theme park rides—and which, without becoming entirely cinematic themselves, nevertheless owe their existence to the fundamental principles of analog cinema. What has been happening since the end of the 1990s is an irreversible obliteration of boundaries between the internal structure of the cinema and its more or less clearly defined external structure (digital video, digital television, online multi-media content and so on). With digital VFX production, filmmaking and re-mastering becoming standard by the 1990s, on the surface analog cinema no longer had anything to conquer as its century of existence was being celebrated. A legitimate form of art in its own right, Hollywood cinema in its digital state is simultaneously similar to and different from its earlier analogic state. This can be correlated, for instance, to Sean Cubitt's observation that because 'cinema so clearly traces a history from mechanical to digital time, . . . the shifting temporalities of the commodity film have neither ceased to change nor mutated into something utterly different in the digital era.'[10] With digital filmmaking firmly in place, the future of the cinema, and so of the Hollywood cinema industry, will be irremediably intertwined with further industrial research and development in corporate technoscience in general and in computer programming and digital imaging in particular.

Bazin's fundamental question 'What is cinema?' after analog cinema's fifty years of existence and Bouquet's question another fifty years later "is this still cinema?" are of even more pertinence now in relation to the digitized state of contemporary Hollywood cinema. The fundamental definition of the cinema has always inevitably involved the process of its re-definition according to the techno-visual, economic, industrial and ideological specificities of the context or period in question, and of its comparison with preceding definitions. Is digital (Hollywood) cinema still "cinema"? Yes, but not according to previous definitions of what the cinema was supposed to be—at least until the 1990s after which, from incorporating and assimilating all other un-cinematic digital media and forms, the cinema as we have known it for a century simply became *impur*.

NOTES

1. Langford, *Post-Classical Hollywood* 199.
2. Using the latest in computer-generated imaging, Industrial Light & Magic re-designed and re-shot all the space sequences, replacing the original X-Wing and TIE fighter miniatures with eye-jolting computer-generated versions. Other sequences were also digitally updated with new or extra CGI "creatures" and skeletal droids, the most prominent digital tweaking being the restoration of a missing scene involving Han Solo and Jabba the Hut.
3. Harbord, 'Digital Film' 24.
4. *Rendez-Vous à Montreal* was made by Nadia and Daniel Thalmann, both assistant professors at the University of Montréal at the time, while their students and collaborators Philippe Bergeron, Pierre Lachapelle, Daniel Langlois and Pierre Robidoux made *Tony de Peltrie*.
5. Tupac Shakur's hologram was produced by James Cameron's visual effects production house Digital Domain in collaboration with hologram-imaging companies AV Concepts and Musion Systems.
6. Tryon, *Reinventing Cinema* 39.
7. Ebert, '*Final Fantasy*' (online).
8. For more details, see *Star Wars: The Old Republic* (online).
9. Tesson, 'Éloge' 4.
10. Cubitt, *The Cinema Effect* 8.

Appendix A: A Time-Line of Landmark (VFX-Intensive) Films, Techno-Scientific Innovations, Financial and Industrial Events That Have Contributed to the Evolution of Digital VFX in the Hollywood Cinema Industry up to the 1990s

[All "events" mentioned in the time-line have occurred in the United States of America, unless otherwise specified]

1869

- J. W. Hyatt invented celluloid.[1]

1879

- Thomas Edison devised an incandescent carbon filament electric light bulb that can last more than 13 hours.[2]

1881

- Thomas Edison established Edison Electric Light Company, the first incandescent lamp factory.[3]

1884

- Paul Nipkow filed a patent in Germany for an "electric telescope" wherein the interaction of selenium and electricity leads to the transformation of light into a modulated electrical current. A year's work later, he filed a master patent for "television."[4]
- George Eastman introduced photographic roll paper.[5]

1886

- George Westinghouse founded Westinghouse Electric Company.[6]

1887

- Edison Phonograph Company was formed.[7]

1888

- After meeting Edward Muybridge and being shown his Zoopraxiscope motion picture machine, Thomas Edison instructed William Dickson and other assistants in his company to develop a Kinetoscope that was intended to be 'an instrument which does for the eye what the phonograph does for the ear.'[8]
- George Eastman introduced the Kodak box roll-film still camera.

1889

- Herman Hollerith invented a machine that utilized punch cards for computation; its main application was for tabulating census results. Hollerith consequently founded the Tabulating Machine Company in 1896, which became International Business Machines (IBM) in 1924.[9]
- George Eastman developed the flexible film roll for still photography.[10]
- Edison General Electric Company and Edison Manufacturing Company were established by Thomas Edison.[11]

1891

- Following Thomas Edison's earlier request for a motion picture machine, William Dickson developed motion picture film and recorded images on it, leading to the emergence of the Kinetoscope (described in the patent application as 'an apparatus for exhibiting photographs of moving objects,' although destined for only one viewer at a time) and of the Kinetograph (basically a camera for recording motion pictures).[12]
- Royal Philips Electronics Inc. (better known later as Philips) was founded by Gerard Philips in Eindhoven, Holland.

1892

- General Electric Company was formed following the merger between Edison General Electric Company and Thomson-Houston Company.[13]

- The Eastman Kodak Company was founded by George Eastman and Henry Strong. It would become the largest supplier of photographic still and motion picture celluloid films worldwide and influence the evolution of picture quality standards in the Hollywood cinema industry and elsewhere by setting the standard of 35mm film for commercial filmmaking and theatrical exhibition.

1893

- Thomas Edison's Kinetoscope motion picture system—as developed by William Dickson—was publicly demonstrated for the first time; applications were submitted the following year to the Patent Office for both the Kinetoscope and the Kinetograph.[14]
- The construction of Thomas Edison's Black Maria outdoor film studio was finished; this was where the earliest Edison motion pictures were recorded.[15]

1894

- The world's first Kinetoscope parlor opened its doors.[16] Although this kind of venue did not yet offer the experience of collective spectatorship, it was nevertheless a significant "ancestor" of the 20th century cinema theater.
- Edison Manufacturing Company began producing and selling films as well as Kinetoscope machines.[17]

1895

- The Lumière camera and projector were introduced in France; compared to others in existence, the Lumière motion picture machine's mechanism allowed reverse film motion.[18] Researchers in photochemistry and manufacturers of photographic equipment by trade, Auguste and Louis Lumière called 'their projector the "Cinématographe," a name reminiscent of Sellers's paddle-wheel machine [1861] and anticipating the universal word for motion pictures—"cinema" (from the Greek *kinema, kinematos,* or "motion").'[19]
- The Lumière brothers in France showed what was the first collection of "short films" in cinema history, such as *La Sortie Des Usines Lumière, Repas de Bébé, Partie de Cartes* and others—essentially recordings of common everyday activity.
- Woodville Latham's Eidoloscope motion picture projector was publicly presented.[20]
- The Phantoscope motion picture projector was publicly demonstrated by Francis Jenkins and Thomas Armat.[21]

- The Bioskop motion picture projector was completed by Max Skladanowsky.[22]
- The Kinetophone was invented, following experiments by Thomas Edison and William Dickson in synchronizing sound with film; it would be commercially introduced in 1913 but eventually abandoned altogether by 1915.[23]
- William Dickson left Edison Manufacturing Company and established American Mutoscope Company in partnership with Herman Casler, Henry Marvin and Elias Koopman. It would later be known as American Mutoscope & Biograph Company and become a major competitor to Thomas Edison in the very early years of the motion picture business.[24]
- The stop-camera technique—which involves the visual transformation of objects in the middle of a shot by stopping the camera and adding or subtracting these objects in question from the scene before starting the camera again—was applied for the first time, by the intermediary of a Kinetograph, during the making of the short film *The Execution of Mary, Queen of Scots* directed by Alfred Clark.[25] This was arguably the first *special effect* in motion picture history.

1896

- Edison Manufacturing Company introduced the Projecting Kinetoscope or Projectoscope. Alongside the Vitascope and Thomas Edison's machines, many different types of projection systems were concurrently being used in American motion picture theaters: Lumière's Cinématographe, Woodville Latham's Eidoloscope (re-engineered to include the innovations of the Vitascope), Birt Acres's Kineopticon and American Mutoscope and Biograph Company's Biograph, among others.[26]
- Raff & Gammon—owners of the Kinetoscope Company and of the rights for the Phantoscope machine (renamed Vitascope and advertised as a Thomas Edison invention)—began the practice of forwarding to the Library of Congress short pieces of positive nitrate celluloid film derived from motion pictures for copyright deposit.[27]
- Thomas Tally opened Electric Theater, the first movie house or cinema theater in the United States[28]
- Alexandre Promio, one of Lumière's itinerant cameramen/exhibitors, is said to have performed the first camera movement—a travelling shot—in *Le Grand Canal à Venise*, by mounting the camera on a moving vehicle (a gondola in this case).[29]
- Georges Méliès in France further refined the stop-camera technique while making *Escamotage d'une Dame chez Robert-Houdin*.[30]

1897

- The cathode ray tube (CRT) was introduced as a result of complex research and experimentation carried out by J. J. Thompson; Sir William Crookes had previously described cathode rays in England in 1878.[31] Less than a century later, this device would have proliferated to the extent of omnipresence—in television sets, computer monitors, and practically any kind of visual display screen.[32]
- R. W. Paul in England designed the first real panning head for a camera tripod in order to film in one single shot the passing processions marking Queen Victoria's "Diamond Jubilee."[33]

1898

- Georges Méliès in France made and showed *La Lune à un Mètre*; with its three scenes, it was arguably the cinema's first multi-shot film.[34]

1899

- American Vitagraph Company (film production and distribution) was founded by James Stuart Blackton.[35]
- James Stuart Blackton and Albert Smith used cut-outs and miniature sets in the making of *The Battle of Santiago.*

1900

- General Electric Company established the first industrial laboratory solely dedicated to scientific research.[36]
- Edison Manufacturing Company was incorporated and subsequently known as Edison Co.[37]
- Edwin S. Porter was hired by Edison Co. to work with motion picture equipment; he would rapidly become not only the company's most famous film director but also a key innovator of early cinema.[38]
- In England, George Albert Smith made *Grandma's Reading Glass* which pioneered the practice of dividing one scene into multiple shots.[39]

1901

- The construction of the first glass-enclosed indoor film studio was completed for Edison Co.[40]

1902

- In *The Twentieth Century Tramp*, Edwin S. Porter pioneered the use of double exposure in separate areas of the frame, created by masking in the camera or printer rather than using a black area built into the décor as Georges Méliès and others had done previously; he would improve on this technique while making *The Great Train Robbery* a year later.[41]
- Georges Méliès's one-reeler VFX-intensive *Le Voyage Dans La Lune* was publicly exhibited.

1903

- Nikola Tesla patented electrical logic circuits referred to as switches or gates—an important milestone in the evolution of computer technology.[42]
- Edwin S. Porter's *The Great Train Robbery* was shown publicly: it contained one of the earliest applications of the optical compositing technique—which would eventually become essential to the future production of VFX in particular—for creating a realistic scene on film.[43]

1905

- Built by vaudeville magnate Harry Davis, the first Nickelodeon motion picture theater was launched in Pittsburgh.[44]
- Edwin S. Porter introduced the technique of single-frame filming by the intermediary of object animation in *How Jones Lost His Roll* and *The Whole Dam Family and the Dam Dog*.[45]

1906

- Lee de Forest invented the thermionic triode valve, pivotal to the future evolution of computer technology. It was the first electronic mechanism capable of comparing two inputs to produce a logical output, hence making possible the application of Boolean algebra.
- Segundo de Chomòn's *Sculpture Moderne*, made at and produced by Pathé Studios in France, was arguably the first motion picture involving the technique of clay animation.[46]
- The technique of single-frame filming was applied to a series of drawings made by James Stuart Blackton to create the first entirely animated motion picture: *Humorous Phases of Funny Faces*.[47]

1907

- Norman O. Dawn made an important contribution to the domain of cinematic VFX when he produced the first glass shots—by painting additions to the scene being photographed on a sheet of glass fixed several feet away in front of the camera; he later improved on the technique by developing the glass matte painting shot in 1911. It is only after 1913 that the technique would be extensively applied in commercial filmmaking.

1908

- The Kinemacolor process emerged. Developed by George Albert Smith in England following an earlier unsuccessful attempt by Edward Turner at a three-color additive process, Kinemacolor was the only genuine motion picture coloring process to be commercialized years before the advent of Technicolor in the 1930s.[48]
- Edison Co shifted its motion picture production activities to its new indoor film studio.[49]
- An association of Edison Licensees and Film Service Association was established by Edison Co. The Motion Picture Patents Company was subsequently founded from this association and includes Biograph licensees, as well as other film production companies (Essanay, American Vitagraph, Pathé, Star-Film, etc.). Known as "the Trust," this association temporarily monopolized and controlled the movie industry by virtue of various agreements of exclusive exploitation being established between film production and distribution companies, film exchanges, theater owners and raw film stock supplier Eastman Kodak.[50]
- D. W. Griffith was hired by American Mutoscope & Biograph Company.[51]

1909

- The ductile tungsten filament (still currently used in incandescent lamps) was developed at General Electric Company's research laboratory.[52] This enabled Westinghouse Electric Company to introduce and market the first continuous tungsten filament light bulb.[53]

1910

- Edison Co. was re-structured into Thomas A. Edison Inc.[54]

1911

- The Home Projecting Kinetoscope, using 21mm film, was launched by Thomas A. Edison Inc. without achieving much success.[55]

1912

- The Bell & Howell continuous film printer was introduced.[56]
- Famous Players (film production and distribution) was founded by Adolph Zukor; following a merger, it later became Famous Players-Lasky Corporation in 1916.[57]

1913

- Fox Film Corporation (theaters; film distribution; film production studio by 1915) was established by William Fox, who had already built a theater chain and initiated his own film distribution concern by 1910.[58]

1914

- James Brautigan, working as cinematographer for Thomas A. Edison Inc. devised a completely *mechanical* motion-control method for repeating camera movements while shooting a scene that required double-exposure. This constituted the earliest form of experimentation with motion-control technology (which would be more extensively developed, through the use of electronics, during the production of VFX for *Star Wars* in the 1970s).[59]
- Paramount Film Corporation (film distribution) was created by William Hodkinson.

1915

- American Mutoscope & Biograph Company ceased its film production operations.
- U.S. Technicolor Motion Pictures Corporation (Technicolor Inc.)—which deals with color film R&D—was founded by Herbert Kalmus and Donald Comstock.[60]
- D. W. Griffith's controversial but spectacular epic *The Birth of a Nation* was released—an important benchmark in the development of (American) cinema.

1917

- The rotoscope camera—a key piece of equipment in the production of two-dimensional animation—was developed by animation pioneer Max Fleischer.[61]
- First National (a theater chain that subsequently funded, produced and distributed feature films) was founded by Thomas Tally and John Williams.[62]

1918

- Thomas A. Edison Inc.'s motion picture studio stopped production; it would be sold to Lincoln & Parker Film Company.[63]
- Matsushita Electric Industrial Company was founded by Konosuke Matsushita in Japan.[64]
- The *Out of the Inkwell* animation series—directed by Dave Fleischer and animated by Max Fleischer—introduced the concept of compositing on film two-dimensional cel animation with three-dimensional live action.

1919

- United Artists Corporation (film production and distribution) was co-founded by Charlie Chaplin, Mary Pickford, Douglas Fairbanks and D. W. Griffith.
- General Electric Company created Radio Corporation of America (RCA) for marketing wireless radio equipment produced by both Westinghouse Electric Company and General Electric Company.[65]

1920

- CBC Sales Corporation (film distribution) was established by Harry Cohn, Joseph Brandt and Jack Cohn.[66]
- Marcus Loew (theaters; film distribution) acquired Metro Pictures Corporation (film production).[67]

1921

- Universal Pictures (film production) was established by Carl Laemmle.[68]

1922

- Henry Luce and Briton Hadden's Time Company was incorporated, becoming Time Inc.[69]

1923

- Vladimir Zworykin patented a complete electrical television system that included a pick-up tube, an electronic camera adapted from the standard cathode ray tube.[70]
- The first issue of *Time* magazine was published by Time Inc.[71]
- The four Warner brothers (Jack, Sam, Albert and Harry) created Warner Bros. Pictures (film production) after working for nearly a decade in the motion picture business.
- Walt Disney, with his brother Roy, launched Disney Brothers Studio.[72]

1924

- The Moviola editing viewer was introduced.
- Columbia Pictures became the new name for CBC Sales Corporation as the company diversified into film production, in addition to its film distribution activities; Columbia Pictures would eventually have its own movie studio by 1926.[73]
- Goldwyn Pictures (film production) was bought by Loew's Inc. then merged with Metro Pictures Corporation (film production) and with Louis B. Mayer Company (film production) to form Metro-Goldwyn-Mayer (MGM) Pictures.[74]
- Music booking agency Music Corporation of America (MCA) was founded by Jules Stein and William Goodheart Jr.

1925

- Warner Bros. purchased American Vitagraph Company.[75]
- Produced by First National, *The Lost World* was released theatrically, featuring innovative VFX work by pioneer effects craftsman Willis O'Brien.

1926

- Eastman Kodak introduced the first duplicating positive and duplicating negative film stocks; improved versions of these would appear in the early 1930s.[76]
- John Logie Baird introduced the first usable television set in England with a picture of 30 vertical lines scanning at 12.5 frames per second.[77]
- General Electric Company, RCA and Westinghouse Electric Company jointly purchased WEAF radio station. It became the anchor station for their newly co-founded National Broadcasting Company (NBC) that immediately started network radio broadcasting; its

first program was relayed to 25 radio stations in 21 American cities.[78]

- Disney Brothers Studio became Walt Disney Studio.

1927

- Philo Farnsworth filed a patent for an electrical pick-up tube (the heart of the camera) described in the application as 'an apparatus for television which comprises means for forming an electrical image.'[79]
- Technicolor Inc. introduced the Technicolor camera, together with an innovative two-strip color process for film printing.[80]
- The United Independent Broadcasters radio network was set up by talent agent Arthur Judson, then taken over by Columbia Phonographic Manufacturing Company (which also owns Columbia Records) and renamed Columbia Phonographic Broadcasting System. It was subsequently sold for $500,000 to William Paley who streamlined the name of the corporation to Columbia Broadcasting System (CBS).
- The CBS Radio network was in operation.
- The Victor Talking Machine Company of Japan Limited, later to become Victor Company of Japan Ltd. (better known as JVC) was founded in Yokohama, Japan.
- Produced by Warner Bros., *The Jazz Singer* was much publicized as the first "talkie" or motion picture with sound (some music and several talking scenes) in cinema history. Using sound-on-film technology developed by RCA Photophone, it was an instant hit with audiences. With the theatrical release of *The Jazz Singer*, Hollywood cinema had officially begun a new techno-industrial phase.[81]
- Paramount Film Corporation absorbed Famous Players-Lasky Corporation.[82]

1928

- Westinghouse Electric Company introduced the Iconoscope, the first television camera tube.[83]
- The sprocketed 35mm optical soundtrack was perfected: it allowed sound components to be edited and accurately synchronized with visual components. This constituted a major step forward from the inflexibility of the Vitaphone sound process previously used.
- Victor Schertzinger's feature film *Redskin* was one of the first demonstrations of the capabilities of the Technicolor camera and the two-strip color film process recently introduced by Technicolor Inc.[84]

- The Dunning process enabled the first successful application of the travelling matte technique whereby the masking effect was generated purely photographically. As such, it created the foundations for the subsequent development of the blue-screen technique that would become essential to future VFX production.[85]
- Produced by Warner Bros., *Lights of New York* was the first "all-talking" motion picture.
- Mickey Mouse and Minnie Mouse were introduced in the silent black-and-white short animation film *Plane Crazy*—produced by Walt Disney for $1,772, directed and animated by Ub Iwerks; it premièred as a sneak preview at a Los Angeles theater.[86]
- Produced by Walt Disney, *Steamboat Willie* was the first animation film (in black-and-white) to feature synchronized sound. It starred Mickey Mouse—with Walt Disney himself providing Mickey's first sound voice—and Minnie Mouse.[87]
- RKO Radio—later known as RKO Pictures—was jointly created by RCA (Radio Corporation of America), General Electric Company and Westinghouse Electric Company through the amalgamation of RCA Photophone (sound technology R&D), Film Booking Office (film production and distribution), and Keith-Albee-Orpheum (theaters)—to exploit commercially RCA Photophone's patents for film sound equipment.[88]

1929

- Two years after the advent of sound in motion pictures, box-office admissions nearly doubled: from 60 million in 1927 to 110 million by 1929.[89]
- Walt Disney Studio became Walt Disney Productions.
- Warner Bros. expanded by acquiring First National and Stanley Company (theaters).[90]

1930

- The sound Moviola editing viewer, used to edit sound film, became available.[91]
- *Fortune* magazine was launched by Time Inc.[92]

1932

- Technicolor Inc. introduced its new three-strip color film process.[93]
- Produced by Disney, *Flowers and Trees* (in the *Silly Symphony* cartoon series, started in 1930) was the first animation film in Technicolor.[94]

1933

- Twentieth Century Pictures (film production) was co-founded by Joseph Schenck and Darryl Zanuck.[95]
- Paramount Film Corporation declared bankruptcy and was eventually re-organized as Paramount Pictures Corporation.
- Produced by RKO, *King Kong* was released in theaters; it contained groundbreaking VFX and animation work by Willis O'Brien. All combinations of animated models with live-action scenes were achieved by means of the background projection technique.

1934

- BASF in Germany was the first industrial corporation to produce magnetic tape for data recording.[96]
- The Federal Communications Commission (FCC) was established—as an upgrading of the Federal Radio Commission of 1927—by the 1934 Communications Act. Directly responsible to U.S. Congress, as an independent government agency the FCC was meant to regulate 'interstate and international communications by radio, television, wire, satellite and cable. The FCC's jurisdiction covered the 50 states, the District of Columbia, and U.S. possessions.'[97]
- The new Technicolor three-color film process was used for the first time in live-action filming during the production of *La Cucaracha*, a short film financed by Technicolor Inc.'s major shareholder Pioneer Pictures.[98]

1935

- The formation of Twentieth Century-Fox was the outcome of a merger between Twentieth Century Pictures and Fox Film Corporation.[99]
- The first feature-length movie in three-color Technicolor, *Becky Sharp*, was released.[100] Produced by Pioneer Pictures, the film itself was not particularly memorable but nevertheless marked yet another important techno-industrial transition for the Hollywood cinema industry after the introduction of sound.

1936

- Eastman Kodak introduced Kodachrome, the first 35mm color film.

1937

- Alan Turing presented his adaptation of the notion of algorithm to the computation of functions and his theoretical conceptualization of a computing machine in his *On Computable Numbers* paper.[101]

1938

- Chester Carlson patented the photocopying process.
- Paramount purchased half ownership of DuMont Laboratories (already involved in extensive television research and experimentation).[102]
- CBS bought American Record Corporation, the parent company of Columbia Records.
- Disney's *Snow White and the Seven Dwarfs* was released; made in Technicolor, it was the first feature-length animation film in color ever made.[103]

1939

- At General Electric Company's Research Laboratory, Katharine Blodgett invented non-reflecting invisible glass, which would become the main prototype in the subsequent development of coatings applied to practically all camera lenses and optical devices in the future.[104]
- As a result of purchasing Hytron Laboratories, CBS became involved in color broadcasting R&D and television set manufacturing.
- In the wake of Alan Turing's groundbreaking work on algorithms and computation, a number of calculating and computing devices—each with varying capabilities and degrees of efficiency and speed—were built during and immediately after World War II:[105]
    - the *Atanasoff-Berry* at Iowa State College, in 1939;[106]
    - the code-breaking *Colossus* in England, in 1941;
    - the *ASCC Mark I* (weighing five tons) at Harvard University with the support of IBM, circa 1943;
    - the U.S. Army's *ENIAC* (Electronic Numerical Integrator Analyzor and Computer) at the University of Pennsylvania, in 1945;
    - the *Pilot ACE* at the National Physical Laboratory in England, in 1947;
    - the *Manchester Mark I* at Manchester University, in 1948;
    - the *BINAC* (Binary Automatic Computer)—the first computing device to operate in real time—in 1949;[107]

- the *UNIVAC* (Universal Automatic Computer)—the first mass-produced computer—in 1950.

1940

- WNBT-TV and WRGB-TV (owned by General Electric Company) made the first U.S. network television broadcast.[108]

1941

- The CBS Television network was in operation.
- CBS-Hytron demonstrated its mechanical color television system: because of its incompatibility with RCA's already established black-and-white standards, the FCC eventually rejected CBS's technology in favor of RCA's.

1942

- The U.S. government reached 'a temporary settlement with the major studios in the so-called Paramount case, and, for the "big five" vertically integrated majors [Twentieth Century-Fox, MGM, Paramount, RKO, Warner Bros] at least, the twin practices of block booking and blind selling [were] either effectively curtailed or banned outright. For the first time, all their products had to be sold to exhibitors individually.'[109]

1943

- American Broadcasting Corporation (ABC) was created out of NBC's existing broadcasting network as a result of an anti-trust lawsuit.[110]

1944

- *The Three Caballeros* was the first feature-length film to sustain a mixture of two-dimensional animation and three-dimensional live-action within the same frame since the introduction of the concept by the *Out of the Inkwell* animation series and Walt Disney's early experiments during the 1920s.[111]

1945

- The concept for the geosynchronous orbit communications satellite was published by science-fiction writer Arthur C. Clarke.[112]
- With its 630TS 10-inch TV set model, RCA re-introduced television; 50,000 sets were made and sold at $375 each.[113]

- Westinghouse Electric Company became Westinghouse Electric Corporation.[114]

1946

- Sony Corporation was founded in Japan by Akio Morita.[115]

1947

- The transistor was invented at Bell Labs by John Bardeen, Walter Brattain and William Shockley, as a replacement of vacuum tubes. It was considered to be the most important breakthrough of the 20th century in the development of the computer and its related technologies.[116]
- Dennis Gabor produced miniature holograms of microphotographs by using a high-pressure mercury lamp.[117]
- Xerox photocopying machines became commercially available.

1948

- The vertically-integrated companies constituting the Hollywood cinema industry lost their monopoly in film distribution, as a result of an anti-trust action brought earlier by theater owners and of the consequent U.S. Supreme Court's "Paramount decrees" compelling them to give up ownership of movie theaters.
- RCA purchased the Farnsworth Radio and Television plant, converting the facility to picture tube production within a few months to satisfy the increasing demand for television sets.[118]
- RKO Radio Pictures Inc. was bought by eccentric multi-millionaire Howard Hughes (aviation).

1949–1950

- Printed circuits for computers were developed. Their obvious advantage was a significant reduction of the computer's bulkiness, as well as faster transmission since the various electronic components—resistors, transistors, capacitors and so on—were closely wired together on a circuit board.[119]
- EastmanColor positive and negative film stocks were introduced.[120]
- NBC made the first coast-to-coast television broadcast, using AT&T's coaxial cable.[121]
- With the assistance of IBM, Matthew Fox developed an experimental pay-television system—using a punch card and scrambled

broadcast signals—called Subscribervision that was consequently tested by RKO.[122]

- Cable television was born in the United States when four radio dealers (George Bright, William McDonald, Robert Tarlton and Rudolph Dubosky) established Panther Valley Television Company in Lansford, Pennsylvania, charging subscribers an installation fee and monthly rental.[123]

1951

- A video recording machine prototype, using magnetic tape, was demonstrated for the first time at Bing Crosby's studio in Los Angeles.[124]
- The first feature-length films shot and printed on Eastman Color positive and negative stocks began to appear.[125]

1952

- The first interactive computer graphics began to appear with the emergence of the TX-2 computer: it possessed 64,000 bytes of memory and displayed information on a video screen.[126]
- Cinerama was demonstrated publicly for the first time with Merian Cooper's documentary *This Is Cinerama.*
- Universal (film production) was taken over by Decca Records (music production).
- Produced by United Artists, *Bwana Devil* was Hollywood cinema's first 3D movie to be released theatrically.[127]

1953

- The first high-speed printer was developed by Remington-Rand.[128]
- An influential factor in the NTSC's ("National Television Systems Committee") decision to establish the U.S. standard for color television at 525 lines of scanning was the efficiency of RCA's recently demonstrated all-electronic color TV system.[129]
- Following the introduction of an improved version of the Eastman Color negative, the use of the three-strip color camera would be phased out as a negative source for Technicolor motion pictures. From 1955 onwards, the name Technicolor only represented 'a laboratory carrying out a unique printing process.'[130]
- The anamorphic CinemaScope process (originally devised by Henri Chrétien in 1928) was revived by Twentieth Century-Fox 'in exactly the same form, with a supplementary lens incorporating a cylindrical element placed in front of an ordinary camera lens.'[131] *The Robe* was the first Hollywood film made and released in CinemaScope—

to become the standard format for all theatrical exhibition two years later.

- Paramount developed VistaVision, its very own widescreen format.[132]
- Due to the overwhelming popularity of Twentieth Century-Fox's CinemaScope, Paramount's VistaVision did not last very long and eventually faded out.
- JVC (radio and TV set production) was acquired by Matsushita Electric Industrial Company.
- Buena Vista Pictures (film distribution) was created as a subsidiary of Disney.[133]

1954

- IBM published Formula Translation (FORTRAN)—the first advanced form of programming language—created by John Backus.[134]
- Production began for RCA's commercial color television model; 31 television stations across the United States were equipped for color broadcasting.[135]
- For the first time in its history, the U.S. motion picture industry released less than 300 films during the year.[136]

1955

- Disneyland was officially inaugurated in Anaheim, California.[137]
- Disney's first live-action film without animated characters, the VFX-intensive *20,000 Leagues under the Sea,* was theatrically released.
- After years of mismanagement, RKO Radio Pictures Inc. was sold by Howard Hughes to the General Tire and Rubber Company for $25 million.

1956

- The magnetic disc for computer data storage was introduced by IBM.[138]
- CBS broadcasted the first videotaped television program: *Douglas Edwards with the News.*[139]
- Following the demise of the studio system and in response to the rapid expansion of the television industry, hundreds of pre-1948 feature films were released by Hollywood studios for broadcasting on American TV networks.[140]

1957

- Morton Heilig filed a patent for the Stereoscopic-Television Apparatus for Individual Use (STAIU) that would influence the development of future cyberspace and virtual reality systems.[141]
- RKO Radio Pictures Inc. was bought from General Tire and Rubber Company by Desilu (TV production) and became solely dedicated to the production of television programs.[142]

1958

- The final breakthrough in the "teething" phase of computer technology was achieved with the Integrated Circuit (basically a silicon chip composed of transistors, resistors and capacitors) made by Jack Kilby from Texas Instruments in collaboration with Fairchild Semiconductor; it was patented the following year.[143]
- Seymour Cray built the first fully transistorized supercomputer, the CDC 1604, for Control Data Corporation. It would strongly influence the development of future generations of supercomputers such as the CDC 6600, the Cray-I (1975), the Cray-II (1983), the Cray-2S (1987), the $20 million Cray Y-MP (1988), and the Cray Y-MP2E (1990). Supercomputers would play a significant role in the experimentation with and production of CG graphics, animation and digital VFX for Hollywood motion pictures, especially from the mid-1980s onwards.
- Motion picture production ended for good at RKO Radio Pictures Inc.; all its assets were progressively sold to other Hollywood movie production companies.

1959

- Grace Hopper created COmmon Business Oriented Language (COBOL).[144]
- MIT introduced the first courses on computing for undergraduates and set up a research department on Artificial Intelligence.[145]
- The U.S. motion picture industry released less than 200 American-made films during the year; there was more than one television set in use per household in the United States at this time.[146]

1960

- Laser (Light Amplification by Stimulated Emission of Radiation) technology was demonstrated by Hughes Aircraft Company—a very influential step towards the future development of compact disc-related technologies.[147]
- Haloid Corporation shipped the first Xerox photocopier.[148]

- In-flight movies were introduced on commercial U.S. air carriers.[149]
- United Artists acquired Ziv Television Programs, the largest TV syndication company and leading programming force outside the principal broadcasting networks during the 1950s.[150]

1961

- Fairchild Semiconductor released the Integrated Circuit commercially.[151] All the essential components having been created and introduced, computer-related technologies would consequently evolve in terms of getting smaller, faster and more sophisticated.
- Morton Heilig patented the Sensorama Simulator; it would influence the development of virtual reality systems.[152]
- RCA provided a demonstration of the first transistorized broadcast-standard videotape recorder.[153]
- The first computer hackers emerged at MIT.[154]
- *Spacewar*, the first popular computer game, was created by Steve Russell and fellow computer hackers at MIT's Project MAC.[155]

1962

- General Electric Company's scientist Bob Hall invented the solid-state laser: an efficient and compact source of highly controlled light that would strongly influence the future development of compact disc players, laser printers and various fiber optic communication devices.[156]
- Ivan Sutherland created Sketchpad, arguably the first real-time interactive computer graphics system; it also happened to include a light pen that allowed the user to draw directly on a cathode ray tube—a precursor of future touch-screen technology.[157]
- United Artists's recently acquired subsidiary Ziv Television Programs became United Artists Television.[158]
- Universal as well as its owner Decca Records both became part of MCA (radio and television production; music production; talent agency; Paramount 1929–1949 film library).

1963

- Digital Equipment Corporation released the PDP-8, considered to be the first mini-computer.[159]
- NASA launched the geostationary satellite Syncom I, designed and produced by Hughes Aircraft Company based on Arthur C. Clarke's original concept; it could transmit television signals. Syncom II was launched later in the year while Syncom III was launched in 1964 and successfully transmitted the Tokyo Olympic

Games to the United States[160] This was but a preliminary stage in the future proliferation of global broadcasting and communication satellites.

- The first computer-generated film WASS (Wavefront Analysis of Spatial Sampling—a term associated with R&D in aircraft landing approach) was produced by Edvard Zajac at Bell Labs. It was meant to simulate the motion of a communication satellite so as to ensure that the same side of the satellite always faced Earth.[161] This was possibly the earliest instance of convergence between computer graphics and motion picture technologies.

1964

- IBM introduced the first word processor, the Integrated Circuit-based IBM 360; it would eventually become a standard for future mainframe computers.[162]
- The sodium light travelling matte process—using a Technicolor beam-splitter camera developed in England in the 1950s—was imported to Hollywood by Disney to produce VFX for *Mary Poppins*.[163]
- Warner Bros. was bought by Seven Arts Productions (film production and distribution); Jack Warner stayed as studio president and retained controlling interest in Warner Bros.

1965

- Sony Corporation introduced the ½-inch CV-2000, promoted as 'the world's first home videocorder'; it used a 7-inch reel that could hold up to an hour of material.[164]
- Super 8mm film became commercially available.[165]
- NBC, CBS and ABC completed their conversion from black-and-white and became all-color television networks.[166]

1966

- IBM introduced its RAMAC 305 computer that had a disc storage system and could hold up to 5 MB of data.[167]
- Created by Gene Roddenberry, the *Star Trek* television series was launched. This would lead to the development of many generically related TV series, the growth of a gigantic international franchise worth billions, and the production of numerous feature-length movies, many of which would enable Hollywood cinema's VFX industry to innovate techno-visually.
- Paramount was acquired by Charles Bluhdorn's Gulf+Western Industries Inc. (sugar; zinc; fertilizer; financial services; real estate;

wire and cable), with the financial assistance of Chase Manhattan Bank.

1967

- The first magnetic floppy disc for data storage was produced by IBM.[168]
- Charles Csuri's *Hummingbird,* made in collaboration with James Shaffer at Ohio State University, was one of the first computer-generated films to incorporate representational figures.[169]
- New Line Cinema (independent film production and distribution) was founded by Robert Shaye.[170]
- United Artists was taken over by Transamerica Corporation (life insurance; computers; airlines; car rental; record companies; real estate and other assets) for $185 million.[171]
- When Seven Arts Productions (film production) acquired for $32 million Jack Warner's controlling interest in Warner Bros., Warner-Seven Arts was formed.

1968

- An early version of hypertext was demonstrated by Doug Engelbart.[172]
- Japanese company Nippon Hoso Kyokai (NHK) started research in High Definition Television or HDTV.[173]
- Intel Corporation (Integrated Electronics Corporation) was co-founded by Robert Noyce and Gordon Moore; this was the beginning of a new industrial phase in the evolution of computers.[174]
- Produced by MGM, *2001: A Space Odyssey* was theatrically released. Production of its groundbreaking VFX sequences—by a team including Douglas Trumbull and John Dykstra—involved the first time application in feature filmmaking of the Slit Scan motion-control technique (originally pioneered by John Whitney Sr. at his Motion Graphics Inc. company).[175]

1969

- Data General Corporation introduced the first 16-bit minicomputer, the Nova, conceived by Edson de Castro.[176]
- The Defense Department's ARPAnet went online—the first network of all networks—enabling UCLA and other institutions to link up, hence establishing the foundations of the future Internet.[177]
- MGM was purchased by billionaire Kirk Kerkorian (real estate; airlines; hotels and casinos).

- Warner-Seven Arts was bought by Kinney National Services Inc. (car parks; car rental; funeral homes; construction) and renamed Warner Communications.

1970

- The IBM 3740 was one of the first computers with a floppy disc device to be released commercially.[178]
- RCA introduced its line of solid-state color TV models.[179]
- The Imax ultra-widescreen system was publicly exhibited for the first time at Expo '70 in Japan. For the Imax system, the image was printed on regular 70mm film stock but ran 'horizontally through the camera and projector without the use of sprocket holes, at an accelerated rate (5.5 feet per second). The film format is ten times the size of an ordinary 35mm frame and three times the size of 70mm, and images are projected on huge screens.'[180]
- Paramount and MCA jointly formed Cinema International Corporation (CIC) to distribute their films theatrically outside the American market.[181]

1971

- Computer memory chips were able to contain 1KB (1024 bits) of information each.
- The 4-bit Intel 4004 microprocessor was introduced by the recently founded Intel Corporation.[182]
- Sony Corporation introduced the U-MATIC format (which utilized ¾ inch color tape) for non-broadcast video application.[183]
- Walt Disney World opened in Orlando, Florida.
- George Lucas's recently formed Lucasfilm Ltd. (film production) was incorporated.

1972

- The Intel 8008 was the first 8-bit microprocessor to be released commercially.[184]
- Philips produced the VLP disc which constituted the first practical laser application to the problem of audio-visual data storage; it held up to 45 minutes of material which could be read by a laser-generated spot.[185]
- Nolan Bushnell founded Atari and created the first commercial video game, *Pong*, which was an immediate success. Two decades later, the business of video and electronic games would be worth billions.[186]

- Magnavox released the Odyssey cartridge-based console game system for home use.[187]
- Charles Dolan sold his Sterling Manhattan Cable company to Time Inc. and convinced them to invest in a new cable pay-TV service; Home Box Office (HBO) was thus created.[188]
- Columbia and Warner merged some of their physical assets and began to jointly operate the old Warner Bros. film studio facilities renamed Burbank Studios.[189]

1973

- Bob Metcalfe at Xerox PARC developed Ethernet.[190]

1974

- Computer memory chips could hold 4KB of information each.
- The Intel 8080 microprocessor was introduced.[191]
- Westar was the first American domestic communications satellite to be launched.[192]
- Electronic mail was invented by Ray Tomlinson; he sent the first email on ARPAnet.[193]
- The new Computer Graphics Lab was created at NYIT with Ed Catmull as director.

1975

- Eastman Kodak's electrical engineer Steven Sasson invented the digital still camera.
- Bill Gates and Paul Allen developed Beginners' All-purpose Symbolic Instruction Code (BASIC)—the first computer language written specifically for a personal computer—and created a software company, Microsoft Corporation.[194]
- Ed Robert's Altair 8800, using the new Intel 8080 microprocessor, was the first minicomputer kit to become available for public sale.[195]
- The first laser printers were launched commercially by IBM.[196]
- Sony Corporation introduced the Betamax video system that used ½ inch tape.[197]
- Dolby motion picture sound system was introduced.[198]
- Industrial Light & Magic was founded by George Lucas to produce VFX for his forthcoming *Star Wars* feature film project.
- Sprocket Systems was created for the sound mixing and editing of *Star Wars*; this was the foundation of what would later become Skywalker Sound (responsible for continuously innovating and establishing new standards in motion picture sound engineering).[199]

- The Dykstraflex, the first electronically commanded motion-control camera system, was developed at Industrial Light & Magic during the production of VFX for *Star Wars*. It was named after special photography VFX supervisor John Dykstra, who used a computer to control the movement of movie cameras. This would influence subsequent VFX cinematography for animating models and miniatures in particular.
- By shifting its programming delivery from microwave to satellite, Time Inc.'s HBO established the first national cable TV network for cable system operators to offer to subscribers at a premium price.[200]

1976

- The TMS9900 was introduced by Texas Instruments as the first 16-bit microprocessor.[201]
- After completing work on a computer circuit board that they called the Apple I computer, Steve Jobs and Stephen Wozniak created Apple Computer Company (which was incorporated a year later).[202]
- One of the first computers to feature a built-in display, the Commodore Pet 2001, was introduced.
- Fairchild introduced Channel F, the first programmable (via plug-in cartridges) home video game system at $170.[203]
- Sony Corporation started the "home video revolution" by commercially releasing the Betamax videotape system in the American consumer market.[204]
- Warner Communications bought Atari from Nolan Bushnell for $26 million.[205]
- Viacom (television production) created the Showtime pay-TV cable channel.[206]
- Ted Turner's Pay-TV channel WTCG was beamed via satellite to cabled homes.[207]

1977

- Computer memory chips held 16KB of information.
- Stephen Wozniak at Apple Inc. created the first industry-standard disc drive.[208]
- Apple Inc. introduced the Apple II, the first personal computer with color graphics.[209]
- The Atari Video Computer System was introduced and later renamed Atari 2600.[210]
- RCA introduced ColorTrak television sets that "think in color."[211]

- Matsushita Electric Industrial Company's subsidiary JVC introduced the VHS (video home system) player-and-recorder to compete with Sony Corporation's Betamax video system.
- George Lucas's *Star Wars* was released theatrically and became a huge box-office hit. Pioneering the use of Dolby Stereo sound, *Star Wars* also helped to 'inaugurate a fundamental transformation in the production, use and reproduction of sound in Hollywood cinema,' other than its innovations in VFX production.[212]

1978

- Intel Corporation commercially released the 8086 and 8088 microprocessors.[213]
- Philips marketed a laserdisc variant for computer data storage.[214]
- Under the supervision of Nicholas Negroponte, the Aspen Project—essentially an experiment in random access computer control—created the foundations of what would in the future become multimedia.[215]
- Taito in Japan developed the *Space Invaders* arcade video game; within the next two years, it would be released and distributed to video game arcades worldwide.[216]
- Cinematronics released the *Space Wars* arcade video game.[217]
- APF Electronics introduced the MP-1000 video game unit.[218]
- Magnavox released the Odyssey 2 cartridge-based console game system.[219]
- MCA introduced the DiscoVision videodisc system—phased out by the mid-1980s.[220]
- The official *Star Wars* Fan Club was created while official *Star Wars*-related toys became available for the first time in stores.[221]
- Following the unprecedented commercial success of *Star Wars*, George Lucas re-structured and expanded Lucasfilm Inc. and Industrial Light & Magic, as well as began pre-production for *Star Wars: The Empire Strikes Back* as a sequel to *Star Wars*.
- CBS partnered with MGM to form MGM-CBS Home Video.
- Orion Pictures Corporation Studio was formed by ex-top executives at United Artists.[222]
- Time Inc. acquired cable operator American Television & Communications (ATC).[223]

1979

- Apple Inc. released the Apple II Plus microcomputer with 48KB of memory and Visicalc, the first spreadsheet program.[224]

- Sony Corporation, in joint-venture agreement with Philips, began to market a five-inch laserdisc—which eventually created the standard for the optical disc storage of digital audio.[225]
- Atari produced the first coin-operated *Asteroids* arcade game machine.[226]
- Upon its release, *Alien* was considered to be one of the first motion pictures to 'use computer-generated images within the main body of the movie as opposed to the title sequence.'[227] In the film, the computer screen information displayed on the bridge of the spacecraft and, in particular, the wire-frame contour map of the planet's surface on which the spacecraft was meant to land, were created by System Simulation, a computer graphics company based in London. *Alien* came to be perceived as a landmark film with regards to science fiction and futuristic imagery, establishing a model and techno-visual standard for many consequent productions of such genre to follow.[228]
- George Lucas created the Computer Research and Development Division, to be headed by Ed Catmull, within Lucasfilm Inc. to develop computer hardware and software oriented towards filmmaking-related activities, such as digital imaging, digital audio post-production, computer-generated animation and so on.
- ESPN and C-SPAN cable networks were established.[229]
- Miramax Films (film distribution) was founded by Bob and Harvey Weinstein.[230]
- Turner Communications Group became Turner Broadcasting System Inc.[231]
- Warner Communications in partnership with American Express (financial services) created The Movie Channel and Nickelodeon as part of its cable television network.[232]

1980

- The Media Lab at MIT started taking shape under the impulse of its director Nicholas Negroponte.[233]
- Ray Tracing was developed at Bell Labs and Cornell University: a method of texturing and lighting computer models.
- A motion-control multi-plane camera called the Automatte was built at Industrial Light & Magic. It was meant to improve the integration of models and matte paintings during the production of VFX for *Star Wars: Return of the Jedi*, as well as to allow more complex camera moves than were previously possible.[234]
- *Star Wars: The Empire Strikes Back* was theatrically released as a sequel to *Star Wars*.
- The construction of George Lucas's Skywalker Ranch began.[235]

- Ted Turner's Cable News Network (CNN)—the first 24-hour all-news cable TV service—premièred with 1.7 million subscribers.[236]
- The USA Network cable television channel was jointly formed by United Artists, Columbia Pictures and Gulf+Western Inc.[237]
- Cable television Premiere Channel was created on the basis of an agreement between Twentieth Century-Fox, Universal, Paramount, Columbia and Getty Oil.[238]
- Twentieth Century-Fox was bought by Marvin Davis (oil exploration; petroleum) for $725 million.[239]

1981

- Computer memory chips could hold 64 KB of information each.[240]
- General Electric produced 3 feet of fused quartz ingots that could be stretched into 25 miles of fiber optic strands; this invention inaugurated the use of fiber optics in communication technologies.[241]
- Nintendo designed and commercially released the *Donkey Kong* video game.[242]
- Bally Corporation's Midway Division made $200 million in machine sales for the *PacMan* video game.[243]
- Microsoft obtained the contract for the provision of an operating system for the IBM PC.[244]
- The introduction of IBM's desktop microcomputer—the IBM PC (which uses Microsoft's DOS operating system)—established an enduring standard for the personal computer market.
- Osborne Computer introduced the Osbourne I, the first portable computer.[245]
- Silicon Graphics Inc. was founded by James Clark. This company would end up producing some of the most powerful and expensive interactive computer workstations on the market and contributing extensively to R&D and innovation in 2D and 3D computer graphics, digital VFX and digital filmmaking in particular.
- The High Definition Television or HDTV standard, with 1,125 lines of scanning, was demonstrated in Japan.[246]
- Warner Communications in partnership with American Express (financial services) launched Music Television or MTV as part of its cable television network.[247]
- During VFX production for *Dragonslayer*, Phil Tippett and Ken Ralston at Industrial Light & Magic developed go-motion photography.[248]
- Featuring VFX mostly produced at Industrial Light & Magic, *Raiders of the Lost Ark* was the first installment of the Indiana Jones movies; it rapidly turned into a box-office smash hit upon theatrical release.

- Richard Edlund and Industrial Light & Magic received a Scientific and Engineering Award for 'the concept and engineering of a beam-splitter optical composite motion picture printer'—a device extensively used in the post-production of many VFX.[249]
- Stuart Ziff and Dennis Muren from Industrial Light & Magic received a Technical Achievement Award for 'the development of a motion picture figure mover for animation photography.'[250]
- Westinghouse acquired cable television operator Teleprompter and merged it with its own cable subsidiary Group W.[251]
- CBS joined with Twentieth Century-Fox to form CBS-Fox Video.
- As a result of Michael Cimino's *Heaven's Gate* exceeding its allocated production budget and being a box-office fiasco, United Artists went bankrupt and was sold by Transamerica Corporation to MGM's owner Kirk Kerkorian for $320 million; MGM-United Artists Entertainment was formed as a result.[252]

1982

- General Electric developed special grades of Lexan resin to be used as coating for audio and video discs which were becoming increasingly popular consumer products.[253]
- Eastman Kodak introduced Datakode film stock that used a 30-frame-per-second standard as opposed to the conventional 24-frame-per-second film stock; Datakode allowed for a motion picture to be recorded on celluloid film using a time code that corresponded to videotape.[254]
- Atari obtained exclusive worldwide rights to market video games based on the motion picture *E.T. The Extra-Terrestrial.*[255]
- RCA introduced the Selectavision videodisc player; it would fade away by the mid-1980s.[256]
- The Hyperion microcomputer was released as the first IBM-compatible portable microcomputer.[257]
- John Warnock founded Adobe Systems.[258] This company later commercialized Photoshop—a groundbreaking computer graphics software for both digital VFX production and the desktop publishing industry.
- The newly founded Mouse Systems Corporation introduced the very first commercial mouse for the IBM PC.[259]
- The Intel 286 microprocessor was released.[260]
- Lucasfilm's THX Sound System was introduced and the Theater Alignment Program was initiated.[261]
- Disney's *Tron* was released. Although a box-office failure, it was an important techno-visual landmark in terms of displaying the possibilities and future potential of computer-generated imaging for the Hollywood cinema industry. MAGI-SynthaVision, Digital Ef-

fects Inc. Robert Able & Associates, Stargate Films and Triple-I (Information International Inc.) were the four main companies with enough computer graphics expertise and resources at the time to produce *Tron*'s 16 minutes of entirely computer-generated visuals, background mattes and VFX—the largest amount ever seen in a Hollywood movie at the time.

- *Star Trek II: The Wrath of Khan* was theatrically released and was a box-office success. It was the first Hollywood feature film to include a continuous sequence of computer-generated digital images and animation: the "Genesis" sequence—the transformation of a dead planet into a verdant world—was the outcome of groundbreaking 3D computer-generated graphics and animation produced at Lucasfilm's Computer Division headed by Ed Catmull.
- During the production of VFX for *Star Trek II: The Wrath of Khan*, the most significant technological development was the Renderman 3D software at Lucasfilm's Computer Division. It would rapidly become an essential work tool in the growing domain of digital VFX production.[262]
- Columbia Pictures was purchased by Coca-Cola Company.
- TriStar Pictures (film distribution) was created as a joint venture between Columbia, CBS and HBO.[263]
- Facing financial problems, Warner Communications downsized and sold some of its assets (such as Atari, Franklin Mint, the New York Cosmos soccer team, Panavision Inc. and Warner Cosmetics).

1983

- The total number of computers being used in the United States was more than ten million units—subsequently in excess of 30 million units by 1986, and more than a hundred million units worldwide by 1989.[264]
- Starcom released *Dragon's Lair* to arcade game centers: it was the first laserdisc-based arcade video game. 'Revenue from arcade games now exceeds theatrical film rentals as the locus of power in the American entertainment industry begins to shift from Hollywood to Silicon Valley.'[265]
- *Star Wars: Return of the Jedi* was released, featuring some computer-generated VFX produced by Lucasfilm's Computer Division (such as the hologram image used in a Rebel strategy meeting sequence).
- Disney launched the Disney Channel for cable television.[266]

1984

- Computer memory chips held 256 KB of information each.[267]

- The Macintosh computer, with mouse and window interface, was released by Apple Inc. (the television advertising campaign comprised a single commercial directed by Ridley Scott and played only once during the NFL Superbowl).[268]
- Apple Inc. commercially released AppleWorks (written by Rupert Lissner): one of the earliest integrated software packages containing modules for database management, spreadsheet calculations, and word processing.[269]
- Essentially an extension of audio-CD technology, the CD-ROM (Read Only Memory) was commercially developed and marketed as a data-storage device by Philips and Sony Corporation.
- Silicon Graphics started shipping out its first 3D graphics workstations.[270]
- Domain Name Server (DNS) was introduced for computer network users.[271]
- Canadian science fiction writer William Gibson's *Neuromancer* novel—wherein the term "cyberspace" was coined—became a best seller.
- The first videotape-based electronic non-linear editing system for film appeared in the shape of the Montage Picture Processor.[272]
- The EditDroid, a laserdisc-based non-linear film editing system was introduced. Developed by Lucasfilm's Computer Division, it used multiple laserdisc players and offered random access as opposed to linear videotape-based systems.[273]
- Distribution company Criterion released *Citizen Kane* and *King Kong* on laserdisc.
- The American cable television industry was de-regulated.[274]
- *The Last Starfighter* was released, featuring 25 minutes of computer-generated scenes created by Digital Productions Inc. and digital VFX contributions from Industrial Light & Magic. Using as model the similarly named Atari game to design the space fight simulation sequences, Digital Productions Inc. devised in the process a new technique of digital scene simulation for motion pictures.[275]
- An entirely computer-generated animation short film of 1.8 minutes long, as such a very important techno-visual breakthrough, *The Adventures of André and Wally B.* was completed at Lucasfilm's Computer Division. 16 computers were used during its making to process data, including four high-speed VAX computers and two Cray supercomputers—one XMP-2 and one XMP-4, the most powerful and expensive computers in the world at the time.[276]
- Touchstone Pictures (film production and distribution) was founded by Michael Eisner and Jeffery Katzenberg as a subsidiary of Disney.[277]

1985

- IBM and Bill Gates's Microsoft signed a joint-development agreement to collaborate on future computer operating systems; this would lead to the development and monopoly of Microsoft Windows as the standard operating system for PCs worldwide.[278]
- Inmos in England built the transputer, a microprocessor with integral memory designed for parallel processing.[279]
- The Intel 386 microprocessor was released.[280]
- Fiber optics were used for the first time to link mainframe computers.[281]
- Advanced LCD video screen technology was launched by Japanese watch company Citizen.[282]
- Thomas Zimmerman and Young Harvill created the DataGlove for VPL Research Inc. as part of developing virtual reality technology.[283]
- Nintendo released the Nintendo Entertainment System, an electronic video game console, together with what would become a standard business model of software licensing for third-party developers.[284] At the time the best-selling domestic gaming console on the market, it would subsequently set the standard—in everything from controller layout to design—for other game consoles to follow.
- The Ediflex, a videotape-based electronic non-linear editing system, was introduced.[285]
- Color System Technologies Inc. introduced the process of electronic colorization for black-and-white film.[286]
- *We Will Rock You: Queen Live in Concert*, a filmed concert of British rock band Queen, was released as the first motion picture made in the Dynavision format that would be short-lived.[287]
- Originally made for television, *Young Sherlock Holmes* was the first feature-length movie containing VFX that were painted directly on the film with a laser.[288] The sequence involving the stained-glass figure on a church window of a sword-wielding knight that comes to life and stalks a priest was a techno-visual breakthrough: it contained the first moving digital character or "synthespian" ever seen in a Hollywood feature film. All digital VFX in *Young Sherlock Holmes* were developed and produced by Lucasfilm's Computer Division and Industrial Light & Magic.
- The Pixar, a laser-based digital image-processing system, was developed during the production of digital VFX for *Young Sherlock Holmes* at Lucasfilm's Computer Division.[289]
- Ed Catmull and his team left Lucasfilm's Computer Division altogether to form a separate company that intended to concentrate on 3D computer animation and its related R&D. The company was

named Pixar after the image-processing computer system originally developed during their tenure at Lucasfilm Inc.[290] Following up from *The Adventures of André and Wally B.* in 1984, Pixar subsequently produced a number of groundbreaking computer-generated 3D animation short films that would influence and lead to the making of the first computer-generated feature-length 3D animation film *Toy Story* in 1995: *Luxo Jr.* (1986), *Red's Dream* (1987), *Tin Toy* (1988), *Knick Knack* (1989), *Light & Heavy* (1991) and *Surprise* (1991)—all supervised/directed by John Lasseter.

- Steve Jobs left Apple Inc. the company he co-founded, with a $100 million settlement. Through a $60 million investment, he acquired a major interest in the newly formed Pixar, later renamed Pixar Animation Studios, and appointed Ed Catmull as its Chief Technical Officer. He also started NeXT Software Inc. that would evolve in the direction of software R&D, design and production.[291]
- Originally conceived by Steve Case and Jim Kimsey, Quantum Computer Services was a new type of company meant to deliver via PC modems online information and other services to consumers; this constituted the foundation of what would later become America Online (AOL)—at a time when the World Wide Web was yet to exist.[292]
- Blockbuster Entertainment Corporation opened its first video rental store; within less than ten years it would have a chain of more than 2,000 stores internationally.[293]
- The ABC television network was bought by Capital Cities Broadcasting for $3.5 billion.[294]
- Twentieth Century-Fox and Metromedia Television (the largest group of independent American television stations) were both acquired by Rupert Murdoch's News Corporation (newspapers; magazines; publishing).

1986

- Grolier's *Academic American Encyclopedia* was the first general public-oriented CD-ROM publication to be released commercially.[295]
- Apple Inc. released the first commercial hypermedia product—the Hypercard.[296]
- The Byte by Byte software company commercially released one of the first 3D modeling programs for computers, the Sculpt 3D.[297]
- IBM laboratories in Zurich discovered high-temperature superconductors that paved the way for superconducting computers in the future.[298]
- TouchVision was the third videotape-based electronic non-linear film editing system to be commercially launched.[299]

- *Howard the Duck* was released and flops at the box-office. It had the merit of techno-visual innovation: a wire-removal computer program developed during the production of its VFX at Industrial Light & Magic.[300]
- A Medal of Commendation for "cinematic pioneering" was presented by the Academy of Motion Picture Arts and Sciences to John Whitney Sr., considered by many as the "father" of computer-generated art.[301]
- For the first time ever, revenues from home video rental and purchase surpassed theater box-office takings for feature films—leading to speculations of the cinema's "death by video."[302]
- The Twentieth Century-Fox Television network started broadcasting.[303]
- RCA Corporation (including NBC) was acquired by one of its original founders, namely General Electric.[304]
- MGM-United Artists was sold by Kirk Kerkorian to Ted Turner (who also purchased both the RKO and pre-1950s Warner Bros. film and television libraries) for $1.5 billion.[305]
- Ted Turner sold MGM-United Artists back to Kirk Kerkorian and to Lorimar Telepictures Corporation for $480 million but kept the MGM, RKO and pre-1950s Warner Bros. film and television libraries. Ted Turner's vast film and television libraries collection would constitute the pillar of Turner Entertainment Company.[306]

1987

- Computer memory chips could hold 1MB of information.
- Texas Instruments released the first AI or "Artificial Intelligence" microprocessor chip.[307]
- Commodore announced the Amiga 500 (followed by the Amiga 1000 a month later); it possessed a 68000 processor and custom chips to support animation, video and audio.[308]
- The Animate 3D software was released for Commodore's Amiga computer.[309]
- The Macintosh II and the Macintosh SE were released by Apple Inc.[310]
- Letraset released ImageStudio, the first commercial, grey-scale image-editing program for the Macintosh computer.[311]
- Philips announced a CD video format called CD-V (Compact Disc Video).[312] This would lead to the later development of the Video Compact Disc (VCD).
- RCA demonstrated DVI (Digital Video Interactive) technology at the second Microsoft CD-ROM conference; although it enabled an hour of full-motion video to be held on compact disc by virtue of signal compression, it would not be commercially successful.[313]

- Developed in Japan in 1985, the first *Super Mario Bros* console game was introduced (and 23 million copies of it were quickly sold) by Nintendo following its recent introduction of the Nintendo Entertainment System video game console.
- *The Witches of Eastwick* was released theatrically, featuring a combination of analog and digital VFX produced by Industrial Light & Magic.
- General Electric sold its RCA Consumer Electronics and General Electric Consumer Electronics divisions to Thomson Consumer Electronics.[314]
- Sumner Redstone (theater chains; hyper malls) bought Viacom Inc. for $3.4 billion.[315]

1988

- NeXT Inc. unveiled its innovative workstation computer: the first of its kind to use erasable optical disks as a primary mass storage device. IBM licensed NeXT's graphics user interface.[316]
- The Animate 4D software for the Commodore Amiga was released.[317]
- Adobe released the Adobe Illustrator 88 for Macintosh computers.[318]
- Tandy announced the Thor CD, an erasable versatile compact disc system for music, video or any digitized data—a precursor of commercial re-writable CD-RWs and DVD-RWs to come.[319]
- Sony Corporation bought Columbia Records from CBS for $2 billion, to be re-named Sony Music Entertainment a few years later.[320]
- Produced by Disney, *Who Framed Roger Rabbit?* showcased Industrial Light & Magic's groundbreaking achievement of compositing optically three-dimensional live action with more than 82,000 frames of hand-drawn, two-dimensional cel-animated cartoon characters, giving the "toons" more credible three-dimensionality in the process.[321]
- Containing more than 350 scenes of complex VFX, fantasy movie *Willow* was moderately successful at the box-office upon its release (although later to become a "cult" movie on video). Its major techno-visual innovation in terms of digital VFX was the technique of morphing (or metamorphosis) as applied to the transformation VFX shot. The Morphing software, created by programmer Doug Smythe, was a digital image-processing breakthrough whereby one image was progressively and seamlessly altered to transform into another image.[322]
- John Knoll—hired by Industrial Light & Magic as a motion-control camera operator in 1986—and his brother Thomas from the Univer-

sity of Michigan developed Photoshop, a draw-and-paint software specifically designed for use on Macintosh computers.[323]

- Rupert Murdoch's News Corporation purchased *TV Guide* for $3 billion from Annenberg Communications.[324]
- Ted Turner created Turner Network Entertainment (TNT), a cable television outlet to capitalize on his extensive film and television libraries of old Hollywood classics.[325]

1989

- Apple Inc. introduced the portable Macintosh computer.[326]
- Intel Corporation developed the 80486 microprocessor, as well as the I860 Reduced Instruction Set Chip (RISC) co-processor chip that operated ten times faster than previous chips. They both had more than a million transistors.[327]
- Nintendo introduced the Gameboy.[328]
- Sega introduced the 16-bit Genesis home video game system.[329]
- The Byte by Byte software company released Animate 4D for Macintosh computers.[330]
- Sony Corporation released the Mavica, the first digital magnetic still camera; it would only become successful with consumers about 8 years later.[331]
- Seymour Cray re-structured his company into two divisions: Cray Research and Cray Computer Corporation (meant to develop a gallium arsenide-based supercomputer).[332]
- *Back to the Future II* was released, revealing a more extensive application by Industrial Light & Magic of the digital wire- and rod-removal software program introduced during the making of *Howard the Duck*.[333]
- *Indiana Jones and the Last Crusade* contained the first all-digital composite shot in a feature film: the "Donovan's Destruction" VFX sequence, produced by Industrial Light & Magic, showed Walter Donovan (played by actor Julian Glover) physically ageing and disintegrating within seconds.[334]
- *The Abyss* was released. The planning and production of its digital VFX involved the first-time application of the Virtus Walk-through program, such as for the sequence of first contact of humans with the non-terrestrial water "pseudopod."[335] It also revealed extensive and innovative use of Photoshop in digital VFX production. The software contributed to the manipulation of digitized photographic data of background plate sets and to the creation of a digital model with which to integrate the film's famous computer-generated non-terrestrial water "pseudopod" as well as to perfecting visually its digital environment.[336]

- The technical specifications for what was soon to become the World Wide Web were released by Tim Berners-Lee from the European Centre for Particle Physics.[337]
- America Online (AOL) was born when Steve Case re-named his online information and services provider company Quantum Computer Services. Upon its launch, AOL online services included e-mail, games, special interest forums, as well as the novel feature of allowing AOL members to communicate through one-on-one, real-time conversations.[338]
- NBC—owned by General Electric—launched CNBC (Consumer News and Business Channel), a 24-hour consumer-oriented news program service on cable television.[339]
- TriStar Pictures became a subsidiary of Columbia.[340]
- Columbia Pictures Entertainment Inc. (Loew's Theaters; Columbia Pictures; Columbia film and television library; TriStar Pictures) was purchased by Sony Corporation for $3.4 billion from Coca-Cola Company. Sony Corporation additionally acquired Guber-Peters Entertainment (film production) for $200 million to manage their new acquisitions.[341]
- Paramount became Paramount Communications as a result of a corporate name change.
- Warner Communications purchased Lorimar Telepictures Corporation.[342]
- Time-Warner is formed as a result of the merger between Warner Communications and Time Inc.: it was the largest information, communication and entertainment entity worldwide at the time.

1990

- Microsoft introduced Windows 3.0.
- Photoshop was commercially marketed by Adobe and, outside its extensive application in film and television digital VFX production, was rapidly adopted by the print and publishing industries.
- Disney's Computer Software division released the Disney Animation Studio for the Commodore Amiga.[343]
- There was no highly reliable input scanner for feature film production until the introduction of the second-generation Tri-linear Multi-spectral High Resolution CCD (Charge Coupled Device) Digital Input Scanner. A joint-venture between Eastman Kodak and George Lucas's Industrial Light & Magic, it could scan a frame of film in roughly every twenty to thirty seconds, an important improvement on the existing in-house Pixar laser scanning image-processing system.[344]
- Symbolics Graphics Centre introduced a new turnkey paint workstation, the PaintAmation. 'This system provided the industry's

most comprehensive set of paint software supported by revolutionary multi-sync videographics hardware. The product offered high-speed painting, sophisticated 2D animation and the unique ability to automate repetitive paint tasks. The PaintAmation System included FrameThrower, a powerful video graphics processor' which allowed 'significant acceleration in paint operations that include soft-matting, rotation, scaling brushes, color space conversions, and shape operations.'[345]

- *The Hunt for Red October* was commercially successful upon its theatrical release. It involved the innovative use of digital technology in the shape of a 'particle-generating system that can imitate natural, random movements such as those of stars, fire, or dust particles,' developed during the production of its VFX at Industrial Light & Magic.[346]
- SkyTV, part of Rupert Murdoch's News Corporation, merged with British Satellite Broadcasting to become BskyB, the largest satellite TV network worldwide.[347]
- Hollywood Pictures (film production) was created as a subsidiary of Disney.[348]
- Universal-MCA was purchased for $6.1 billion by Matsushita Electric Industrial Company (consumer electronics; information and communications technologies; professional audio-video equipment).[349]
- Kirk Kerkorian sold MGM-United Artists for $1.3 billion to Giancarlo Paretti of Pathé Communications Company (with the financial backing of Crédit Lyonnais Bank Nederland) based in Europe.[350]

1991

- Computer memory chips could hold 4 MB of information each.
- Apple Inc. introduced QuickTime digital audio/video technology, to be used for the integration of dynamic media for Macintosh computers.[351]
- The Macintosh PowerBook portable computer, which could also function as a desktop computer, was introduced by Apple Inc.[352]
- Microsoft released DOS 5.0.[353]
- The Federal Trade Commission began to investigate Microsoft's business practices.
- Nintendo released the 16-bit Super Nintendo Entertainment System.[354]
- Philips launched the CD-I (Interactive) format; it offered full motion and full resolution by the intermediary of video compression.[355]

- TV series *The Young Indiana Jones Chronicles* was one of the first television productions to include entirely digital matte paintings.[356]
- *Terminator 2: Judgment Day* was released. A combination of available state-of-the-art digital technologies and innovative software/hardware development, the movie's most techno-visually groundbreaking VFX was the "liquid metal" T-1000 cyborg. Its production required the use of a hardware set-up of some 35 high-powered Silicon Graphics workstations and of the newly developed Body Sock and Make Sticky softwares—both created through in-house R&D development at Industrial Light & Magic.[357]
- Toshiba partnered with Time-Warner, with support from Matsushita Electric, Hitachi, Mitsubishi Electric, Pioneer, Thomson and JVC, to develop a new optical disc storage format. This would eventually lead to Digital Video Disc (DVD) technology.
- MGM-United Artists was renamed Metro-Goldwyn-Mayer by Crédit Lyonnais Bank Nederland (banks; finance; real estate) after taking over management from Giancarlo Paretti and Pathé Communications Company.

1992

- Microsoft introduced Windows 3.1.[358]
- SuperMac Technology shipped the DigitalFilm video production system for Macintosh.[359]
- Eastman Kodak commercially released the Photo CD.[360]
- Sony Corporation introduced the MiniDisc (MD) player that used smaller compact discs.[361]
- Time-Warner acquired all the stock of cable systems operator American Television and Communications (ATC) and, in combination with its own Warner Cable, created the Time-Warner Cable Group that served over 7.1 million subscribers.[362]
- *Death Becomes Her* was released and achieved box-office success. Many new programs were tested and used extensively for the first time during the production of its digital VFX by Industrial Light & Magic:
    - SoftImage, an animating software package that allowed CG artists to create 3D match-moves for any live-action plate and to weave into shots complex and hand-held camera moves;[363]
    - Colorburst, a painting and digital rotoscoping package that also included the Matador software, a major improvement over previous matte-generating systems;[364]
    - C-Bal, a program developed in-house at Industrial Light & Magic to improve digital blue-screen matte extraction and

allow CG artists to adjust more easily the color saturation of elements and final composites;

- MM2, a 2D match-move tool than could accomplish digitally what was once an analog VFX obtained from applying the camera pin-blocking technique.[365] This software animated or tracked 2D images, allowing elements to be re-photographed using a match-move program so that all elements, and the background plate, could move and interlock together.

- The film's complex digital VFX requirements—especially the challenge of duplicating the look of human skin digitally and convincingly—led the "Mac Squad" at Industrial Light & Magic to make the transition from Macintosh computers to higher-end Silicon Graphics hardware.[366]
- Doug Smythe from Industrial Light & Magic and Tom Brigham (who had conducted his own pioneering morphing work at MIT in the early 1980s) were both honored with a Technical Achievement Award from the Academy of Motion Picture Arts and Sciences for advancing cinema technology with the Morph computer software program, introduced during the production of digital VFX for *Willow* at Industrial Light & Magic.[367]
- Doug Smythe and George Joblove from Industrial Light & Magic received a Scientific and Engineering Award for the 'concept and development of the digital motion picture re-touching system for removing visible rigging and dirt/damage artifacts from original motion picture imagery'—introduced during the making of *Howard the Duck* in 1986 and more extensively applied during the post-production of *Back to the Future II* in 1989.[368]
- For *The Young Indiana Jones Chronicles* TV series, all VFX (including digital compositing, matte paintings and replication of actors) as well as the entire post-production process were completed at Industrial Light & Magic using only digital video technology.

1993

- Computer memory chips could hold 16 MB of information each.
- Microsoft launched Windows NT commercially.
- Intel Corporation released the Pentium processor that allowed computers to 'incorporate "real world" data such as speech, sound, handwriting and photographic images more easily.'[369] In being consequently adopted by most computer manufacturers, it became the standard chip for the computer industry; it used 32-bit registers, with a 64-bit data bus and incorporated 3.1 million transistors.[370]

- Apple Inc. introduced the AV Macintosh systems that integrated video and telecommunication technologies on the desktop for the first time.[371]
- The CD-R (Recordable) format was launched.[372]
- Voyager (a multi-media distribution company) launched Richard Lester's *A Hard Day's Night* on multi-media CD-ROM for Macintosh computers: it was the first digital CD to contain an entire feature-length movie.[373]
- *Jurassic Park* was massively successful at the box-office worldwide upon its theatrical release. Its groundbreaking complex visuals were conceived and produced by Industrial Light & Magic (for full-motion dinosaurs) and by Stan Winston Studio (for live-action dinosaurs), amongst others. During the production of digital VFX for *Jurassic Park*, some major innovations emerged:
    - the DID or Direct Input Device;[374]
    - the Viewpaint program, developed at Industrial Light & Magic—which revolutionized how CG artists created surface-texture maps by allowing the surface of CG dinosaurs to be painted as if it were a real sculpture;[375]
    - Enveloping, an in-house tool developed at Industrial Light & Magic specifically to enhance the animation of CG dinosaurs.[376]
- To fully exploit the technological convergence of the digital medium in the entertainment, motion picture, information and communication industries at large, George Lucas re-structured his business empire into: Lucasfilm Inc., Lucas Digital Ltd., Skywalker Sound, Industrial Light & Magic, Industrial Light & Magic Commercial Productions, LucasArts Entertainment, Lucas Learning and Lucas Licensing.[377]
- George Lucas's Industrial Light & Magic formed a strategic alliance with Silicon Graphics Images: the Joint Environment for Digital Imaging (JEDI)—to cooperate in the research, development and installation of high-powered workstations.[378]
- Miramax Pictures (film production and distribution) was acquired by Disney.[379]
- Rupert Murdoch's television networks gained access to a potential audience of 3 billion people when he acquired the control of Star TV, a satellite system that could reach most Asian countries between Israel and Taiwan, including India and China.[380]
- Ted Turner acquired New Line Cinema (film distribution and production), its subsidiary Fine Line Features (film production), and Castle Rock Entertainment (film production).[381]

- Paramount Communications was acquired by Sumner Redstone's Viacom Inc. (MTV; Nickelodeon; Showtime; television stations; syndicated television companies).

1994

- Aldus and Adobe merged to form the Aldus-Adobe software company.[382]
- Apple Inc. launched the Power Macintosh computer that involved the first-time use of the Reduced Instruction Set Chip (RISC) microprocessor—designed by Apple, Motorola and IBM—in the personal computer environment.[383]
- Sony Corporation introduced the 32-bit home video game system PlayStation in Japan.[384]
- The AOL community (1 million members) was linked to the Internet for the first time; AOL.com—AOL's Internet portal—was announced.[385]
- Digital Satellite System (DSS) direct-to-home broadcasting started with the introduction by Thomson Consumer Electronics (a subsidiary of Thomson Multimedia Inc.) of RCA's Satellite System, featuring DIRECTV programming.[386]
- The third-generation scanner image-processing system, developed at Industrial Light & Magic, became operational.[387] Electronic and digital processes were beginning to completely replace long-standing photochemical optical processes in Hollywood filmmaking.
- Time-Warner Cable Group received a special Emmy Award for its pioneer engineering work on fiber optic transmission.[388]
- Westinghouse established a joint venture in radio and television operations with CBS.[389]
- Wayne Huizenga's Blockbuster Video Entertainment, the largest American video retail chain, was bought by Viacom.[390]
- DreamWorks SKG was officially launched. It was the much publicized "dream team" joint-venture movie production company between filmmaker Steven Spielberg, ex-Disney CEO Jeffrey Katzenberg and record producer David Geffen.[391]

1995

- Computer memory chips could hold 64 MB of information.
- The Pentium Pro processor was released by Intel Corporation.[392]
- Microsoft released Windows 95 commercially.
- Apple Inc. shipped QuickTime VR, bringing virtual reality to Macintosh and Windows-based personal computers.[393]
- Sega introduced the Saturn video game system.[394]

- Sony Corporation commercially released worldwide its 32-bit game home video game system PlayStation.[395]
- Sony Corporation and Philips introduced the single-sided Multimedia Compact Disc (MMCD), with dual-layer option and a storage capacity of 3.7 GB.
- Toshiba and Time-Warner introduced the single-layer Super Density Disc (SD), with double-sided option and a storage capacity of 5 GB.
- To avoid a format war, Sony Corporation, Philips, Toshiba and Time-Warner eventually reached a mutual agreement about the standard of what would become the most common type of Digital Video Disc (DVD): a single-layer single-sided high-density CD with a storage capacity of 4.7 GB.
- The *Star Wars* trilogy in the original format was commercially released on VHS tape for the last time: the first home video title to feature Lucasfilm's THX Digital Mastering Process.[396]
- Doug Smythe, Lincoln Hu, and Douglas Kay received a Technical Achievement Award from the Academy of Motion Picture Arts and Science for their 'pioneering efforts in the creation of Industrial Light & Magic's digital film compositing system.'[397]
- After signing a three-movie deal with Disney, Pixar Animation Studios released the $30 million *Toy Story*: it rapidly became the top box-office hit of the year. Its many years of production involved a variety of new powerful and sophisticated digital modeling, rendering, and animation tools and engines, such as:
    - Autodesk Inc.'s Windows NT-capable 3D Studio MAX animation software (and its Biped 3D plug-in extension) that enabled the animation of two-legged skinned characters;
    - Charybdis Enterprises's MythOS graphics development engine, for creating titles with dazzling 3D graphics, fluid animation, and realistic, multi-dimensional visuals. MythOS consisted of a set of C++ libraries that provided a foundation for creating a wide range of versatile software products and it was optimized for graphics-intensive, real-time 3D applications;
    - Diamond Multimedia's Edge 3D, a chip-based graphic accelerator;
    - Number Nine Visual Technology's 9FX Reality, a new graphics board.[398]
- *Toy Story* was the first feature-length Hollywood motion picture to contain entirely computer-generated 3D animation, involving the use of very expensive and powerful Silicon Graphics and Sun Microsystems interactive workstations.[399]

- AOL International began to launch its online services outside the United States: AOL Germany, AOL UK, AOL Canada and AOL France.[400]
- NBC (owned by General Electric) sold a 50 percent stake to Microsoft in one of its cable TV channels, with the aim of jointly creating an additional news network to challenge CNN's monopoly; the outcome is MSNBC—the first combined cable TV network and online website—it would begin broadcasting and netcasting less than a year later.[401]
- Steven Spielberg, Jeffery Katzenberg and David Geffen announced construction of studio facilities for the recently launched DreamWorks SKG; this was the first soundstage studio to be built for a Hollywood major since the demise of the Hollywood studio system more than 50 years ago.[402]
- 80 percent of Universal-MCA was sold by Matsushita Industrial Electric Company to Seagram (Canadian liquor giant).
- CBS was purchased by Westinghouse for $5.4 billion.[403]
- Disney acquired Capital Cities (owner of ABC Television network) for $19 billion.[404]
- Ted Turner sold his entire media empire to Time-Warner for $7.5 billion.[405]

1996

- Nintendo released its 64-bit game system, the Nintendo 64.[406]
- Cray Research—responsible for the production of Cray supercomputers—merged with Silicon Graphics.[407]
- A decade after being created by Steve Jobs with some of the profits received upon leaving Apple Inc., NeXT Software Inc. was acquired by Apple Inc.; Jobs returned to the helm of the company he originally co-founded.[408]
- Aiming to liberalize restrictions previously placed on telecommunications companies, the 1996 U.S. Telecommunications Act removed the limitations to radio and television station ownership as well as allowed cable and telephone companies to compete in each other's industries and markets, amongst other things.[409]
- The technological convergence of broadcast television and multicast Internet was on the way with the introduction of the WebTV receiver, which coincided with MSNBC on cable TV (24-hour news and information) and MSNBC on the Internet (interactive online news service) being launched by NBC in partnership with Microsoft.[410]
- Westinghouse bought Infinity Broadcasting.[411]

- News Corporation bough out New World Communications for $2.48 billion; Rupert Murdoch was now reaching 40 percent of all American television viewers.
- Creative Artists Agency and Intel Multimedia started off a partnership by opening the CAA-Intel Media Lab, as an attempt to break Hollywood's filmed entertainment monopoly.[412]
- Metro-Goldwyn-Meyer was sold for $1.3 billion by Crédit Lyonnais Bank Nederland to a business group led by Kirk Kerkorian and including MGM's chairman Frank Mancuso, as well as the Australian Seven Network Group Ltd. broadcasting company (partly owned by Rupert Murdoch's News Corporation).[413]

1997

- Intel Corporation released the 7.5 million-transistor Pentium II processor, specifically designed to process more efficiently digital graphics, audio and video data.[414]
- The sale of motion pictures on DVD began.[415]
- The digitally re-mastered 20th anniversary "special edition" of *Star Wars*—later followed by its two sequels—was theatrically released worldwide. Becoming instant box-office hits, the digital re-masters contained enhanced picture quality, scenes and VFX, as well as a few minutes of previously unseen footage.
- Directed by James Cameron and allegedly produced for a record $200 million, *Titanic* quickly made cinema history upon its theatrical release. It became so popular (and faced so little competition) that its billion dollar box-office earnings worldwide made it the top-grossing movie ever at the time.
- Westinghouse became CBS Corporation after putting its remaining industrial divisions up for sale.[416]
- The new MGM acquired the 2,200-film library of Metromedia Television Group (owned by News Corporation) as well as Orion Pictures (film production).[417]

1998

- Intel released the Pentium II Xeon processor.[418]
- Apple Inc. (with Steve Jobs as Interim CEO) introduced the iMac computer.
- *Antz* was the second all computer-generated 3D animation motion picture to be made and released since *Toy Story* in 1995, but without the involvement of Pixar Animation Studios. *Antz* became a box-office success. Produced by DreamWorks SKG, its digital images were mostly supplied by Pacific Data Images while 3D character

and environment digitizing and modeling were done by Viewpoint Datalabs International.

- Pixar Animation Studios followed up its movie deal with Disney by releasing *A Bug's Life*. Made with a budget of $45 million, *A Bug's Life* was Pixar's second entirely computer-generated 3D animation movie in three years, preceding the forthcoming release of their *Toy Story 2*. It made about $33 million on its opening weekend and went on to be a smash hit.
- Twentieth Century-Fox acquired the rights to distribute George Lucas's highly anticipated forthcoming *Star Wars* prequel trilogy.
- Digital television broadcasting started but due to an oversight, cable systems were not ready to carry the signals. HDTV sets still cost $10,000. The first network digital broadcast on the ABC TV network featured Disney's 1996 remake of *101 Dalmatians*.[419]
- Microsoft, Alcatel, NEC and RCA's DIRECTV took equity positions in Thomson Multimedia (owner of Thomson Consumer Electronics) to develop new products.[420]
- America Online (AOL) acquired Netscape for $4 billion.[421]
- Disney purchased a major stake ($475 million) in online services provider Infoseek.[422]
- John Malone's Tele-Communications Inc. (TCI)—the second largest cable provider in the United States—acquired *TV Guide* from Rupert Murdoch's News Corporation for $2 billion.[423]
- Twentieth Century-Fox's offshore studios in Sydney, Australia were officially opened.

1999

- According to American Web magazine *eMarketer*, there were 130 million Internet users worldwide, an estimated figure expected to increase to 350 million by 2003—60 percent of whom would be based outside the United States[424]
- Time-Warner Cable Group held two video-on-demand tests in Texas and Florida.[425]
- Time-Warner and AOL (with nearly 30 million members worldwide) announced their future merger; Time-Warner-AOL would eventually become a single entity in 2001.[426]
- Metro-Goldwyn-Meyer purchased the Polygram film library from Universal.[427]
- CBS and Sumner Redstone's Viacom Inc. announced a $37 billion merger.[428]
- Intel released three processors in the same year (albeit not simultaneously): the Celeron, the Pentium III and the Pentium III Xeon.[429]
- Personal Video Recorders, which used digital hard discs, become commercially available.[430]

- As a sequel to *Toy Story*, the entirely computer-generated 3D animation movie *Toy Story 2* by Pixar Animation Studios—originally intended for direct-to-video release—rapidly became a box-office hit upon its theatrical release, making $81.1 million after only two weeks of exhibition.[431] It was the first feature film in motion picture history to be entirely created, mastered and exhibited digitally, and the first Hollywood movie to be commercially released in both analog and digital formats.
- Directed by George Lucas, *Star Wars: Episode I—The Phantom Menace* made $207 million in its first two weeks of theatrical release.[432]
- Two different digital motion picture display systems were tested with the *digital* exhibition of *Star Wars: Episode I—The Phantom Menace* in a few cinema theaters. AMC's Burbank 14 featured 'a Texas Instruments prototype digital projector which includes a chip with over one million tiny mirrors,' while Pacific's Winnetka used 'a liquid crystal light valve system from Hughes-JVC'; in both instances, the film was played out from a computer.[433] As such, *Star Wars: Episode I—The Phantom Menace* became the first feature-length live-action Hollywood movie to be theatrically exhibited in digital.

NOTES

1. 'Westinghouse Electric History' (online).
2. 'Timeline for Inventing Entertainment' (online).
3. 'Timeline for Inventing Entertainment' (online).
4. Winston, *Misunderstanding Media* 8–9.

Prior to becoming known as "television," it was called 'telephotography,' 'telescopy,' and 'teleautography' according to Brian Winston. 'As late as 1911, a British patent official opened a new file on the matter as a branch of facsimile telegraphy, even though he called it television, *a term first coined independently by Persky in 1900*.' Winston, *Media, Technology and Society* 94. Emphasis added.

5. Monaco, *How to Read a Film—The World of Movies* 571.
6. 'Westinghouse Electric History' (online).
7. 'Timeline for Inventing Entertainment' (online).
8. 'Timeline for Inventing Entertainment' (online).
9. 'Chronology of Scientific Developments' 1729.
10. Monaco, *How to Read a Film—The World of Movies* 571.
11. 'Timeline for Inventing Entertainment' (online).
12. Following A. R. Fulton's explanations, William Dickson

> cemented together sheets of emulsion-covered celluloid to form a strip half an inch wide. Then, because this area proved too narrow, he substituted a one-and-one-half-inch strip, which allowed for one-inch pictures and additional space for perforations along the edge. The perforations enabled the teeth of a locking device to hold the strip of sheets steady as it moved, by a stop-motion device, through the camera. . . . From the negative, Dickson made a positive print which he placed in a box-like structure, about four feet high and two feet square, containing a battery-run motor. Propelled by the motor, the strip ran on a loop between an electric lamp and a shutter.

The pictures were visible by flashes under a magnifying lens as the viewer looked through a slit in the top of the box.

Fulton, 'The Machine' 23–24.

13. In more than a century of existence, General Electric would have considerably expanded its range of products, activities and operations—domestic appliances, aviation, medical and industrial systems, plastics, transportation, broadcasting, communications, real estate and insurance—to the extent of becoming one of the largest industrial corporations worldwide. Furthermore, it would be the only company to remain continuously listed in the Dow Jones Industrial Index since the original 1896 Index. 'General Electric History' (online).

14. 'Timeline for Inventing Entertainment' (online).

15. 'Timeline for Inventing Entertainment' (online).

16. 'Timeline for Inventing Entertainment' (online).

17. 'Timeline for Inventing Entertainment' (online).

18. Salt, *Film Style* 41.

A. R. Fulton points out that the Lumière brothers had been experimenting with other available motion picture devices before producing their own. When the Kinetoscope was demonstrated in France for the first time in 1894, only a few months later after its introduction in the United States, they found out that the continuous motion in the Kinetoscope would not do for a projection machine. Accordingly, they built a stop-motion device. They also built a camera, which differed from Edison's Kinetograph in the speed at which the film was fed through it, that is, in the number of pictures, or frames, it recorded each second. Whereas the Kinetograph took forty-eight frames per second, the Lumières decided on sixteen as the proper rate. By early 1895, they had completed both projector and camera and taken some pictures, and on March 22, at their factory in Lyons, they demonstrated their accomplishment.

Fulton, 'The Machine' 26–27.

19. Fulton, 'The Machine' 27.

20. 'Timeline for Inventing Entertainment' (online).

21. 'Timeline for Inventing Entertainment' (online).

22. Monaco, *How to Read a Film—The World of Movies* 571.

23. The Kinetophone essentially operated according to the loose synchronization of a Kinetoscope image with a cylinder phonograph.

'Timeline for Inventing Entertainment' (online).

24. 'Timeline for Inventing Entertainment' (online).

25. Salt, *Film Style* 46.

26. 'Timeline for Inventing Entertainment' (online).

27. 'Timeline for Inventing Entertainment' (online).

28. Vaz, *Industrial Light & Magic* 3.

29. Salt, *Film Style* 42.

30. Salt, *Film Style* 46.

31. Winston, *Media, Technology and Society* 90.

The cathode ray tube's invention has been attributed elsewhere to F. Braun in Germany.

See, 'Chronology of Scientific Developments' 1729.

32. Paradoxically the CRT would remain confined to the laboratory for a number of years as 'a tool of advanced physics and nothing more.'

See Winston, *Media, Technology and Society* 90.

33. Salt, *Film Style* 42.

34. Salt, *Film Style* 44.

35. Monaco, *How to Read a Film—The World of Movies* 572.

36. 'General Electric History' (online).

37. 'Timeline for Inventing Entertainment' (online).

38. 'Timeline for Inventing Entertainment' (online).

39. Salt, *Film Style* 60.

40. 'Timeline for Inventing Entertainment' (online).

41. Salt, *Film Style* 72.
42. 'A Chronology of Computer History' (online).
43. Two robbers hold a telegraph operator at gunpoint while a train can be seen outside the telegraph office window; the train was composited into the shot using the in-camera matte technique. *The Great Train Robbery* also contained the first "emblematic" (close-up) shot: Justus D. Barnes, as one of the outlaws, pointing a gun directly at the camera and, by extension, at the movie spectator. The film would be remade by Sigmund Lubin in 1904.
44. Showing movies all day long, Nickelodeons attracted a large and varied clientele including women and children. Although there would be about 8,000 of them in 1908, the Nickelodeon boom would start declining by the end of 1907 as a result of entrepreneurs becoming more interested in the construction of larger movie theaters with greater seating capacities.
'Timeline for Inventing Entertainment' (online).
45. In these two films, according to Barry Salt, 'cut-out letters are made to move about to form words by shifting them a small amount between each single frame exposure, so introducing at one stroke what was to be the standard filmic animation technique.'
Salt, *Film Style* 72.
46. Salt, *Film Style* 130.
The film contained figures of birds, people and so on, made in modelling clay which gradually metamorphose into one another, apparently without human intervention, by the use of small changes made to them between the exposures of a succession of single frames—as such one of the "ancestors" of digital morphing.
47. Salt, *Film Style* 72.
48. Salt, *Film Style* 102.
49. 'Timeline for Inventing Entertainment' (online).
50. Blandford, Grant & Hillier, *The Film Studies Dictionary* 154.
51. D. W. Griffith made more than four hundred films (one- and two-reelers) for American Mutoscope & Biograph Company until his departure in 1913.
Blandford, Grant & Hillier, *The Film Studies Dictionary* 23.
52. 'General Electric History' (online).
53. 'Westinghouse Electric History' (online).
54. 'Timeline for Inventing Entertainment' (online).
55. 'Timeline for Inventing Entertainment' (online).
56. Salt, *Film Style* 128.
57. Blandford, Grant & Hillier, *The Film Studies Dictionary* 90.
58. Allen, 'William Fox Presents' 129.
59. See Lee, 'Motion Control' 60; Finch, *Special Effects.*
60. Blandford, Grant & Hillier, *The Film Studies Dictionary* 237.
61. According to Mark Vaz:

> The Fleischer rotoscope allows animators to project film images one frame at a time onto a surface where the outline of the desired image can be traced by hand. By tracing live-action footage, realistic movements could be created. . . . Rotoscoping also allows for the creation of hand-drawn travelling mattes, a laborious, time-consuming method in which the frame-by-frame drawn silhouettes punch out "matte windows" in background plates that can later be filled with film images conforming perfectly to the exact shape of the window—a process utilised to composite the spaceship into the star fields shots of *2001: A Space Odyssey* for instance.

Vaz, *Industrial Light & Magic* 98.
62. Blandford, Grant & Hillier, *The Film Studies Dictionary* 100.
63. 'Timeline for Inventing Entertainment' (online).
64. Wasko, *Hollywood in the Age* 64.
65. 'About RCA' (online).

66. The three letters stood for the names of the three men who created the company. Buscombe, 'Notes on Columbia Pictures' 27.
67. Blandford, Grant & Hillier, *The Film Studies Dictionary* 147.
68. Blandford, Grant & Hillier, *The Film Studies Dictionary* 250.
69. 'AOL-Time-Warner Timeline' (online).
70. Winston, *Misunderstanding Media* 56.
71. 'AOL-Time-Warner Timeline' (online).
72. The first contract of the Disney brothers was to produce a dozen short animated cartoons to constitute the *Alice's Wonderland* series, which would be followed by another contract in 1927 for a series of short cartoons for "Oswald the Rabbit."
Hollister, 'Genius at Work' 25.
73. Buscombe, 'Notes on Columbia Pictures' 27.
74. Blandford, Grant & Hillier, *The Film Studies Dictionary* 115.
75. Blandford, Grant & Hillier, *The Film Studies Dictionary* 260.
76. According to Barry Salt, by

> making a print of an original negative on duplicating positive stock, and then printing this onto duplicating negative stock, it was now possible for the first time to create a duplicate negative that had almost as good definition and as low contrast as the original negative, and so capable of being used in its place either to make duplicate prints, or optical effects in the printer without much loss of quality.

Salt, *Film Style* 222.
77. By 1928, The Baird Television Development Company would be 'building televisors (or receivers) for public domestic sale' in the United Kingdom.
Winston, *Media, Technology and Society* 95.
78. 'About RCA' (online).
79. According to Brian Winston, Philo Farnsworth's invention operated on the basis of very different principles to those of Vladimir Zworykin's earlier camera and produced motion pictures of better definition and sharper contrast. 'Called an image dissector, it had the advantages of offering a more stable picture [and] used neither a scanning spot nor the storage principle but worked by translating the image into a pattern of electrons which were then passed across an aperture.'
Winston, *Media, Technology and Society* 108.
80. Salt, *Film Style* 234.
81. Kristin Thompson argues however that the implementation of the innovative sound process 'began with Warner Bros.'s Vitaphone shorts and *Don Juan* in August of 1926, and was well under way by the time of *The Jazz Singer*'s première in October of 1927.'
Thompson, *Exporting Entertainment* 10.
For a more detailed account of the invention, application and diffusion of sound in the American cinema, see: Gomery, *The Coming of Sound*; Neale, *Cinema and Technology*.
82. Blandford, Grant & Hillier, *The Film Studies Dictionary* 90.
83. 'Westinghouse Electric History' (online).
84. Salt, *Film Style* 237.
85. According to Barry Salt, the process then required that

> the background scene that had to be combined into one shot with the foreground action was shot first in the ordinary way. . . . The actors whose actions were to be combined with the background scene then performed in front of a blue backing sheet illuminated with white light, while being filmed with a camera loaded with a double layer of film passing through it. This double layer of film was made up of the Dunning plate in front of the unexposed panchromatic negative on which the final combined image was to be produced. The stated effect of doing all this was that the blue light from the backing sheet printed a negative image of the background from the red positive, which was opaque to the blue light in the appropriate

> places, but no image from the background was produced where the actors blocked the blue light from the lens.

Salt, *Film Style* 234.

86. Walt Disney subsequently applied for a trademark with the U.S. Patent Office for the use of the "Mickey Mouse" character and name in motion pictures.
See Polsson, *Chronology of Walt Disney* (online).
Mickey Mouse would again feature, a few months later, in another silent black-and-white short animation film *The Gallopin' Gaucho*.
See Hollister, 'Genius at Work' 25.

87. *Steamboat Willie* was not actually drawn by Walt Disney himself but by his studio's chief animator and eventually long-time collaborator, Ub Iwerks. First made as a silent film, it was subsequently re-worked into a sound film—following the official introduction of sound technology during the previous year—and finally delivered to theatrical exhibitors in July 1928. See Hollister, 'Genius at Work' 26.
Over the next few years, Disney would introduce a number of enduring cartoon characters such as Pluto in 1930, Goofy in 1932 and Donald Duck in 1934, among others.
See Polsson, *Chronology of Walt Disney* (online).

88. 'About RCA' (online).

89. Monaco, *How to Read a Film—The World of Movies* 575.

90. Blandford, Grant & Hillier, *The Film Studies Dictionary* 100.

91. The sound Moviola is a 'simple adaptation of the silent Moviola, with a continuously turning sprocket drive pulling the soundtrack film under a photoelectric sound head identical to that in a sound projector, the whole unit being mounted beside the standard Moviola picture head, and driven from it in synchronism by a rigid shaft drive. . . . The soundtrack can be moved slowly by hand under the sound head, and the exact position of any part of a sound identified.'
Salt, *Film Style* 281.

92. After *Time* and *Fortune*, the Time Inc. empire continued to expand through the launching of dozens of other magazines which included, amongst others: *Life* (1936), *Sports Illustrated* (1954), *Money* (1972), *People* (1974), *Entertainment Weekly* (1990), *In Style* (1994) and *Teen People* (1998). Time Inc.'s book division Time-Life Inc. would be created in 1961.
See 'AOL-Time-Warner Timeline' (online).

93. It was a substantial improvement upon the previous Technicolor two-strip color film process by virtue of an additional blue component being included in what had previously been a red and green system only. Neupert, 'Painting a Plausible World' 106.
According to Edward Branigan, prior to this technological innovation, 80-90 percent of American films were only *tinted* in some manner. See Branigan, 'Color and Cinema' 127.

94. Hollister, 'Genius at Work' 27.
'Disney became the first studio to exploit commercially Technicolor's three-color cinematography process. . . . As a result of this *aesthetic* innovation during the years of 1932 to 1935, the *Silly Symphony* cartoon series became an unlikely prototype for Technicolor's subsequent application to live-action feature films.'
See Neupert, 'Painting a Plausible World' 107.

95. Blandford, Grant & Hillier, *The Film Studies Dictionary* 247.

96. Winston, *Misunderstanding Media* 88.

97. *Federal Communications Commission* (online).
According to James Monaco,

> the Communications Act of 1934 recognizes the interdependency of telephone, telegraph, and radio (and television) broadcasting but treats the older media differently from the new radio industry. Telephone and telegraph are seen as natural monopolies and designated as "common carriers" that must furnish service as requested at rates governed by an organization

to be called the Federal Communications Commission. Broadcasting, however, is considered a competitive activity. The principle of public ownership of the airwaves is recognized: the FCC will issue limited licenses to broadcasters and govern the nature of their activities.

Monaco, *How to Read a Film—The World of Movies* 576.

98. Salt, *Film Style* 274.

99. Blandford, Grant & Hillier, *The Film Studies Dictionary* 103.

100. The three-color system was made possible, according to Barry Salt,

> by the introduction of the new special Technicolor three-strip camera, which had a 45 degree split-cube prism behind the lens to produce two images, one of the green part of the spectrum on a panchromatic film directly behind the prism block, and another deviated by 90 degrees onto a bi-pack of two films with their emulsions in contact in another gate, to record the blue image and the red image respectively. In this second gate carrying the bi-pack, the light passed through the transparent base of the first film before forming an image in the blue-sensitive emulsion and then through a red filter layer coated onto its surface before forming the red image in the panchromatic emulsion on the second film facing the first.

Salt, *Film Style* 274.

101. Winston, *Misunderstanding Media* 109.

Computers work by manipulating numbers—they do not understand anything other than 0s and 1s fed into microprocessors and electronic components in the form of changes in voltage, normally five volts for 1 and zero volts for 0. The computer "looks" at a series of voltage pulses representing binary numbers and uses extremely simple logic to deal with them. And such decisions made by the computer are based on the theory of symbolic logic, developed circa 1847 by George Boole while working on his "Universal Engine" and published as *The Mathematical Analysis of Logic*. The signal involved for determining decisions taken by the computer does not necessarily have to be voltage, but what really matters is the principle that a numerical value has to be assigned to either the presence or absence of an input signal. So, computing decisions are based on Boolean algebra and involve comparing two pieces of data and using the logic operators "and," "or" and "not" to determine what to do next.

102. Wasko, *Hollywood in the Age* 12.

103. Hollister, 'Genius at Work' 36.

104. 'General Electric History' (online).

105. According to Timothy Binkley, before the advent of electric, electronic and digital computers, the word "computer" apparently described a person whose job was to perform numerical calculations; it is only in the 1940s that it has come to refer to actual machines.

See Binkley, 'Reconfiguring Culture' 122.

106. It is not easy to pinpoint exactly which one of the 1940s prototypes should be considered the *first computer*, but the *Atanasoff-Berry* will be ruled in a court of law in 1973 to be the first automatic digital computer.

107. Computation in "real time" means that humans can interact "live" with the machine, obliging it to function according to human terms.

108. 'About RCA' (online).

109. Kerr, 'Joseph H. Lewis' 52.

110. Monaco, *How to Read a Film—The World of Movies* 577.

111. Burton-Carvajal, 'Surprise Package' 139.

112. Within the next two decades, not only would it be realised but it would also become the essential driving force of the rapidly growing broadcasting and communications industries with which the Hollywood cinema industry has constantly maintained a symbiotic relationship.

113. 'About RCA' (online).

114. 'Westinghouse Electric History' (online).

115. Monaco, *How to Read a Film—The World of Movies* 578.

116. Brian Winston indicates, however, that 'the transistor was not crucial to the development and growth of the computer,' since it

> was not, of itself, a significant invention but rather the signpost to one. It took 21 years to get from the transistor to the invention—the microprocessor; 21 years in which the computer industry played very little part, except towards the very end. . . . IBM, for example, never marketed a commercially successful fully transistorised computer, and valve-based computers were still being shipped to customers a decade after the transistor was commercially available.

Winston, *Misunderstanding Media* 181.

117. Winston, *Misunderstanding Media* 101.

118. 'About RCA' (Online).

119. 'Chronology of Scientific Developments' 1729.

120. Monaco, *How to Read a Film—The World of Movies* 579.

121. Monaco, *How to Read a Film—The World of Movies* 579.

122. Wasko, *Hollywood in the Age* 12.

123. See Winston, *Media, Technology and Society* 307–10.

124. Wasko, *Hollywood in the Age* 116.

125. Salt, *Film Style* 310.

126. Hafner & Lyon, *Where Wizards Stay Up* 32–33.

127. Tino Balio explains that

> the invention of Milton Gunzberg, the three-dimensional process, which he dubbed Natural Vision, required two projectors to exhibit and Polaroid glasses for the audience. This stereoscopic method, as opposed to the panoramic method of Cinerama, was regarded by the trade as the true third dimensional experience. . . . In the wake of *Bwana Devil*, other studios jumped onto the 3D bandwagon, producing pictures in Naturescope, Paravision, Tri-Opticon, in addition to Natural Vision and other stereoscopic techniques. (Requirements for equipping a theater for 3D were minimal: a high-intensity reflective screen—or the old one painted aluminium—and an interlocking system for the projectors.) A string of 3D hits [followed], such as Warner's *House of Wax*, Paramount's *Sangaree* and Universal's *It Came from Outer Space*.

Because of the unwillingness of producers to invest in and experiment further with the creative potential of such an innovation, 3D cinema would have eventually fizzled out by early 1954, although later re-emerging in better shape during the 2000s.
See Balio, *United Artists* 49–51.

128. 'A Chronology of Computer History' (online).

129. 'About RCA' (online).

130. This process involved 'printing by dye transfer via three positive relief matrices held in contact with a "blank" emulsion on three successive register pin bel machines, one for each of the positive colors.'
Salt, *Film Style* 311.

131. The image consequently produced on the film included twice the horizontal field while leaving the vertical field unchanged.
Salt, *Film Style* 316.

132. Tana Wollen explains that 'VistaVision widens the image, not through an anamorphic lens but by increasing the size of the film negative frame in the camera, running it horizontally, exposing a frame with eight sprocket holes.'
Wollen, 'The Bigger the Better' 14.
As for Barry Salt, the sole advantage of VistaVision over ordinary non-anamorphic widescreen photography is that the larger area on the negative produces a sharper image in the final print. This is because most of the loss in image sharpness occurs at

the negative stage, since the negative emulsion is always far grainier than the positive emulsion.
See Salt, *Film Style* 320.

133. Prior to the foundation of Buena Vista, the exhibition and circulation of Disney movies depended entirely upon distribution agreements with other Hollywood studios: Columbia (1929–1931), United Artists (1931–1936) and RKO (1936–1954).
See Gomery, 'Disney's Business History' 77.

134. 'A Chronology of Computer History' (online).

135. 'About RCA' (online).

136. Cheatwood, 'The Tarzan Films' 179.

137. Built for $17 million, Disneyland was actually composed of separate themed "lands" (such as Fantasyland, Frontierland, Adventureland, and Tomorrowland); it would already be generating gross revenues of $10 million within a year of operation. Other Disneylands will later be opened outside the United States: in Tokyo, Japan (1983) and Paris, France (1992).
See Grover, *The Disney Touch* 8.

138. Monaco, *How to Read a Film—The World of Movies* 580.

139. Winston, *Misunderstanding Media* 90.

140. Monaco, *How to Read a Film—The World of Movies* 580.

141. Coyle, 'The Genesis of Virtual Reality' 151.

142. Desilu was owned by TV stars Lucille Ball and Desi Arnaz, both from the very popular *I Love Lucy* series.

143. Winston, *Misunderstanding Media* 200.

144. Pennings, *History of Information Technology* (online).

145. Haddon, 'Interactive Games' 125.

146. Cheatwood, 'The Tarzan Films' 179.

147. Pennings, *History of Information Technology* (online).

148. Monaco, *How to Read a Film—The World of Movies* 581.

149. Monaco, *How to Read a Film—The World of Movies* 581.

150. Balio, *United Artists* 107-8.

151. Polsson, *Chronology of Personal Computers* (online).

152. Coyle, 'The Genesis of Virtual Reality' 151.

153. Winston, *Misunderstanding Media* 91.

154. Stewart Brand writes that hackers basically invented themselves and '[w]ithout specifications they would just start programming, quick and dirty. They did the first computer graphics, the first word-processing, the first computer games, the first time-sharing.'
Brand, *The Media Lab* 56. See also Levy, *Hackers*.

155. Haddon, 'Interactive Games' 126.

156. 'General Electric History' (online).

157. Binkley, 'Reconfiguring Culture' 110.

158. Balio, *United Artists* 110.

159. Polsson, *Chronology of Personal Computers* (online).

160. Winston, *Media, Technology and Society* 287.

161. Binkley, 'Reconfiguring Culture' 106.

162. Pennings, *History of Information Technology* (online).

163. This process provided much better results compared to any previous travelling matte system, even more so after the introduction of Eastman Kodak's Color Reversal Intermediate film stock in 1968. See Salt, *Film Style* 339.

164. Wasko, *Hollywood in the Age* 119.

165. Monaco, *How to Read a Film—The World of Movies* 582.

166. Monaco, *How to Read a Film—The World of Movies* 582.

167. Polsson, *Chronology of Personal Computers* (online).

168. Polsson, *Chronology of Personal Computers* (online).

169. Binkley, 'Reconfiguring Culture' 107.

170. Monaco, *How to Read a Film—The World of Movies* 583.

171. Balio, *United Artists* 303–305.
172. Pennings, *History of Information Technology* (online).
173. Renaud, 'Towards Higher Definition Television' 47.
174. Polsson, *Chronology of Personal Computers* (online).
175. Slit Scan is a technical process that essentially enables any photograph or hand-drawn item in animation to be put 'in striking perspective, as if you were looking at it from a great distance in space or in a desert.'

Myers, 'Oscar computes' 68.

176. 'A Chronology of Computer History' (online).
177. Monaco, *How to Read a Film—The World of Movies* 584.

ARPA stands for 'Advanced Research Projects Agency' and was originally formed by President Eisenhower in 1958 for the development of U.S. military defence-related projects.

For more details, see Hafner & Lyon, *Where Wizards Stay Up*.

178. Pennings, *History of Information Technology* (online).
179. 'About RCA' (online).
180. Wasko, *Hollywood in the Age* 182.
181. Balio, *United Artists* 317.
182. According to Intel's official definition, a microprocessor

> is an integrated circuit on a tiny silicon chip that contains thousands or millions of tiny on/off switches, known as transistors. The transistors are laid out along microscopic lines made of superfine traces of aluminium that store or manipulate data. These circuits manipulate data in certain patterns, patterns that can be programmed by software to make machines do many useful tasks. One of the biggest tasks microprocessors perform is acting as the brains inside a personal computer. . . . Computers are not the only way in which microprocessors are used. Microprocessors also help many devices—the telephone, thermostat, car or a traffic light for instance—to remember and adjust to incoming information.

'Intel's History' (online).

183. Pennings, *History of Information Technology* (online).
184. 'Intel's History' (online).
185. Winston, *Misunderstanding Media* 97.
186. Polsson, *Chronology of Personal Computers* (online).
187. Monaco, *How to Read a Film—The World of Movies* 585.
188. HBO would take many years to fully grow commercially. With about only 57,000 subscribers in 1974, it would have attracted 12 million of them by the early 1980s, especially after going live by satellite for the Ali-Frazier heavyweight boxing title fight in the Philippines. By 1995, HBO would be servicing 25 million subscribers, that is 25 percent of all American households. See Winston, *Media, Technology and Society* 312–13.
189. According to Tino Balio, 'to reduce operating costs, personnel reductions were made in the executive, production, technical, and clerical departments. Columbia and Warner would not only share studio space, but also rent the facilities to third-party motion picture and television production companies—a practice also followed by Paramount, MGM and Fox.'

Balio, *United Artists* 317.

190. 'In an Ethernet, machines talk to each other like humans: when more than one tries to talk at the same time, someone will randomly go first, while the rest wait for it to finish.'

See Hafner & Lyon, *Where Wizards Stay Up* 237–38.

191. 'Intel's History' (online).
192. Monaco, *How to Read a Film—The World of Movies* 585.
193. Pennings, *History of Information Technology* (online).
194. Polsson, *Chronology of Personal Computers* (online).

195. Polsson, *Chronology of Personal Computers* (online).
196. Pennings, *History of Information Technology* (online).
197. Monaco, *How to Read a Film—The World of Movies* 585.
198. Monaco, *How to Read a Film—The World of Movies* 585.
199. *Lucasfilm* (online).
200. Wasko, *Hollywood in the Age* 75.
201. Polsson, *Chronology of Personal Computers* (online).
202. Polsson, *Chronology of Personal Computers* (online).
203. Polsson, *Chronology of Video Game Systems* (online).
204. Balio, 'Introduction to Part II'—*Hollywood in the Age of Television* 266.
205. Polsson, *Chronology of Video Game Systems* (online).
206. Wasko, *Hollywood in the Age* 78.
207. 'AOL-Time-Warner Timeline' (online).
208. Winston, *Misunderstanding Media* 219.
209. Polsson, *Chronology of Personal Computers* (online).
210. Polsson, *Chronology of Video Game Systems* (online).
211. 'About RCA' (online).
212. Neale & Smith, 'Introduction'—*Contemporary Hollywood Cinema* xix.
213. 'Intel's History' (online).
214. Winston, *Misunderstanding Media* 100.
215. The Aspen Project was a prototype for the Defence Department made up of a set of videodiscs allowing 'the user to tour Aspen [in Colorado], virtually driving down every street, choosing which way to turn at every corner.'
Monaco, *How to Read a Film—The World of Movies* 587.
216. Polsson, *Chronology of Video Game Systems* (online).
217. Polsson, *Chronology of Personal Computers* (online).
218. Polsson, *Chronology of Video Game Systems* (online).
219. Polsson, *Chronology of Video Game Systems* (online).
220. Jointly-developed by MCA and Philips, '"DiscoVision" utilises a reflective laser optical—that is a laserdisc—system that offers a top-quality picture, stereophonic sound, random access, and freeze frame, among other features.'
Balio, 'Introduction to Part II' 269.
221. *Lucasfilm* (online).
222. Monaco, *How to Read a Film—The World of Movies* 586.
223. 'AOL-Time-Warner Timeline' (online).
224. Monaco, *How to Read a Film—The World of Movies* 587.
225. Winston, *Misunderstanding Media* 100.
226. Polsson, *Chronology of Video Game Systems* (online).
227. Baker, 'Computer Technology' 33.
228. As confirmed by Kevin Baker, 'the sounds [simple machinery "pings," "beeps" and "bloops"], colors (mostly green) and images were very contrived in the sequence from *Alien* and this characterization of computing was to be exploited further in later movies.'
Baker, 'Computer Technology' 34.
229. Monaco, *How to Read a Film—The World of Movies* 587.
230. Monaco, *How to Read a Film—The World of Movies* 587.
231. 'AOL-Time-Warner Timeline' (online).
In spite of its mix of old movies and TV shows, by 1982 WTBS would nevertheless be delivering cable television programming to 26 million homes in the Unites States.
See Winston, *Media, Technology and Society* 313.
232. Winston, *Misunderstanding Media* 290.
233. Brand, *The Media Lab* 142.
234. Vaz, *Industrial Light & Magic* 92.
235. *Lucasfilm* (online).
236. Winston, *Media, Technology and Society* 312.
237. Wasko, *Hollywood in the Age* 78.

238. Wasko, *Hollywood in the Age* 78.
239. Grover, *The Disney Touch* 16.
240. Monaco, *How to Read a Film—The World of Movies* 588.
241. 'General Electric History' (online).
242. Polsson, *Chronology of Video Game Systems* (online).
243. Polsson, *Chronology of Video Game Systems* (online).
244. Monaco, *How to Read a Film—The World of Movies* 588.
245. 'A Chronology of Computer History' (online).
246. Brand, *The Media Lab* 72.
247. Wasko, *Hollywood in the Age* 78.
248. Smith, *Industrial Light & Magic* 93.
249. Vaz, *Industrial Light & Magic* 308.
250. Vaz, *Industrial Light & Magic* 308.
251. 'Westinghouse Electric History' (online).
252. Balio, *United Artists* 341.
253. 'General Electric History' (online).
254. Wasko, *Hollywood in the Age* 27.
255. Polsson, *Chronology of Video Game Systems* (online).
256. RCA's biggest gamble in consumer electronics since color television, Selectavision 'utilizes the capacitance electronic disc—that is, a "needle-in-the-groove"—system that's simple to play. One merely had to insert the disc with the plastic sleeve still on, remove the sleeve, and press a button. This simplicity of design was supposed to differentiate Selectavision from the more complicated-to-operate VCR and attract a different buyer.'
Balio, 'Introduction to Part II' 269.
257. Polsson, *Chronology of Personal Computers* (online).
258. Polsson, *Chronology of Personal Computers* (online).
259. Polsson, *Chronology of Personal Computers* (online).
260. 'Intel's History' (online).
261. *Lucasfilm* (online).
262. Vaz, *Industrial Light & Magic* 111.
263. Monaco, *How to Read a Film—The World of Movies* 589.
264. 'A Chronology of Computer History' (online).
265. Monaco, *How to Read a Film—The World of Movies* 589.
266. Wasko, *Hollywood in the Age* 78.
267. Monaco, *How to Read a Film—The World of Movies* 589.
268. Polsson, *Chronology of Personal Computers* (online).
269. Polsson, *Chronology of Personal Computers* (online).
270. Polsson, *Chronology of Personal Computers* (online).
271. Pennings, *History of Information Technology* (online).
272. Ohanian & Phillips, *Digital Filmmaking* 145.
273. Ohanian & Phillips, *Digital Filmmaking* 146.
274. Monaco, *How to Read a Film—The World of Movies* 590.
275. Digital Productions Inc. was then using, amongst other things, one of the most powerful computers of the time: a Cray X-MP supercomputer (which handles 320 million instructions per second or 160 million floating point calculations) with Prevue software, high-speed VAX computers from Digital Equipment Corporation and a IMI555 from Interactive Machines Inc.
See Myers, 'Oscar Computes' 69.
276. At the time, scanners required 16 hours to scan and store *one* minute of film, while it took 720 disks to store a single two-hour movie.
Vaz, *Industrial Light & Magic* 111.
277. Monaco, *How to Read a Film—The World of Movies* 590.
278. Polsson, *Chronology of Personal Computers* (online).
279. 'Chronology of Scientific Developments' 1729.
280. 'Intel's History' (online).

281. 'Chronology of Scientific Developments' 1729.
282. Coyle, 'The Genesis of Virtual Reality' 154.
283. Coyle, 'The Genesis of Virtual Reality' 157.
284. The Nintendo Entertainment System used a 6502 processor and generated images with 256 x 240 resolution in 16 colors. See Polsson, *Chronology of Video Game Systems* (online).
285. Ohanian & Phillips, *Digital Filmmaking* 145.
286. Monaco, *How to Read a Film—The World of Movies* 590.
Although a film's electronic conversion from black-and-white to color could cost between $1,500 and $3,000, the investment could be recovered profitably via sales to syndicated television, cable channels and videocassette stores. See Wasko, *Hollywood in the Age* 35.
287. Wollen, 'The Bigger the Better' 25.
288. Smith, *Industrial Light & Magic* 212.
289. Vaz, *Industrial Light & Magic* 111.
290. Vaz, *Industrial Light & Magic* 111.
291. Streisand, 'The Patience of Jobs' 83.
292. 'AOL-Time-Warner Timeline' (online).
293. Wasko, *Hollywood in the Age* 153.
294. Monaco, *How to Read a Film—The World of Movies* 590.
295. Brand, *The Media Lab* 22.
296. Pennings, *History of Information Technology* (online).
297. Polsson, *Chronology of Personal Computers* (online).
298. 'Chronology of Scientific Developments' 1729.
299. Ohanian & Phillips, *Digital Filmmaking* 145.
300. In the wire-removal process, scenes containing rigs and flying wires—used in conjunction with the blue-screen technique to create the effect of flying objects during shooting—are scanned into a computer and the pixels on either side of the rigs are manipulated to meld the surrounding colors and effectively erase the rigs and wires. 'This process also requires a separate program to add film grain to the pristine digital image, which helps match the original photography.'
See Vaz, *Industrial Light & Magic* 115.
301. Myers, 'Oscar Computes' 68.
302. Brand, *The Media Lab* 29.
303. Monaco, *How to Read a Film—The World of Movies* 591.
304. 'About RCA' (online).
305. Balio, *United Artists* 343.
306. Balio, *United Artists* 343. Color System Technologies Inc. would consequently undertake the electronic colorization (a technique it introduced during the previous year) of many black-and-white Hollywood cinema "classics"—such as *Casablanca*, *The Maltese Falcon* and others—solely owned by Ted Turner.
307. 'A Chronology of Computer History' (online).
308. Polsson, *Chronology of Personal Computers* (online).
309. Polsson, *Chronology of Personal Computers* (online).
310. 'A Chronology of Computer History' (online).
311. Polsson, *Chronology of Personal Computers* (online).
312. Brand, *The Media Lab* 23.
313. Brand, *The Media Lab* 23.
314. 'About RCA' (online).
315. Monaco, *How to Read a Film—The World of Movies* 591.
316. 'A Chronology of Computer History' (online).
317. Polsson, *Chronology of Personal Computers* (online).
318. Polsson, *Chronology of Personal Computers* (online).
319. Polsson, *Chronology of Personal Computers* (online).
320. Monaco, *How to Read a Film—The World of Movies* 591.
321. Vaz, *Industrial Light & Magic* 125.

The film was a major success, generating $154 million at the American box-office and $174 million overseas. See Grover, *The Disney Touch* 121.

322. According to Mark Vaz, it was all done at the new computer graphics department of Industrial Light & Magic, and it took nearly two years until competitors in the visual effects industry could effectively duplicate the technique.
See Vaz, *Industrial Light & Magic* 114 & 135.

323. Vaz, *Industrial Light & Magic* 159–60.

324. Monaco, *How to Read a Film—The World of Movies* 591.

325. Monaco, *How to Read a Film—The World of Movies* 591.

326. 'A Chronology of Computer History' (online).

327. 'A Chronology of Computer History' (online).

328. Pennings, *History of Information Technology* (online).

329. This 16-bit device was based on an 8 MHz 68EC000 processor; graphics were produced with 320 x 224 resolution in 32 colors.
See Polsson, *Chronology of Video Game Systems* (online).

330. Polsson, *Chronology of Personal Computers* (online).

331. Monaco, *How to Read a Film—The World of Movies* 592.

332. 'A Chronology of Computer History' (online).

333. Vaz, *Industrial Light & Magic* 304.

334. Unlike *Willow*, in which specific elements were digitally manipulated first then optically composited later, all the main elements of the "Donovan's Destruction" sequence in *Indiana Jones and the Last Crusade* were manipulated and composited solely in the digital realm. A few months later, two more all-digital composites would be accomplished in the non-terrestrial water "pseudopod" sequence of *The Abyss*.
See Vaz, *Industrial Light & Magic* 135 & 138.

335. Wasko, *Hollywood in the Age* 22.

336. During the production of *The Abyss*, the VFX crew at Industrial Light & Magic was faced with the 'challenge of imitating the elemental mystery of water to create a life-size, "pseudopod" creature composed entirely of seawater.' The team led by Dennis Muren, John Knoll, and CG artists Mark Dippé and Jay Riddle created the alien creature using powerful Silicon Graphics workstations enhanced with a proprietary modelling and animation software package from Canadian company Alias.
See Vaz, *Industrial Light & Magic* 117, 193 & 195.

337. Monaco, *How to Read a Film—The World of Movies* 592.

338. 'AOL-Time-Warner Timeline' (online).

339. 'General Electric History' (online).

340. Wasko, *Hollywood in the Age* 81.

341. Monaco, *How to Read a Film—The World of Movies* 591.

342. Wasko, *Hollywood in the Age* 50.

343. Polsson, *Chronology of Personal Computers* (online).
This particular software package included four modules—Pencil Test, Exposure Sheet, Ink & Paint, and a supervisor program called DAS—'to re-create the traditional process of cel animation used in all Disney cartoons.'
Anzovin, 'Disney Animation Studio' 78.
The Disney Animation Studio software package was meant to allow any animator or filmmaker to produce Disney-style cartoons without requiring prior training in this particular style and technique. The "side-effect" of such a tool however was its standardization of the Disney style; more cartoons made in the same way inevitably reduced opportunities for alternative techno-stylistic experimentation and development in the field.

344. For the CCD scanner to work, light has to strike 'an assembly of CCD chips, which function as image-recording devices, converting the information to an array of binary digits that provide meaningful information to a computer. The tri-linear array of the ILM/Kodak scanner is able to provide high resolution at a maximum of eight thousand pixels in a horizontal direction.' The CCD scanner immediately contributed to the big task of digitally manipulating and compositing matte paintings with live-

action elements (which required 15 Macintosh computer disks, each holding about 600 megabytes of information) in the 35 seconds-long final sequence of *Die Hard II: Die Harder* for instance—the camera pulls back to reveal a snowy runaway crowded with planes, ambulances, military and police vehicles, and dozens of bystanders. See Vaz, *Industrial Light & Magic* 115 & 210.
The development team included Industrial Light & Magic's digital experts Lincoln Hu, Joshua Pines and Jeffrey Light, optical experts Mike MacKenzie and Mike Bolles, VFX supervisor Scott Squires, and researchers Mike Davis and Glenn Kennel from Eastman Kodak.

345. 'What's New' 18 & 20.
346. Vaz, *Industrial Light & Magic* 116.
347. Monaco, *How to Read a Film—The World of Movies* 592.
348. Wasko, *Hollywood in the Age* 53.
349. Monaco, *How to Read a Film—The World of Movies* 592.
350. Wasko, *Hollywood in the Age* 65.
351. Polsson, *Chronology of Personal Computers* (online).
352. Polsson, *Chronology of Personal Computers* (online).
353. 'A Chronology of Computer History' (online).
354. The 65816 processor of the Super Nintendo Entertainment System was a 16-bit version of the previous 6502 processor, but the new system could not use game cartridges from the original Nintendo system; it produced graphics in 512 x 448 resolution with 256 colors.
See Polsson, *Chronology of Video Game Systems* (online).
355. Pennings, *History of Information Technology* (online).
356. *Lucasfilm* (online).
357. Body Sock allows for the edges of a digitally modelled surface to be smoothed out and blended together, while the Make Sticky software allows scanned-in, two-dimensional background plates to be projected and to stick to 3D computer models as they are animated.
See Vaz, *Industrial Light & Magic* 205–6.
358. 'A Chronology of Computer History' (online).
359. Polsson, *Chronology of Personal Computers* (online).
360. Unlike Sony Corporation's digital Mavica still camera, introduced a few years earlier, Kodak's Photo CD enabled the company and consumers alike to maintain their 'huge investment in chemical film technology while still enjoying the benefits of digital recording.'
Monaco, *How to Read a Film—The World of Movies* 593.
361. Monaco, *How to Read a Film—The World of Movies* 593.
362. Wasko, *Hollywood in the Age* 82.
363. SoftImage 3D—developed at first for Silicon Graphics systems, and later applied to PCs through Windows NT 3.51—has been subsequently used, mainly at Industrial Light & Magic, to add 3D animation to *Jurassic Park, The Mask* and *Casper,* among other digital VFX-intensive films.
364. The advantage of the Matador software was that instead of each outline having to be drawn by hand, rotoscopers could put down points and digitally generate curves and lines for key frames. The computer automatically produced the frames in between. The rotoscoper then went in for the fine-tuning of each frame as needed.
See Vaz, *Industrial Light & Magic* 223.
365. 'The pin-block technique, developed for the photochemical optical process, involves re-photographing composite elements under an animation stand by utilizing an automated shuttle to position and register the film.'
Vaz, *Industrial Light & Magic* 314.
366. Vaz, *Industrial Light & Magic* 205–6 & 212.
367. Vaz, *Industrial Light & Magic* 135.
368. Vaz, *Industrial Light & Magic* 308.
369. 'Intel's History' (online).

370. Polsson, *Chronology of Personal Computers* (online).
371. Polsson, *Chronology of Personal Computers* (online).
372. Pennings, *History of Information Technology* (online).
373. Rosebush, 'Digital Video' 127.
374. The Direct Input Device was basically

> a motion-capture system linking traditional stop-motion talent with computer graphics technology. The DIDs were created as individual dinosaur skeletal armatures equipped with a system of encoders at individual pivot points that could work with Silicon Graphics workstations. . . . When manipulated in stop-motion style, the encoders would translate the manipulations to three-dimensional wire-frame models that were visible on the computer monitors. The DID is composed of three basic components: the physical armature, a controller box, and the computer software that enable the manipulations to interface with the digital realm. . . . Of the fifty-two CG shots on *Jurassic Park,* fifteen would be animated with the DID.

Vaz, *Industrial Light & Magic* 76.

375. Vaz, *Industrial Light & Magic* 220.
376. Used in combination with the Body Sock software, developed earlier for the liquid metal cyborg of *Terminator 2: Judgment Day*, Enveloping allowed animators to make computer-generated flesh movements with all the realism of actual organic skin moving against muscle and bone. See Vaz, *Industrial Light & Magic* 220–21.
377. Vaz, *Industrial Light & Magic* 237. See also: *Lucas Digital Ltd.* (online); *Lucasfilm* (online).
378. Vaz, *Industrial Light & Magic* 119.
379. Wasko, *Hollywood in the Age* 53.
380. Monaco, *How to Read a Film—The World of Movies* 593.
381. Monaco, *How to Read a Film—The World of Movies* 593.
382. 'A Chronology of Computer History' (online).
383. Pennings, *History of Information Technology* (online).
384. Playstation featured a 34 MHz processor and integrated special audio and graphics functions; graphics resolution was 640 x 480 in 24-bit color.
385. 'AOL-Time-Warner Timeline' (online).
386. 'About RCA' (online).
387. This particular scanner employed the same tri-linear CCD imaging technology as the second-generation CCD scanner introduced in 1989 except that, as Lincoln Hu explains, 'We introduced new electronic, optical, mechanical, and illumination designs to improve scan speed, image quality, and flexibility. We made our components to be easily serviceable and interchangeable between 35mm and 65mm formats. We can switch to a mechanical assembly to accommodate the Imax format in two hours, after calibrating and realigning the system for 70mm.'

Quoted in Vaz, *Industrial Light & Magic* 278.

This new scanner would be used for the first time during the production of Ben Burtt's documentary *Special Effects* in 1996, the first IMAX format (70mm and 15-perforation film frames) motion picture undertaken by Industrial Light & Magic.

388. 'AOL-Time-Warner Timeline' (online).
389. 'Westinghouse Electric History' (online).
390. Monaco, *How to Read a Film—The World of Movies* 594.
391. By the end of the 1990s, DreamWorks SKG would already be on the verge of turning into a large entertainment empire and influential corporate "player" in the Hollywood cinema industry landscape.
392. Using 5.5 million transistors and packaged together with a second speed-enhancing cache memory chip, it was 'designed to fuel 32-bit server and workstation applications, enabling fast computer-aided design, mechanical engineering and scientific computation.'

See 'Intel's History' (online).

393. Polsson, *Chronology of Personal Computers* (online).

394. Saturn featured a 68000 processor for audio, a pair of SH7064 processors for game play like the 32X, an expansion slot for future additions, as well as a SH7034 processor for the CD-ROM drive; graphics resolution was 640 x 224 in 24-bit color. See Polsson, *Chronology of Video Game Systems* (online).

395. Polsson, *Chronology of Video Game Systems* (online).

396. *Lucasfilm* (online).

397. Vaz, *Industrial Light & Magic* 308.

398. Nathans, 'Beyond *Toy Story*' 18.

399. Vaz, *Industrial Light & Magic* 111.
According to *Toy Story*'s official website, the film's technical accomplishments included, among other things:

> 79 minutes and 114,240 frames of computer animation, 800,000 machine hours, 4.5 million lines of code (270Mb), film frames totalling 160 billion pixels, 600 billion bytes or 1200 CD-ROMs full of uncompressed data, 2 Terabytes or 2 trillion bytes of data for the total storage of all film information. Amidst all the hi-tech equipment involved, the "polys" or pencil tests were apparently shot on a Kodak camera built in 1912.

See *Toy Story* (online).

400. 'AOL-Time-Warner Timeline' (online).

401. Monaco, *How to Read a Film—The World of Movies* 595.

402. Monaco, *How to Read a Film—The World of Movies* 595.

403. Monaco, *How to Read a Film—The World of Movies* 595.

404. Monaco, *How to Read a Film—The World of Movies* 594.

405. Monaco, *How to Read a Film—The World of Movies* 595.

406. Polsson, *Chronology of Personal Computers* (online).

407. Polsson, *Chronology of Personal Computers* (online).

408. Monaco, *How to Read a Film—The World of Movies* 597.

409. Because the U.S. congress misunderstood the dynamics of large telecommunication corporations, the Telecommunications Act of 1996 essentially gave them 'a licence to raid and consolidate when it intended, rather, to increase competition.'
Monaco, *How to Read a Film—The World of Movies* 596.

410. 'Computer and television manufacturers settle their battle over the new digital television standard by agreeing to differ: computers will use progressive scanning while TV sets will offer both progressive and interlaced scanning.'
Monaco, *How to Read a Film—The World of Movies* 597.
MSNBC premièred with 22 million subscribers; NBC Asia & Europe and CNBC Asia & Europe were also launched around the same time. See 'General Electric History' (online).

411. 'Westinghouse Electric History' (online).

412. See Krantz, 'Hollywood Gets Wired' 58.

413. This was the third time that Kerkorian would be involved in the purchase/sale of Metro-Goldwyn-Mayer. See Monaco, *How to Read a Film—The World of Movies* 596.

414. According to Intel, 'PC users can now capture, edit and share digital photos with friends and family via the Internet; edit and add text, music or between-scene transitions to home movies; and, with a video phone, send video over standard phone lines and the Internet.'
See 'Intel's History' (online).

415. Monaco, *How to Read a Film—The World of Movies* 597.

416. 'Westinghouse Electric History' (online).

417. Monaco, *How to Read a Film—The World of Movies* 597.

418. It featured 'technical innovations specifically designed for workstations and servers that utilize demanding business applications such as Internet services, corporate data warehousing, digital content creation, and electronic and mechanical design

automation. Systems based on the processor can be configured to scale to four or eight processors and beyond.'
'Intel's History' (online).

419. Monaco, *How to Read a Film—The World of Movies* 599.
420. 'About RCA' (online).
421. Monaco, *How to Read a Film—The World of Movies* 599.
422. Monaco, *How to Read a Film—The World of Movies* 599.
423. Monaco, *How to Read a Film—The World of Movies* 598.
424. See *eMarketer* (online).
425. 'AOL-Time-Warner Timeline' (online).
426. 'AOL-Time-Warner Timeline' (online).
427. Monaco, *How to Read a Film—The World of Movies* 600.
428. Monaco, *How to Read a Film—The World of Movies* 601.
429. The cheaper Celeron processor was designed for the value PC market segment. The Pentium III processor, featuring 70 new instructions—Internet streaming SIMD extensions—dramatically enhanced 'the performance of advanced imaging, 3D, streaming audio, video and speech recognition applications. It was designed to significantly enhance Internet experiences, allowing users to do such things as browse through realistic online museums and stores and download high-quality video. The processor incorporates 9.5 million transistors, and was introduced using 0.25-micron technology.' As for the Pentium III Xeon processor, it provided additional capabilities for e-Commerce applications and advanced business computing. Designed for systems with multi-processor configurations, it incorporated the Pentium III processor's 70 SIMD instructions, which enhanced multi-media and streaming video applications. See 'Intel's History' (online).
430. Monaco, *How to Read a Film—The World of Movies* 600.
431. Texas Instruments's DLP Cinema projector prototype—the first of a variety of innovative exhibition technologies developed for future theater installations—was used for the theatrical digital exhibition of *Toy Story 2* which had six digital "playdates": at El Capitan Theater in Hollywood, at Edwards Irvine Spectrum (Southern California), at AMC Burbank, at AMC Van Ness in San Francisco, at Cinemark Legacy in Plano, Texas, and at AMC Pleasure Island in Orlando, Florida.
432. Released in May 1999, *Star Wars: Episode I—The Phantom Menace* would end up making $500 million by the end of that year at the U.S. box-office alone. In January 2000, Twentieth Century-Fox and George Lucas scheduled the film's re-release in 755 American theaters.
433. See 'Digital Film Debuts' 9.

Appendix B: Top 100 All-Time Domestic Grosses (2012)

Adjusted for ticket price inflation.
[Source: http://boxofficemojo.com/alltime/adjusted.htm]

	Title	*Adjusted Gross*	*Unadjusted Gross*	*U.S. Theatrical Release*
1	*Gone with the Wind*	$1,604,234,300	$198,676,459	**1939**
2	*Star Wars*	$1,414,269,600	$460,998,007	**1977**
3	*The Sound of Music*	$1,130,778,100	$158,671,368	**1965**
4	*E.T.: The Extra-Terrestrial*	$1,126,323,400	$435,110,554	**1982**
5	*The Ten Commandments*	$1,040,140,000	$65,500,000	**1956**
6	*Titanic*	$1,019,066,300	$600,788,188	**1997**
7	*Jaws*	$1,016,945,800	$260,000,000	**1975**
8	*Doctor Zhivago*	$985,635,500	$111,721,910	**1965**
9	*The Exorcist*	$878,157,400	$232,906,145	**1973**
10	*Snow White and the Seven Dwarfs*	$865,460,000	$184,925,486	**1937**
11	*101 Dalmatians*	$793,343,000	$144,880,014	**1961**
12	*The Empire Strikes Back*	$779,554,300	$290,475,067	**1980**
13	*Ben-Hur*	$778,120,000	$74,000,000	**1959**
14	*Avatar*	$772,206,800	$760,507,625	**2009**
15	*Return of the Jedi*	$746,831,700	$309,306,177	**1983**
16	*The Sting*	$707,794,300	$156,000,000	**1973**

17	*The Lion King*	$707,461,100	$422,782,411	**1994**
18	*Raiders of the Lost Ark*	$699,846,300	$242,374,454	**1981**
19	*Jurassic Park*	$684,474,400	$357,067,947	**1993**
20	*The Graduate*	$679,436,900	$104,901,839	**1967**
21	*Star Wars: Episode I - The Phantom Menace*	$673,516,900	$431,088,301	**1999**
22	*Fantasia*	$659,365,200	$76,408,097	**1941**
23	*The Godfather*	$626,645,200	$134,966,411	**1972**
24	*Forrest Gump*	$623,652,200	$329,694,499	**1994**
25	*Mary Poppins*	$620,763,600	$102,272,727	**1964**
26	*Grease*	$611,135,400	$188,755,690	**1978**
27	*Thunderball*	$593,912,000	$63,595,658	**1965**
28	*The Dark Knight*	$589,799,700	$533,345,358	**2008**
29	*The Jungle Book*	$585,018,500	$141,843,612	**1967**
30	*Sleeping Beauty*	$577,047,900	$51,600,000	**1959**
31	*Shrek 2*	$564,144,400	$441,226,247	**2004**
32	*Ghostbusters*	$561,601,100	$238,632,124	**1984**
33	*Butch Cassidy and the Sundance Kid*	$560,229,400	$102,308,889	**1969**
34	*Love Story*	$555,785,300	$106,397,186	**1970**
35	*Spider-Man*	$551,708,900	$403,706,375	**2002**
36	*Independence Day*	$549,994,900	$306,169,268	**1996**
37	*Home Alone*	$537,809,500	$285,761,243	**1990**
38	*Pinocchio*	$535,182,500	$84,254,167	**1940**
39	*Cleopatra (1963)*	$533,436,700	$57,777,778	**1963**
40	*Beverly Hills Cop*	$533,171,100	$234,760,478	**1984**
41	*Goldfinger*	$526,422,000	$51,081,062	**1964**
42	*Airport*	$524,923,600	$100,489,151	**1970**
43	*American Graffiti*	$521,771,400	$115,000,000	**1973**

44	*The Robe*	$519,709,100	$36,000,000	**1953**
45	*Pirates of the Caribbean: Dead Man's Chest*	$513,149,200	$423,315,812	**2006**
46	*Around the World in 80 Days*	$513,046,200	$42,000,000	**1956**
47	*Bambi*	$505,876,800	$102,247,150	**1942**
48	*Blazing Saddles*	$502,026,500	$119,500,000	**1974**
49	*Batman*	$499,859,700	$251,188,924	**1989**
50	*The Bells of St. Mary's*	$498,196,100	$21,333,333	**1945**
51	*The Lord of the Rings: The Return of the King*	$489,419,200	$377,845,905	**2003**
52	*The Towering Inferno*	$487,322,800	$116,000,000	**1974**
53	*Spider-Man 2*	$477,660,500	$373,585,825	**2004**
54	*My Fair Lady*	$476,400,000	$72,000,000	**1964**
55	*The Greatest Show on Earth*	$476,400,000	$36,000,000	**1952**
56	*National Lampoon's Animal House*	$475,528,700	$141,600,000	**1978**
57	*The Passion of the Christ*	$474,056,400	$370,782,930	**2004**
58	*Star Wars: Episode III - Revenge of the Sith*	$471,037,200	$380,270,577	**2005**
59	*Back to the Future*	$468,862,200	$210,609,762	**1985**
60	*The Lord of the Rings: The Two Towers*	$458,338,700	$342,551,365	**2002**
61	*The Sixth Sense*	$457,178,300	$293,506,292	**1999**
62	*Superman*	$455,423,500	$134,218,018	**1978**
63	*Tootsie*	$451,817,300	$177,200,000	**1982**

64	*Smokey and the Bandit*	$451,253,400	$126,737,428	**1977**
65	*Finding Nemo*	$447,319,500	$339,714,978	**2003**
66	*West Side Story*	$444,404,100	$43,656,822	**1961**
67	*Harry Potter and the Sorcerer's Stone*	$443,949,100	$317,575,550	**2001**
68	*Lady and the Tramp*	$442,535,000	$93,602,326	**1955**
69	*Close Encounters of the Third Kind*	$441,269,800	$132,088,635	**1977**
70	*Lawrence of Arabia*	$439,745,400	$44,824,144	**1962**
71	*The Rocky Horror Picture Show*	$437,251,200	$112,892,319	**1975**
72	*Rocky*	$437,017,400	$117,235,147	**1976**
73	*The Best Years of Our Lives*	$436,700,000	$23,650,000	**1946**
74	*The Poseidon Adventure*	$435,921,600	$84,563,118	**1972**
75	*The Lord of the Rings: The Fellowship of the Ring*	$435,106,300	$315,544,750	**2001**
76	*Twister*	$434,223,700	$241,721,524	**1996**
77	*Men in Black*	$433,656,400	$250,690,539	**1997**
78	*The Bridge on the River Kwai*	$431,936,000	$27,200,000	**1957**
79	*Transformers: Revenge of the Fallen*	$427,973,400	$402,111,870	**2009**
80	*It's a Mad, Mad, Mad, Mad World*	$427,770,800	$46,332,858	**1963**
81	*Swiss Family Robinson*	$427,235,500	$40,356,000	**1960**
82	*One Flew Over the Cuckoo's Nest*	$426,261,700	$108,981,275	**1975**

83	*M.A.S.H.*	$426,252,600	$81,600,000	**1970**
84	*Indiana Jones and the Temple of Doom*	$425,050,600	$179,870,271	**1984**
85	*Star Wars: Episode II - Attack of the Clones*	$424,539,700	$310,676,740	**2002**
86	*Mrs. Doubtfire*	$418,314,000	$219,195,243	**1993**
87	*Aladdin*	$416,392,200	$217,350,219	**1992**
88	*Toy Story 3*	$414,482,900	$415,004,880	**2010**
89	*Ghost*	$408,633,900	$217,631,306	**1990**
90	*Duel in the Sun*	$405,102,000	$20,408,163	**1946**
91	*Pirates of the Caribbean: The Curse of the Black Pearl*	$402,152,200	$305,413,918	**2003**
92	*House of Wax*	$401,223,400	$23,750,000	**1953**
93	*Rear Window*	$399,816,700	$36,764,313	**1954**
94	*The Lost World: Jurassic Park*	$396,285,000	$229,086,679	**1997**
95	*Indiana Jones and the Last Crusade*	$392,367,000	$197,171,806	**1989**
96	*Spider-Man 3*	$388,379,400	$336,530,303	**2007**
97	*Terminator 2: Judgment Day*	$386,331,600	$204,843,345	**1991**
98	*Sergeant York*	$382,098,100	$16,361,885	**1941**
99	*How the Grinch Stole Christmas*	$381,968,200	$260,044,825	**2000**
100	*Harry Potter and the Deathly Hallows Part 2*	$381,011,200	$381,011,219	**2011**

Appendix C: A Basic Glossary of Analog and Digital Terminologies and Key Professions in the Hollywood Cinema VFX Industry

Aerial Image Printer

The Aerial Image Printer is a versatile device most commonly used in the production of VFX shots. It can allow several filmed images to be combined at once, using multiple projectors and prismatic beam splitters; in successive passes of the camera, scores of images can be fitted together into matted spaces. If a picture is projected on a white sheet and then the sheet is removed, the "image" is still there in space, as long as the projector remains on—even though it has become invisible to the naked eye. Should someone peer through another lens at the exact spot where the white sheet was, he/she will see the image again, just as clear and well focused as if it were projected on a white sheet. It is at this exact plane, where the image is aerially in focus, that a second film is placed: thus the two images are combined. The optical printer's camera is pointed and focused on the combined images. The camera is then capable of re-photographing the two pieces of film as if they were one. [See also: *Beam Splitter, Matte, Matte Painting, Optical Printer*]

Algorithm

In relation to video compression software, Algorithm refers to formula that is applied to compress or decompress video. Algorithm enables complicated operations to be carried out precisely, such as the digit-by-digit multiplication of large numbers. Algorithms are fundamental to the programming of computerized motion-control systems as well as to the kind of programming involved in digital filmmaking and in the creation and production of CGA and digital VFX. [See also: *CGA, CGI, Computer Graphics, Graphic Design, Motion-Control, Motion-Control Photography, Programming*]

Analog-to-Digital Converter

The Analog-to-Digital Converter is a device that applies digital sampling to transform an analog signal into a digital version of that signal. It is commonly utilized when the film production process combines analog and digital technologies/media. [See also: *Digital-to-Analog Converter*, *Sampling*]

Anamorphic

The term Anamorphic implies unequal scaling in vertical and horizontal dimensions. Anamorphic films were originally introduced to American audiences as Cinemascope, though they are more commonly seen under the Panavision brand name. When an Anamorphic lens is used on a film camera, the developed filmic image appears squeezed in the horizontal axis, making it half as wide as it should be. A circle becomes an upright oval; a square becomes a rectangle with its top and bottom half the size of its height. These distorted images are particularly difficult to deal with in the creation of optical VFX, so that technicians usually use a non-anamorphic camera system (such as VistaVision) when they composite images. If a film is to be released anamorphically, the images are transferred to that particular medium at the very last stage. [See also: *Compositing*, *Optical Compositing*, *Optical Printer*]

Animatics

Animatics constitute a low-grade animated version of VFX shots used to temporarily fill in unfinished scenes in the work print—hence conveying a sense of what the desired outcome would look like before the completion of actual VFX shots in their finalized form.

Animation Camera

The Animation Camera is a type of film camera that is mounted above or beside a table on which animation artwork is photographed. The table centers the artwork, moves it right or left, up or down, illuminates it from above or below (in the case of translucent animation cels). With a built-in computing mechanism, the Animation Camera can do many of these operations automatically: it is controlled by a mechanism that exposes one frame at a time, allowing the animation operator to position the next cel. The camera's movement between exposures, advancing the film, is engineered to provide a very steady image. Animation cameras are supported on vertical or horizontal stands. Modern animation stands closely resemble the motion-control devices used in photographing models and miniatures for VFX shots. In the technique of aerial image animation for instance, a filmed scene is projected onto a field where artwork (animated mattes or painted titles) is placed. Usually the image is projected from below so that the artwork blocks it. The art is painted on clear sheets of plastic so that the filmed scene shows through where there is no paint. [See also: *Aerial Image Printer, Cel, Cel Animation, Matte, Matte Painting, Miniatures, Models, Motion-Control, Motion-Control Photography*]

Animatronics

The term Animatronics was coined at Disney to describe Disneyland robots. The term is commonly used in the domain of VFX production to describe the utilization of electrical or mechanical components and systems involved in the replication and simulation of the movements of live creatures (fictional creations or existing human or animal life forms). Animatronics often involves puppets whose movements are directed by electronic, mechanical or radio-controlled gadgets. [See also: *Puppet*]

Armature

Mostly used during the production of VFX shots, an Armature is a machined-steel skeleton that has ball joints and that is encased in a (sometimes radio-controlled) stop-motion puppet. The Armature allows the puppet to maintain a particular posture while it is photographed, either in motion or as a still. [See also: *Stop-motion Animation, Puppet*]

Beam Splitter

The Beam Splitter is a device containing a partially silvered mirror that reflects some of the light striking it and transmits the rest. Beam Splitters can also be cube-shaped, formed by joining two prisms. Both the silvered and cubic devices are used extensively in the production of VFX shots as they are the means by which an image can be sent off in two directions or several images can be combined simultaneously.

Bicubic Patch

As a component of computer graphics programming, a Bicubic Patch is defined by a set of control points arranged in a 4 x 4 matrix also known as the surface's control hull. The control points approximate the surface in some way. [See also: *Computer Graphics, Programming*]

Blue-screen

Blue-screen commonly refers to a VFX technique that involves filming moving objects or characters in front of a brightly illuminated blue screen (a green screen can also be used). Following the process of optical re-photography, what is obtained is a film of objects or characters positioned against a black background as well as of their silhouettes against a clear-white background. With these two elements, objects or characters can potentially be placed into the background of any scene. The silhouette is used to make a moving black space in the background environment, and the image of the objects or characters is inserted into that black space. After the film is developed, it is re-photographed twice: once with a filter that turns all the blue parts to black and leaves the objects or characters unaltered, and then with a filter that turns all the blue to white and blocks out all other colors, so that objects or characters become silhouettes. The key to the Blue-screen technique is that everything from the foreground to the background has to stay in focus, or mattes cannot be pulled off it: this also means that much more depth of focus is usually required. [See also: *Matte, Matte Painting*]

Blue Spill

While resorting to the blue-screen technique, if light from the blue screen illuminates the object and is seen by the camera, it is called a Blue Spill. If uncorrected during photography, these contaminated areas of Blue Spill may become transparent during the optical compositing process and create an undesirable matte line wherever the blue light has fallen on the object. This condition usually occurs when the subject is glossy or white or placed at an undesirable angle to reflect light, and certain lenses are more susceptible to Blue Spill than others. Special lighting techniques can eliminate or minimize Blue Spill problems. [See also: *Blue-screen, Compositing, Optical Compositing, Optical Printer, Matte, Matte Painting*]

Bump Mapping

In the context of digital VFX production Bump Mapping is a multi-texture technique that, for the purpose of lighting calculations, permits a texture to control which direction a computer-generated surface is facing. Bump Mapping enables a very realistic look to be obtained for complex surfaces (such as tree bark or rough concrete) through the incorporation of accurate lighting as well as coloring details. Bump Mapping is also commonly applied in the production of graphics for video games. [See also: *CGI, Computer Graphics, Graphic Design, Programming*]

CAD

CAD, or Computer-Aided Design, involves the use of a computer in such a way that a designer can see his/her design immediately on a visual display or monitor as well as the consequences of changing it. This might involve showing a perspective view of a complicated 3D object; often the point of view can be moved, giving the impression that the object is being rotated. More sophisticated CAD systems allow for binocular vision, and the designer may even create the impression of walking about inside his/her proposed design. CAD is an important aspect of set design and VFX model making in filmmaking, as well as of engineering practice in diverse areas as architecture, automobile design and electronics. [See also: *Computer Graphics, Graphic Design*]

Cel

An essential component of traditional analog 2D animation, a Cel is essentially a clear acetate or plastic sheet onto which characters and objects are painted or drawn over a background. Whether it has been hand-inked or xerographed, the character's or object's outline is applied to the front of the Cel. The colors are hand-painted directly onto the back of the Cel to get rid of brushstrokes. To reduce undesired glare effects, black paint is sometimes applied to large areas of the Cel's front. [See also: *Cel Animation*]

Cel Animation

When many cels are photographed together in succession under an animation camera, the illusion of movement or animation of characters and objects painted or drawn on each cel is obtained in the completed film. A character or object is drawn in each cel with incremental changes, so that after photography and upon projection, the sequence obtained displays the illusion of motion. [See also: *Animation Camera, Cel*]

CGA

CGA, or Computer-Generated Animation, is usually resorted to in filmmaking for the animation of particularly complex moving objects or characters in the production of extensive VFX sequences. After being supplied with a primary image or model, the computer contributes to its animation by providing a set of tools that will enable it to be moved, displaced, rotated or distorted in a variety of ways. The computer power needed, especially for rendering, tends to be so massive that the kind of machines involved are often used to animate only, and very rarely to create the primary images or visual models that are meant to be animated (which is more a function pertaining to the CGI process). [See also: *CGI, Computer Graphics, Graphic Design, Models, Rendering*]

CGI

An intrinsic aspect of digital VFX production, CGI means Computer Graphic Imaging. The first step in the CGI process is to build a virtual skeleton of the object or character desired. By establishing a number of points and connecting them together, a wire-frame skeleton can be obtained; or it is also possible to assemble virtually together various geometrical shapes and figures—spheres, circles, cubes, rectangles and so on—until the desired skeleton is obtained. Unless meant to be static, the skeleton has to be animated by the intermediary of various softwares according to the characteristics and patterns of motion associated with or meant to be displayed by the desired object or character. Since the animated object or character is ultimately meant to be fitted into the film, not only the motion ascribed but also the size and volume of the skeleton have to be re-adjusted in each frame to synchronize with any movement of the camera in the shot. The skeleton has then to be "dressed up" following the characteristics of the material or substance it is supposed to simulate or resemble—skin, wood, water, metal and so on. This can be carried out by scanning photographs of the desired material or substance so as to store into the computer a databank of textures, colors, tones and so on which can be cut and pasted onto the skeleton; this kind of databank can also be accessed by resorting to available softwares, hence bypassing the necessity to scan photographs. Then the object or character is "polished" in accordance with the intensity and position of light sources in terms of how and from where it is hit by light. These last two steps in the CGI process are quite straightforward in the case of inanimate, lifeless objects or characters. But if motion is involved, "dressing up" and "polishing" have to be re-worked for every single frame because the superficial characteristics of characters and objects as well as the way they are hit by light are inevitably altered by the movement ascribed to them. By coordinating points of movement of an object or character, a computer can be programmed to produce the multiple drawings needed to show motion between those points as well as to adjust and polish the external appearance of the object or character according to the type of motion it has undergone. The CGI process generally requires close collaboration between animators, VFX cinematographers and programmers. [See also: *Computer Graphics, Graphic Design*]

Character Animation

Character Animation involves the process of bringing computer-generated characters virtually to life on screen: more than just imparting the sense of motion to a character, the Character Animation of human or nonhuman entities requires showing its personality as well as specific patterns of movement in particular environments. As such, Character Animation relates to force, resistance and balance: on screen, a believable force should appear to be triggering and stopping a computer-generated character's movement in relation to a believable path of action. Timing and acting are particularly important in the process of Character Animation for the performance of the character onscreen to be believed—in terms how it might think, feel and react to its environment and to others.[1]

Character Animator

Usually part of a VFX production or CGA team, a Character Animator uses both traditional animation skills and computer graphics systems (such as high-end animation software packages like "SoftImage" and "Maya") to bring to life and create motion for computer-modeled characters or objects. A Character Animator executes direction received from the Animation Supervisor or from the VFX Supervisor to express in visual terms believable movement, personality and action as desired for computer-generated characters, creatures and objects. Knowledge of or expertise in computer software alone is not sufficient: the Character Animator typically needs to be knowledgeable in stop-motion or cel animation as well as in principles of traditional animation, in 3D modeling, in compositional design, and in principles of acting and film production. [See also: *CGA, Character Animation, Cel, Computer Graphics, Graphic Design, Models, Stop-motion Animation, VFX Supervisor*]

Chroma Key

Chroma Key is a basic form of video VFX. Most commonly used in television production, it is a process that requires overlaying one video signal over another video signal and replacing the first signal's range of colors with those of the second signal. In common Chroma Key applications, a foreground picture that includes an object or character positioned against one key single-color background is usually photographed first. Another picture is subsequently inserted to replace the first foreground picture's key color.

Colorization

Colorization involves the application of an electronic process to inject artificial colors into a monochromatic image, hence turning a "colorless" (usually black & white) foundation to a wide range of colors. The "Color Systems Technology" electronic process, for example, allows just that.

Compositing

Compositing involves the process of combining two or more separate visual elements (still and moving pictures), through the use of an Aerial Image Printer, for instance, to create a new VFX shot altogether. [See also: *Aerial Image Printer*]

Compositor

Working closely with VFX supervisors, CGI supervisors, Matte Painters and Rotoscope Artists, the Compositor has to seamlessly combine all of a shot's many layers or elements: by compositing the numerous separate stills, live-action shots and computer-generated components that all together will constitute a complete shot. Often the last one working on a shot before it is included in the overall film, the Compositor carries out such tasks as matte extractions, painting, color continuity, tracking, stabilization, and rotoscoping. While needing to be familiar with various computer softwares and platforms, the Compositor typically has good graphic design and artistic skills, graphic artist experience, solid understanding of composition and color, and adequate camera operator skills. [See also: *Compositing, Computer Graphics, Graphic Design, Optical Compositing, Matte, Matte Painting, Rotoscoping, Rotoscope Artist, VFX Supervisor*]

Computer Graphics

William Fetter coined the term Computer Graphics in 1960 as a way of describing novel methods in design that he had developed while working as a Graphics Designer for aviation company Boeing. In very broad terms, Computer Graphics involve visual data manipulation and representation through the application of specialized computer programs. Other than contributing to the production of industrial output in many CAD application areas, Computer Graphics are also used for purely artistic purposes: designs may be recognizable transformations of existing works of art or of photographs, through the use of the Photoshop software for instance. The term CG is commonly used for Computer Graphics in the digital cinematic VFX world to refer to both hardware and software required to make it possible for presenting and manipulating designs via a computer terminal with graphics capability, that is, one on which it is possible to draw lines and curves. [See also: *CAD, Graphic Design, Photoshop*]

Contact Printer

The Contact Printer is a motion picture printing machine that places original processed film in emulsion-to-emulsion contact with unexposed duplicating film. Light passes through the original and exposes the undeveloped film. Contact printers are most commonly used for printing large quantities of release prints for theatrical exhibition, where microscopically precise registration of images is not so important. Most VFX work, however, requires prints in perfect registration, otherwise one part of an image will shift and ride against another.

"Creature" Developer

Usually with experience in, amongst other things, character animation and in 3D modeling of organic models, the "Creature" Developer is part of a VFX production team and executes directions received from VFX Supervisors. The "Creature" Developer ensures that both real and fictional non-human models maintain on the screen an anatomically correct and sculpturally detailed form while they move through animated motions. To achieve this goal, the "Creature" Developer needs to be proficient in: (i) connecting rendered surfaces of models to animation controls, (ii) creating procedurally animated simulations of hair, cloth, muscles and flesh, and (iii) ensuring that models advance in good form through the animation and render pipeline of a VFX shot.[2] [See also: *Character Animation, Character Animator, Models, VFX Supervisor*]

CRT

The CRT, or Cathode Ray Tube, is basically the picture tube of a television set. A TV screen is the front surface of a large picture tube, properly called a Cathode Ray Tube. Streams of electrons from the cathode of the tube bombard the phosphorous screen in front, stimulating it to glow. The electrons are deflected one way and another by magnets, thus forming the lines and thousand of dots that make up a television picture. The CRT is also the means for displaying computer-generated images. When computers are programmed to generate VFX images, these are usually drawn on a Cathode Ray Tube and then photographed for film. Laser beams have replaced cathode rays, giving images with higher resolution and greater color clarity. [See also: *CGI*]

Digital-to-Analog Converter

The Digital-to-Analog Converter is a device that changes the representation of a numerical value from a digital encoding (in which it might be manipulated by a digital computer) to a particular value of some variable physical quantity (usually an electrical potential) to interact with other physical processes—such as regulating automated sections in an industrial plant, or controlling the movement of a computerized stop-motion VFX camera. [See also: *Analog-to-Digital Converter*]

Digital Images

Digital Images are entirely created with the use of a computer and projected on a cathode ray tube for filming, or scanned directly on the film with a laser.

Digital Image Processing

Digital Image Processing is a computer-based process, usually resorted to during film post-production, which allows the virtual manipulation of color, contrast, saturation, sharpness and shape of filmic images. [See also: *Post-production*]

Digital Matte Artist

Working closely with the VFX supervisor and the Director of Photography, the Digital Matte Artist creates digital set extensions and virtual environments for VFX shots by applying digital paint and techniques of photo manipulation in order to create various kinds of backdrops that are meant look realistic and fool the eye. The Digital Matte Artist usually also has the responsibility to design, create and incorporate into matte paintings computer-generated 3D elements and animation, as well as to ensure that planned concept design requirements are fulfilled. The Digital Matte Artist typically has a fine arts background; sound understanding of photography, industrial design or architecture; good knowledge of cameras, lenses, photography and perspective; and the ability to reproduce photo-real imagery with digital tools.[3] [See also: *Matte, Matte Artist, VFX Director of Photography, VFX Supervisor*]

Digital Printer

The Digital Printer can read, encode and store a film shot in digital form so that it can subsequently be manipulated and "printed" onto celluloid film.

DVE

Most commonly used in television and video production, DVE, or Digital Video Effects, are enabled by effects generators that apply digital signal processing to produce 2D and 3D effects.

Effects Animation

Effects Animation is a technique that allows such things as lightning rays, laser swords and flying pixies, amongst a wide range of VFX elements, to be photographed directly from artwork one frame at a time with an animation camera. Most Effects Animation, however, tends to have another function: to camouflage or block out unwanted visual elements that are the by-products of other VFX processes (wires, rigging, blue spill and so on). [See also: *Animation Camera, Blue Spill*]

Film Stock

The motion picture industry usually defines Film Stock in terms of lines per millimeter of film surface. The limiting resolutions of chemical emulsions can be determined when the emulsion is used to photograph a chart upon which standardized blocks or lines of black have been printed. Certain types of Film Stock are more favored than others for the production of VFX shots; for instance, Eastman Kodak developed the 5295 Film Stock to cater for blue-screen shots in particular, although optical processing houses often insist on the low-speed Eastman 5247 Film Stock whenever shots involve optical VFX or even just a few VFX elements. [See also: *Blue-screen*]

Fractal Graphics

An essential aspect of CGI, Fractal Graphics constitute a technique for creating computer images of objects that exist in nature—by turning the objects' natural curves into mathematical formulae which allow the image to be subsequently constructed. Fractal refers to the fact than an object actually looks like a smaller version of itself. For instance, a branch of a tree may look like an entire tree or a section of a mountain may look a lot like a mountain: the "shapeness" of the object is independent of its size. Rivers, coastlines, plants and a number of natural phenomena share this quality. Although nature has a lot of randomness too, the computer is also capable of generating randomness. The combination of these two notions produces a simple method for generating very complex shapes such as mountains and plants with very little information. [See also: *CAD, CGA, CGI, Computer Graphics, Graphic Design*]

Front Projection

Also known as the "Alekan-Gerand Process" and the "Scotchlite Process," the Front Projection process is commonly applied during VFX production, especially when photographing matte paintings. It involves a highly reflective screen being placed behind the subject to be photographed. A projector is positioned next to the camera with its optical axis at 90° to the camera. A beam splitter reflects the projected image to the screen while allowing the camera to see through; thus the camera sees the projected image on the highly reflective screen with the subject in front of it. The screen is so reflective that only a very dim projection light is used;

consequently, the foreground objects do not reflect the projected image. [See also: *Beam Splitter, Matte, Matte Painting*]

Garbage Mattes

There are often large sections of VFX shots in particular that are not used and need to be covered up with Garbage Mattes: these block out "garbage" elements in shots, such as unwanted lights, cables, stands and so on. Garbage Mattes are rapidly made with pieces of black paper placed above a light box and photographed with an animation camera. [See also: *Animation Camera, Matte, Matte Painting*]

Generation

In the context of the analog video- or celluloid-based master print of a movie, Generation refers to each duplicate print successively made that separates a copy from its original master recording. A duplicate made directly from the original master constitutes a second Generation copy, a duplicate of a duplicate of the original master constitutes a third Generation copy, and so on. Each additional Generation obtained implies an increasing loss of details for the image—as illustrated by sync loss and by fluctuations in color range, grain and contrast. Limitations in the video magnetic tape's frequency response and electronic circuitry imperfections, for instance, are common reasons for Generation loss when duplicates are made from master recordings. To minimize the deterioration of picture quality, it is therefore best to limit the number of Generations in making the final print of a movie for the purpose of its exhibition. Generation loss is unimportant for digital video and film formats, however, because each duplicate is an exact copy of the original master.

Glass Matte Painting

As a further development of the glass shot, in Glass Matte Painting a sheet of glass is set up in front of the camera but it is not specially shielded or lit. A matte or mask of opaque black paint is applied to the glass so as to obscure the unwanted areas of the scene in front. Next the scene is filmed with the action taking place in the areas still visible through the parts of the glass that are not blacked out, and further lengths of test footage are exposed in the same way. Back at the studios one of the test sections, but not the main negative, is developed, and then threaded in the gate of a camera which is set up in front of an art board on an easel. Light is shone through the back of one frame of the test film to project an image of the test film onto the white art board. Then the artist is free to slowly build up painted additions to the scene, checking all the while for matching, and he finally blacks out the parts of the board where the filmed parts of the scene fall. The resulting painting is then filmed as a second exposure on the undeveloped negative after a series of test exposures and developments have been made using the other undeveloped test sections. In this way a correctly combined scene can be obtained on one negative after it has been developed.[4] [See also: *Glass Shot, Matte, Matte Painting*]

Glass Shot

A Glass Shot requires a painting to be made on a sheet of glass and then placed between the camera and whatever is meant to be filmed. Almost anything can be added to a scene by means of a Glass Shot—clouds in the sky, a castle on a mountain and so on. The Glass Shot is one of the earliest VFX techniques of the cinema, first used by Norman Dawn in 1907.

Go-motion Animation

As a refinement of the stop-motion animation technique (that can be observed in the work of pioneer VFX practitioners Willis O'Brien and Ray Harryhausen for instance), Go-motion Animation was developed and first used during the production of VFX for *Dragonslayer* in 1980. For the Go-motion Animation technique, blue rods are attached to the major body parts that would move when a puppet is in motion. These rods are then connected to "stepper" motors whose movements are controlled by a computer. As the puppet is animated, the programmer at the computer records the exact position of the rods, creating "programmed" movements for the puppet, based on the animator's design. Its actions can then be replayed and filmed in a more traditional manner while the camera shutter is open so that the camera can expose the movement as it occurred, thereby filming the blur that had been missing in traditional stop-motion cinematography.[5] The stroboscopic jerky movement often associated with stop-motion animated characters is largely eliminated as a result of smooth, almost lifelike movements being obtained with the application of the Go-motion Animation technique. [See also: *Stop-motion Animation, Puppet*]

Graphic Design

Originally referring to design required for media intended for commercial printing purposes—such as newspapers, books and posters—Graphic Design has eventually penetrated practically all domains that involve some form of visual production like gaming, cinema, the Internet, and so on. Other than the creation of images and patterns, Graphic Design also often involves layout, titling, scaling and coloring in such diverse media as video games, films, websites, magazines, and so on. [See also: *Computer Graphics*]

HDTV

HDTV, or High-Definition Television, is a techno-visual standard for producing high resolution video, which involves doubling the common NTSC standard of 525 lines per picture to 1050 lines as well as augmenting the screen aspect ratio from 12:9 to 16:9. The HDTV standard enables a TV screen to be more similarly shaped to a movie theater screen and image quality that approaches 35mm film photography.

Hologram

A Hologram is a photographic record of the difference between a reference beam of light and the pattern of interference created in that beam by an object. It allows for a true stereoscopic representation of the object to be recreated—true, that is, in the sense that the Hologram allows the eye to look round the sides of the object just as it could in reality. This is unlike a 3D system using colored filters and double images to create an illusion of depth where every person sees the same illusion and where head movement will not reveal new facets of the scene.[6]

Image Resolution

Image Resolution refers to how refined or how coarse a digitized image is. Image Resolution is measured in Dots Per Inch or DPI. High resolution ("hi-res") and low resolution ("low-res") respectively describe amelioration and degradation in image quality, caused by increasing and reducing the number of dots per square inch for a particular image.

Interpolation

Interpolation is a process that involves the averaging of pixel information when a digital image needs to be scaled. When an image is increased in size or scaled up, the necessary pixels additionally needed to constitute the larger image are created from averaging the smaller image's pixels. When an image is reduced in size or scaled down, its resulting superfluous pixels are averaged in order for a new single pixel to be created. [See also: *Pixels*]

Layering

In film post-production, Layering involves the processing of the many video layers—stills, live-action shots and computer-generated components—that all together will constitute a complete shot. Each layer of the intended complete shot can be manipulated (re-sizing, cropping, re-positioning, etc.) to expose the next lower layer of the shot's source video. [See also: *Post-production*]

Linear Editing

The Linear Editing of moving pictures involves the utilization of analog media, such as magnetic tape, wherein any recorded audio-visual material can only be accessed in a specific order. From the tape's beginning, the viewing of scene 9 requires going from scene 1 through scene 8, for instance. [See also: *Non-Linear Editing*]

Making Sticky

The Making Sticky software, as used in digital VFX production, allows advanced visual manipulation and involves the technique of grafting the characteristics of one image onto another; it allows scanned-in, two-dimensional background plates to be projected and to stick to three-dimensional computer models as they are animated. [See also: *Models*]

Maquette

A Maquette is a miniature form or model of an object or character that is often used during the design process of VFX shots to display possibilities and during VFX production as a scale model of a final product.

Matte

Most commonly used in the production of VFX shots, a Matte is a solid color signal—adjustable for hue, luminance, and chrominance—which works as a mask that partially covers celluloid film as it travels through a camera. The Matte keeps part of a shot black (the sky, the view through a window or the shape of a person) and whichever part of the shot blacked out by a Matte is usually filled in with another image. Some Mattes are made of black tape or of opaque cardboard, while other Mattes are painted on a glass sheet placed in front of the camera. More sophisticated approaches to Matting are accomplished by using one film to mask another film, as is the case for Traveling Matte processes for example. [See also: *Digital Matte Artist, Matte Artist, Matte Painting, Travelling Matte*]

Matte Box

The Matte Box is a bellows-like device that attaches to the camera and shades the lens. It is also used to hold filters and support objects in VFX work when parts of a scene are being matted out. [See also: *Matte, Matte Artist, Matte Painting*]

Matte Painting

Matte Painting has always been a central technique to the production of VFX shots. It usually entails the making of large realistic paintings (especially of fantastic or spectacular backgrounds) that are shot as separate components and subsequently composited with live-action scenes. [See also: *Compositing, Optical Compositing, Optical Printer, Matte*]

Matte Artist

The Matte Artist is responsible for making the large, realistic matte paintings that will be shot and composited with live-action scenes. [See also: *Compositing, Optical Compositing, Digital Matte Artist, Matte, Matte Painting*]

Matte Photography Camera Operator

The Matte Photography Camera Operator has the responsibility for shooting matte paintings adequately so that they can be properly composited with live-action scenes during post-production. [See also: *Compositing, Optical Compositing, Optical Printer, Matte, Matte Painting, Post-production*]

Miniatures

Miniatures are small-sized objects that are photographed in such a way as to appear larger than they really are in a particular shot. Miniatures of spaceships, buildings, land vehicles, airplanes and so on are made when it is not prudent or possible to film live the required full-sized objects. The art of photographing Miniatures is a highly specialized skill and involves the application of a whole range of lighting and photographic techniques. When skillfully photographed, Miniatures can be very effective used in the production of VFX shots.

Miniature Model-Maker

The Miniature Model-Maker is responsible for the construction of miniature objects and characters that may be inanimate or animated during Miniature Photography as part of VFX shots. [See also: *Miniature Photography*]

Miniature Photography

In VFX production, Miniature Photography is the process of filming inanimate or animated miniature models that are later composited with live-action scenes. [See also: *Compositing, Optical Compositing, Optical Printer, Miniatures, Models*]

Models

Models of animated characters or creatures are handcrafted and filmed as miniatures or life-size, to be later incorporated into the film, against matte paintings or outdoor shots as background. [See also: *Matte, Matte Painting, Miniatures*]

Modeler

Working closely with Art Directors and VFX Supervisors, the Modeler's main job is to convert sculpted maquettes and 2D concept art from VFX designers into highly detailed and topologically accurate 3D wireframe models. After a skeleton is installed and the skin is created for the wireframe model, the Modeler will move on to sculpt specific muscle tension shapes and necessary facial expressions as may be required. [See also: *Maquette, Models, VFX Designers, VFX Supervisor*]

Model Sculptor

The Model Sculptor is responsible for sculpting three-dimensional characters, objects and creatures of all sorts from paper sketches—smaller versions at first during pre-production to facilitate visualization—to full-scale sculptures used during VFX photography. [See also: *Models, Modelers*]

Morphing

As a computer graphics technique, Morphing (short for 'metamorphosis') allows film and computer-generated images to be melded together through the seamless digital compositing of one image being progressively altered to transform into another image. [See also: *CGI, CGA, Compositing, Computer Graphics*]

Motion-Capture

Requiring the use of complex computer softwares, Motion-Capture is a digital VFX technique that involves the capturing/recording of live performances from actors and the application of those recorded live performances to computer-generated digital characters. The animation of a digital character occurs through the mapping of the captured/recorded real-life actions of a human or animal figure onto the digital character. This is generally accomplished by placing motion sensors on various parts of the real-life body of the human or animal figure performing and by capturing/recording, from multiple camera angles, the choreography or movement performed.[7]

Motion-Control

Most commonly used during VFX production, Motion-Control is a process wherein a computerized system is used to move cameras and models together in pre-arranged patterns. Because they are computer-controlled, the model and the camera can repeat their movements over and over again: for visually creating multiple objects from only one model, for adding lights to models, and so on. The computer can direct the camera shutter to remain open while the camera is moving, so that the image blurs slightly, just as actual movie frames of fast-moving objects are blurred, so that the final effect appears as real live action. [See also: *Models*]

Motion-Control Photography

An important aspect of VFX production, Motion-Control Photography involves the combined control of camera movement, the positioning of the model to be photographed and how it is lit, in synchronization with the movement of the filmstrip passing through the camera. For feature films, one second of action on screen can generally be broken into twenty-four single frames running sequentially. If a camera movement is meant to be two seconds long on screen, the camera will need to move for forty-eight frames; through Motion-Control Photography, what is recorded electronically is the changing positions of the motors controlling the movement of everything—cameras and models in the system—on frame one through to frame forty-eight. With a positional map for each frame being established, the camera movement can be carried out as often as required since the camera and models' positions would track according to the map exactly based on specific frame count. [See also: *Models, Motion-Control*]

Multi-Texturing

Multi-Texturing refers to the mapping of more than one texture at a time on a computer-generated shape or polygon in the production of digital VFX.

Non-Linear Editing

The Non-Linear Editing of moving pictures involves the utilization of computer-controlled digital media—characterized by fast retrieval and random access—such as hard disks or CD-ROMs. Hence, any recorded audio-visual material does not have to be accessed in any specific order, contrary to linear editing, For instance, to view scene 7 from the beginning of the tape, one can go directly to that scene without going from scene 1 through scene 6, since Non-Linear Editing allows for quick, random access to any point on the hard disk (commonly within 20–40 milliseconds). [See also: *Linear Editing*]

Optical Compositing

Optical Compositing is the process through which a finished VFX shot is obtained from the combination of two or more images on an optical printer. An optical composite can be as simple as a flash of lightning in the sky or as complicated as fifty individual spaceships in a space battle. The most common application of optical compositing during VFX production is to place actors, photographed separately on a stage, into imaginary environments that merge actual live shots with stationary matte paintings. [See also: *Compositing, Optical Printer, Matte, Matte Painting*]

Optical Printer

The Optical Printer is a camera that re-photographs and combines separate images of motion picture film. Optical Printers consist of a projector and a camera facing each other, with a lens that focuses the projector image on the film of the camera. More advanced Optical Printers are electronically controlled and computerized, allowing strings of complex commands to be carried out fast and with greater accuracy. [See also: *Compositing, Optical Compositing*]

Photomosaics

Photomosaics is a 2D graphic design technique originally developed at MIT and first applied in *The Truman Show* in 1998 for the production of the main poster to promote the film. The technique is fundamentally based on a computer program that "arranges" a large number of different photographs to fit the template of one specific photograph. [See also: *Computer Graphics, Graphic Design*]

Photoshop

Originally developed for the publishing industry to prepare images for printing on four-color offset presses, the Photoshop software is also used during VFX post-production to edit digitized moving pictures. Photoshop edits digitized pictures, allowing computer graphics artists to perform image-processing functions like rotation, re-sizing and color correction. Because scratches, dust, or other problems in the image can also be painted out, the use of Photoshop in film post-production allows flawed shots to be corrected and re-inserted into the film which constitutes a cost- and time-effective advantage over the analog image-processing method of photo-chemical optical printing to fix flawed shots. And the Photoshop software is versatile enough to be used for texture-map editing in computer graphics, to create digital matte paintings, and to design visual concepts in the art department. [See also: *Computer Graphics, Digital Matte Artist, Graphic Design, Matte, Matte Painting, Post-production*]

Pixels

The term Pixel, as an abbreviated form for 'picture element,' refers to the minimum element of raster display that can be represented as a point with a specified intensity or color level. Pixels are the elements of a picture that make up the image on a Cathode Ray Tube, like the "dots" of a TV picture. Television picture resolution is measured by how many Pixels are utilized in the image's formation. For instance, American commercial television is broadcast with 150,000 Pixels per screen image while in most of Europe television is broadcast with 210,000 Pixels per screen (except in France where television resolution is 440,000 Pixels per screen). A computer-generated image's resolution is directly related to the density of the Pixels that constitute it. Computer-generated images of less than 12,000,000 Pixels, however, cannot pass for motion picture-originated shots. [See also: *CGI, CRT, Raster Display*]

Pre-Visualization

Pre-Visualization is carried out during the early stages of a movie's development and usually involves computerized storyboarding and the use of computer-generated virtual sets and digital characters. Back in the early days of cinema, VFX pioneer Georges Méliès worked out his fantastic sets and mechanical VFX through the use of drawings, paintings and detailed schematics before a shot was actually recorded by the camera. Pre-Visualization is particularly crucial in the planning and execution of complex and expensive VFX shots. Since for digital VFX shots, live actors are often likely to be performing opposite computer-generated characters, objects or backgrounds (later to be digitally composited into the movie during post-production), it becomes very important to have a strong Pre-Visualization of scenes and sequences early on during a movie's production. Testing the viability of sets enables the avoidance of potential time- and money-consuming problems such as re-shoots, color corrections, injuries and so on. Scene mock-ups involving digital characters and virtual sets are handy in providing reasonably accurate Pre-Visualizations of scenes and sequences long before shooting starts. [See also: *Post-production*]

Post-production

In the context of filmmaking, the Post-production process refers to all work carried out after raw footage and audio elements have been recorded during a motion picture's shooting or production phase. Editing, VFX insertion, image manipulation and audio mixing are some of the things commonly undertaken during the post-production process.

Process Camera

Most commonly used in the production of VFX shots, the Process Camera is a special type of camera designed to deliver rock-steady images by means of a pilot-pin movement. Shots produced by a Process Camera are usually intended as an element in an optical composite or for the purpose of rear- or front-screen projection during the production of certain types of VFX. [See also: *Compositing, Optical Compositing, Optical Printer*]

Programming

Programming, in the domain of computers, is the process of constructing detailed sets of instructions to make a computer perform specific tasks. This process involves the choice or invention of: a suitable algorithm; decisions about the representation and organization of a particular type of data; actions to be taken if a machine malfunction is detected or the program is presented with inconsistent data. The programmer must also take into account the structure of the computer on which the program is to be run as well as the nature of its operating system and its processing and storage capacity. [See also: *Algorithm*]

Prosthetics

Prosthetics refers to synthetic replacements or additions for body parts through the application, blending and coloring of form-shifting materials—such as rubber, gelatin, foam latex, plastic—on a performer's skin. Skills like sculpting, mold-making, painting, make-up and hair application come into play when prosthetic appliances are developed and applied in the creation of a character.

Puppet

In the filmmaking context, a Puppet usually refers to a replica of a human, animal, or other imaginary character that is animated by the manipulations of a puppeteer. The movement of puppets may be controlled by means of string, rod, hand, cable, joystick, electrical linear actuators, hydraulic cylinder, or any other electronic radio-control devices.

Pyrotechnics

A major component of analog VFX production, Pyrotechnics refers to the art, craft and science of creating and manipulating fireworks.

Quad Printer

Used to process VFX shots in particular, the Quad Printer, first built in 1979 at Industrial & Magic, consists of four projectors and one camera with an anamorphic lens. Two of the projectors are at right angles to the camera and use a cubic beam splitter to reflect the images they project. In 1982, the two right-angle projectors were removed and a camera with a non-anamorphic lens was put in their place. By reversing the beam splitter, the printer has the capability of filming in anamorphic or non-anamorphic film formats. The electronic and optical features of this particular printer have made it the standard for all optical printers subsequently developed. [See also: *Anamorphic, Beam Splitter*]

Raster Display

The common television set is a Raster Display. Raster describes the pattern of horizontal scanning lines that are traced by the electron beam of a video monitor and constitute a video image.

Raster Graphics

Raster Graphics, also known as bit-mapped graphics, involve images that are defined as the pixels in a column-and-row format. [See also: *Computer Graphics, Graphic Design, Pixel, Raster Display*]

Ray Tracing

A computer graphics technique, Ray Tracing involves following every ray of light that falls on a scene, tracing it from its point of origin to some part of the scene, bouncing back and forth within the picture, then finally bouncing back to the image plane and to the viewers. The technique makes possible such effects as transparency, the refraction of images, mirrored surfaces, and optically exact shadows. [See also: *CGA*, *CGI*, *Computer Graphics*, *Graphic Design*]

Real-Time

In the context of computing, Real-Time implies a mode of operation according to which data is received/processed and the outcomes delivered so rapidly as to appear instantaneous. [See also: *Programming*]

Rear-screen Projection

Rear-screen Projection is one of the oldest VFX techniques whereby background scenes are projected onto a screen from behind, while actors perform in front of the screen for the camera. Miniatures may also be shot in front of a rear screen, though the most common use has been to supply the moving images seen outside car windows on screen, for instance. [See also: *Miniatures*]

Rendering

One of the most expensive and time-consuming steps in computer graphics and digital VFX production, Rendering is the process of converting 3D environments and models into a frame-buffer for display on a color raster monitor; high-quality graphics and digital VFX usually require an immense amount of computer power. [See also: *Computer Graphics*, *Graphic Design*, *Models*]

R&D

In the filmmaking context, R&D or "Research & Development" implies Research through the application of scientific knowledge toward the discovery or modification of inventions followed by Development whereby scientific discoveries and inventions are transformed into "new" techniques or products for commercial gain.

Rotoscoping

Rotoscoping is a process for the frame-by-frame re-touching of images in live-action film sequences that essentially involves the blowing up and tracing of individual frames, one at a time, onto cels. With live action being turned into animation (when the cels are re-photographed), VFX elements can subsequently be added to live action—such as inexistent objects being inserted into the hands of real people, and so on. [See also: *Cel*]

Rotoscope Artist

Working very closely with compositors, Rotoscope Artists handle the rotoscoping process by manipulating, inserting and removing isolated elements in composited sequences through the use of both hand-painting and procedural methods. Rotoscope Artists also perform plate clean up as well as additional tasks like wire-removal and blue-screen extraction. Rotoscope Artists usually possess a sound knowledge of compositing and paint animation techniques, and they typically have a background in graphic design, drawing and animation. [See also: *Compositing, Compositor, Graphic Design, Optical Compositing, Optical Printer, Matte, Matte Painting, Rotoscoping*]

Sampling

Sampling is the first step that takes place when an analog signal is to be converted to digital. The analog signal is measured at regular intervals—called samples—and the values obtained are subsequently encoded to produce a digital representation of the analog signal concerned.

Scanner

A scanner is device used to "read" images and convert the visual data obtained into a digital form that can be stored and manipulated by a computer.

Servo

A Servo is a rotary actuator made up of a circuit board, a potentiometer, a motor and a gear train: it is the "muscle" that powers motion-control photography and many animatronics devices. An input apparatus transmits an electronic position signal to the Servo which subsequently "sends" the output arm in the directed position, and whatever is connected to the output arm (a pulley, lever or cable) will then also move. [See also: *Animatronics, Motion-Control, Motion-Control Photography*]

Slit-scan Photography

After being originally used in static photography to deform or blur images, Slit-scan Photography was eventually perfected to enable the creation of psychedelic colors and spectacular animations in VFX shots. The fundamental principle of the mechanical process of Slit-scan Photography has to do with the camera's movement in relation to a light source and in combination with long exposure time. The process of Slit-scan Photography involves the insertion of a moveable slide—into which a slit has been cut—between the subject to be photographed and the camera.

First, an abstract colored design is painted on a transparent support that is subsequently laid down on the glass of a backlighting table and covered with opaque masking into which one or multiple slits have been cut. Positioned on top of a vertical ramp and de-centered in relation to the light slits, the camera takes a single photograph while going down the ramp. At the top of the ramp when it is far away, the camera records a precise image of the light slit, and this image becomes progressively bigger and eventually shifts itself out of frame: what is obtained is a light trail that joins with the screen's edge. The above steps are repeated for each image while the opaque masking is lightly peeled back: variations in colors and in the light stream position are hence produced, thus creating an animation. Since the kind of animation produced by means of Slit-scan Photography can only be carried out image by image, the process is very expensive and time-consuming: a minimum of 240 adjustments is required to produce a 10-second animation sequence, for example.[8]

Spline Algorithm

Commonly applied in digital animation and VFX production, Spline Algorithm is a mathematical procedure for tracing, for each frame of a moving picture, the paths or positions of moving objects from point to point and at key points in space and time. [See also: *Algorithm*]

Split-screen

Split-screen is an analog VFX technique whereby two or more images, going on at the same time on screen, are usually (but not always) separated by a clearly visible line. Thus, half the frame might be matted off and the remaining half shot, then the film rewound in the camera, the opposite side matted off, and a second scene filmed on the other half of the frame. On a more sophisticated level, multiple split screens are a common part of the VFX technicians' repertoire of techniques for seamlessly combining different images. [See also: *Matte, Matte Painting*]

Stop-motion Animation

Sometimes called "stop action," Stop-motion Animation is one of the oldest techniques of analog VFX. It is used for filming an inanimate object, one frame at a time, whereby it is gradually moved between exposures. Stop-motion Animation enables stationary miniatures, models, puppets and objects to be "brought to life" on screen. [See also: *Miniatures, Models, Puppet*]

Synthespians

A Synthespian is an animated figure that is created from the real-life actions of a human or animal figure being mapped onto a computer-generated character following the process of motion-capture. The motion-capture data obtained is subsequently applied to the computer-generated character. Although a good starting point is usually a couple of powerful computers endowed with 3D graphics cards and possessing memory capacities large enough for the intake and manipulation of massive amounts of data as well as for the rendering of desired models, the kind of computer graphics–related software needed for the job ultimately depend on the complexity of the Synthespian wanted. These softwares are often available commercially, but if the job is groundbreaking or exceptionally big and complicated, specific programs and softwares sometimes have to be specially written to enable the creation of the Synthespian in question. [See also: *Computer Graphics, Graphic Design, Models, Motion-Capture, Rendering*]

Technical Director

The Technical Director collaborates with VFX designers and CGI supervisors to produce the desired on-screen look for computer-generated scenes, characters and objects. To bring together the multiple elements that constitute a complete VFX shot, the Technical Director works closely with other VFX personnel such as Character Animators, Rotoscope Artists and Compositors. Possessing reasonably strong computer programming abilities, visual arts skills, adequate knowledge of graphic design and of filmmaking techniques, the Technical Director is usually responsible for lighting, shading, rendering, some compositing, and for creating the motion dynamics and look of VFX simulations of water, smoke, fire and hair, for instance.[9] [See also: *CGI, Character Animator, Compositing, Computer Graphics, Graphic Design, Programming, Rendering, Rotoscoping, Rotoscope Artist, VFX Designer, VFX Supervisor*]

Texture Mapping

In the context of digital VFX production, Texture Mapping describes the way surface texture, color or detail is added to a graphic or to a 3D model to enhance the illusion of reality of computer-generated images. The application of a texture map to a shape's surface is a pivotal process in the production of computer-generated images for VFX shots. Because it is often difficult to make accurately a computer-generated model of an object that is very complex, the solution is often to apply/"map" the characteristics of a photograph or painting of the real object onto the computer-generated surface of the object in question. [See also: *CGI, Computer Graphics, Models*]

Traveling Matte

A technique used in optical printing, Traveling Matte involves moving photographic silhouettes, either obscuring or revealing images during re-photography. Into the black shapes left by these photographic silhouettes, other images are placed—people, vehicles, anything that needs to be seen moving. There are many systems—such as rotoscoping and front-back alternate lighting of a model—that can be used for creating perfectly registered Traveling Mattes, but the one most commonly applied throughout the VFX industry is usually the blue-screen process. [See also: *Blue-screen, Compositing, Matte, Matte Painting, Optical Compositing, Optical Printer, Rotoscoping*]

VFX Designer

On the basis of script specifications and in relation to time and money constraints, the VFX Designer devises the possibilities of how a VFX sequence should be filmed—in terms of camera positioning, camera movement, the choice, look and placement of objects, creatures and characters, and so on—before the sequence is actually realized in production and post-production. CGI-related technologies have contributed enormously to speeding up the designing of VFX sequences during the pre-production stage of filmmaking. [See also: *CGA, CGI, Computer Graphics, Graphic Design, Post-production*]

VFX Director of Photography

The production of VFX sequences is the result of close collaboration and understanding between the VFX production unit and the cinematography crew. The work approach and shooting style adopted by the motion picture's director of cinematography can either make things very simple or very difficult for the VFX photography crew and blue-screen crew, for instance. The blending of miniature, live action and matte painting VFX shots with other shots in the overall motion picture requires a lot of balancing to determine that contrast, color, composition and so on in VFX shots will later match the rest of the film. The VFX Director of Photography hence has the responsibility of shooting as well as supervising the photography of miniatures, models and matte paintings, in such a way that these VFX shots will later blend inconspicuously into the director of cinematography's live-action shots. [See also: *Blue-screen, Matte, Matte Painting, Miniatures, Models*]

VFX Editor

The VFX Editor is responsible for creating and editing 3D animations, graphics, and other elements for VFX shots, often working with CG artists and under the overall supervision of a VFX Supervisor. [See also: *Computer Graphics, Graphic Design, VFX Supervisor*]

VFX Supervisor

The VFX Supervisor oversees and manages all the work carried out by the different VFX units and crews assembled to produce VFX shots for a specific film project.

NOTES

1. See *LucasArts* and *Lucas Digital* (online).
2. See *LucasArts* and *Lucas Digital* (online).
3. See *LucasArts* and *Lucas Digital* (online).
4. See Salt, *Film Style* 159–60.
5. See Smith, *Industrial Light & Magic* 93.
6. See Winston, *Misunderstanding Media* 101.
7. See Ohanian & Phillips, *Digital Filmmaking* 68.
8. See Smith, *Industrial Light & Magic*.
9. See *LucasArts* and *Lucas Digital* (online).

Works Cited

'A Chronology of Computer History.' *HCS Virtual Computer History Museum*. http://www.cyberstreet.com/hcs/museum/chron.htm.
Abbott, Lenwood. *Special Effects: Wire, Tape and Rubber Band Style*. Hollywood: ASC P. 1984.
'About RCA.' *RCA.com*. Radio Corporation of America/Thomson Multimedia Inc. http://www.rca.com/.
'Academy Awards Database.' Academy of Motion Picture Arts and Sciences (AMPAS). http://www.oscars.org.
Ackermann, Robert. *Data, Instruments, Theory*. Princeton: Princeton UP. 1985.
Adams, Walter. *The Structure of American Industry*. New York: Macmillan. 1977.
'Advertisement for *Cinefex* subscription.' *American Cinematographer*, July 1990, 71:7. 25.
Agel, Jerome. *The Making of Kubrick's 2001*. New York: 1970.
Ain't-It-Cool-News. http://www.aint-it-cool-news.com/.
Allen, Jeanne. 'The Industrial Context of Film Technology: Standardisation and Patents,' in *The Cinematic Apparatus*, eds. Teresa de Laurentis and Stephen Heath. London: Macmillan. 1985. 28–39.
Allen, Michael. 'From *Bwana Devil* to *Batman Forever*: Technology in contemporary Hollywood cinema,' in *Contemporary Hollywood Cinema*, eds. Steve Neale and Murray Smith. London: Routledge. 1998. 109–30.
Allen, Robert, and Douglas Gomery. *Film History: Theory and Practice*. New York: Knopf. 1985.
Allen, Robert. 'William Fox Presents *Sunrise*,' in *The Studio System*, ed. Janet Staiger. New Brunswick: Rutgers UP. 1995. 127–39.
Alloway, Lawrence. 'On the Iconography of the Movies.' *Movie*, 7, 1963. 4–6.
Amos, Lindsay. 'Reality Bites—Dean Cundey.' *Cinema Papers*, 114, February 1997. 27–30.
Andrew, Dudley. *Concepts in Film Theory*. New York: Oxford UP. 1984.
Anzovin, Steve. 'Disney Animation Studio (evaluation).' *Compute*, 14:7, August 1992. 78–80.
'AOL-Time-Warner Timeline.' *AOL-Time-Warner Inc*. http://www.aoltimewarner.com/about/timeline.html
Apodaca, Anthony and Larry Gritz. *Advanced Renderman: Creating CGI for Motion Pictures*. San Francisco: Morgan Kaufmann. 1999.
Archer, Steve. *Willis O'Brien: Special Effects Genius*. Jefferson: McFarland. 1993.
Arnheim, Rudolf. *Film als Kunst*. Berlin: Ernst Rowohlt Verlag. 1932.
Arnheim, Rudolf. *Film as Art*. Berkeley: U of California P. 1957.
Arnold, Alan. *Once Upon a Galaxy: A Journal of the Making of The Empire Strikes Back*. London: Sphere Books. 1980.
Aumont, Jacques. *Aesthetics of Film*, trans. Richard Neupert. Austin: U of Texas P. 1992.
Baker, Kevin. 'Computer Technology and Special Effects in Contemporary Cinema,' in *Future Visions: New Technologies of the Screen*, eds. Philip Hayward and Tana Wollen. London: BFI. 1993. 31–45.
Balàzs, Béla. *Theory of Film: Character and Growth of a New Art* (1952), trans. Edith Bone. London: Dover. 1970.
Balio, Tino. '"A major presence in all of the world's important markets": the globalization of Hollywood in the 1990s,' in *Contemporary Hollywood Cinema*, eds. Steve Neale and Murray Smith. London: Routledge. 1998. 58–73.

Balio, Tino. 'Hollywood Production Trends in the Era of Globalisation, 1990–99.' In *Genre and Contemporary Hollywood*, ed. Steve Neale. London: BFI. 2002. 165–84.
Balio, Tino. 'Introduction to Part II,' in *Hollywood in the Age of Television*, ed. Tino Balio. Boston: Unwin Hyman. 1990. 259–96.
Balio, Tino. 'Introduction,' in *Grand Design: Hollywood as a Modern Business Enterprise 1930–1939*, ed. Tino Balio. New York: Scribner's. 1993. 1–12.
Balio, Tino. 'Surviving the Great Depression,' in *Grand Design: Hollywood as a Modern Business Enterprise 1930–1939*, ed. Tino Balio. New York: Scribner's. 1993. 13–36.
Balio, Tino. *United Artists: the Company that Changed the Film Industry*. London: U of Wisconsin P. 1987.
Ballard, Edward. *Man and Technology*. Pittsburgh: Duquesne UP. 1978.
Barbaroux, Évangéline. 'Mondialisation: l'état du cinéma vu par 50 cinéastes de la planète.' *Cahiers du Cinéma*, 557, May 2001. 47–75.
Barr, Charles. 'CinemaScope: Before and After,' in *Film Theory and Criticism*, eds. Gerald Mast and Marshall Cohen. New York: Oxford UP. 1974. 140–68.
Barrett, Neil. *The State of Cybernation: Cultural, Political and Economic Implications of the Internet*. London: Kogan Page. 1996.
Barrett, William. *The Illusion of Technique*. New York: Anchor P. 1978.
Barthes, Roland. 'Upon Leaving the Movie Theatre,' in *Apparatus: The Cinematographic Apparatus, Selected Writings*, ed. Theresa Hak Kyung Cha. New York: Tanam.1980. 1–4.
Baudrillard, Jean. *Seduction*, trans. Brian Singer. London: Macmillan. 1990.
Baudrillard, Jean. *Simulations*, trans. Paul Foss, Paul Patton and Philip Beitchman. New York: Semiotext(e). 1983.
Baudry, Jean-Louis. 'Ideological Effects of the Basic Cinematographic Apparatus,' trans. Allan Williams. *Film Quarterly*, 28:2, winter 1974-75. 39–47.
Baudry, Jean-Louis. 'The Apparatus: Metapsychological Approaches to the Impression of Reality in Cinema,' trans. Jean Andrews and Bernard Augst. *Camera Obscura*, 1, fall 1976. 104–28.
Baxter, John. *George Lucas*. London: HarperCollins. 1999.
Baxter, John. *Steven Spielberg – the Unauthorised Biography*. London: Harper. 1996.
Bazin, André. *What is Cinema? (vol.I)*, trans. Hugh Gray. Berkeley: U of California P. 1967.
Bazin, André. *What is Cinema? (vol.II)*, trans. Hugh Gray. Berkeley: U of California P. 1971.
Beacham, Frank. 'Movies for the Future: Story-Telling with Computers.'*American Cinematographer*, 6:4, April 1995. 36–48.
Beckett, John. 'The Case for the Multi-Market Company' – lecture presented at the Financial Analysts Seminar, University of Chicago, Rockford, Illinois, 22nd August 1969.
Belton, John. 'CinemaScope: the Economics of Technology.' *Velvet Light Trap*, 21, summer 1985.
Benjamin, Walter. 'The Work of Art in the Age of Mechanical Reproduction,' in *Illuminations*, trans. Harry Zohn, ed. Hannah Arendt. London: Jonathan Cape. 1970. 219–53.
Berardinelli, James. '*Close Encounters of the Third Kind*.' *Reelviews Online*. http://www.reelviews.net/movies/c/ce3k.html
Berardinelli, James. '*Independence Day*.' *Reelviews Online*. http://www.reelviews.net/php_review_template.php?identifier=132
Berardinelli, James. '*Stargate*.' *Reelviews Online*. http://www.reelviews.net/php_review_template.php?identifier=93
Berardinelli, James. '*Starship Troopers*.' *Reelviews Online*. http://www.reelviews.net/movies/s/starship.html
Berardinelli, James. '*Star Wars: Episode I – The Phantom Menace*.' *Reelviews Online*.http://www.reelviews.net/php_review_template.php?identifier=171

Binkley, Timothy. 'Reconfiguring Culture,' in *Future Visions: New Technologies of the Screen*, eds. Philip Hayward and Tana Wollen. London: BFI. 1993. 92–122.
Biskind, Peter. 'Blockbuster: The Last Crusade,' in *Seeing Through Movies*, ed. Mark Miller. New York: Pantheon. 1990. 112–49.
Bizony, Piers. *2001: Filming the Future*. London: Aurum. 1994.
Blake, Edith. *On Location on Martha's Vineyard: The Making of the Movie Jaws*. New Orleans: Lower Cape. 1975.
Blanchard, Walter. 'Unseen Camera Aces II: Linwood Dunn, ASC.' *American Cinematographer*, 24:7, July 1943. 268.
Blandford, Steve, Barry Grant and Jim Hillier. *The Film Studies Dictionary*. London: Arnold. 2001.
Bohn, Thomas and Richard Stromgren. *Light and Shadows*. Sherman Oaks: Alfred. 1978.
Bordwell, David and Kristin Thompson. *Film Art: an Introduction* (fourth edition). New York: McGraw-Hill. 1993.
Bordwell, David, Janet Staiger and Kristin Thompson. *The Classical Hollywood Cinema: Film Style and Mode of Production to 1960*. New York: Columbia UP. 1985.
Borgmann, Albert. *Technology and the Character of Contemporary Life*. Chicago: U Chicago P. 1984.
Bouquet, Stéphane. 'Est-ce encore du cinéma?' *Cahiers du Cinéma* (hors-série), April 2000. 20–21.
Brakhage, Stan. 'Metaphors on Vision.' *Film Culture*, 30, fall 1963. 30–35.
Brand, Stewart. *The Media Lab: Inventing the Future at MIT*. New York: Penguin. 1988.
Branigan, Edward. 'Color and Cinema: Problems in the Writing of History,' in *The Hollywood Film Industry – a Reader*, ed. Paul Kerr. New York: Routledge.1986. 121–47.
Braudy, Leo and Marshall Cohen. 'Preface,' in *Film Theory and Criticism: Introductory Readings*, eds. Leo Braudy and Marshall Cohen. New York: Oxford UP. 1999. xv–xix.
Brewster, Bill. 'Film,' in Exploring Reality, eds. Michael Leahy, Dan Cohn-Sherbok and Michael Irwin. London: Allen & Unwin. 1987. 145–67.
Brooker,Will. *Star Wars*. London: Palgrave Macmillan/BFI. 2009.
Brookey, Robert. *Hollywood Gamers: Digital Convergence in the Film and Video Game Industries*. Bloomington: Indiana UP. 2010.
Brown, William. 'Man Without a Movie Camera – Movies Without Men: Towards a Posthumanist Cinema?,' in *Film Theory and Contemporary Hollywood Movies*, ed. Warren Buckland. New York: Routledge. 2009. 66–85.
Brown, Douglas and Tanya Krzywinska. 'Movie-Games and Game-Movies: Towards an Aesthetics of Transmediality,' in *Film Theory and Contemporary Hollywood Movies*, ed. Warren Buckland. New York: Routledge. 2009. 85–102.
Bruck, Connie. *Master of the Game: How Steve Ross Rode the Light Fantastic From Undertaker to Creator of the Largest Media Conglomerate in the World*. New York: Simon & Schuster. 1994.
Buckland, Warren. 'A close encounter with *Raiders of the Lost Ark*: Notes on narrative aspects of the New Hollywood blockbuster,' in *Contemporary Hollywood Cinema*, eds. Steve Neale and Murray Smith. London: Routledge. 1998. 166–77.
Bukatman, Scott. *Blade Runner*. London: BFI Publishing. 1997.
Bulleid, H.A.V. *Special Effects Cinematography*. London: Fountain. 1954.
Burton-Carvajal, Julianne. '"Surprise Package": Looking Southward with Disney,' in *Disney Discourse: Producing the Magic Kingdom*, ed. Eric Smoodin. London: Routledge. 1994. 131–47.
Buscombe, Edward. 'Notes on Columbia Pictures Corporation 1926–41' (1975), in *The Studio System*, ed. Janet Staiger. New Brunswick: Rutgers UP. 1995. 17–36.
Butler, David. *Fantasy Cinema: Impossible Worlds On Screen*. London: Wallflower. 2009.
Canudo, Ricciotto. *L'Usine Aux Images*, ed. Jean-Paul Morel. Paris: Séguier/Arte. 1995.
Carroll, John. *Toward a Structural Psychology of Cinema*. New York: Mouton. 1980.
Castells, Manuel. *End of Millenium*. Malden: Blackwell. 1998.

Champlin, Charles. *George Lucas – the Creative Impulse*. New York: Abrams. 1992.
Chanan, Michael. 'The Treats of Trickery,' in *Cinema: the Beginnings and the Future*, ed. Christopher Williams. London: U of Westminster P. 1996. 117–22.
Cheatwood, Derral. 'The Tarzan Films: An Analysis of Determinants of Maintenance and Change in Conventions,' in *The Studio System*, ed. Janet Staiger. New Brunswick: Rutgers UP. 1995. 163–83.
Chisholm, Brad. 'Widescreen Technologies.' *Velvet Light Trap*, 21, summer 1985. 67–74.
Christopherson, Susan. 'Behind the Scenes: How Transnational Firms are Constructing a New International Division of Labor in Media Work' (2006), in *The Contemporary Hollywood Reader*, ed. Toby Miller. New York: Routledge. 2009. 185–204.
'Chronology of Scientific Developments,' in *The Oxford Encyclopaedic English Dictionary*, eds. Joyce Hawkins and Robert Allen. Oxford: Clarendon. 1991. 1728–1729.
Cinephilia.com Blog. http://www.cinephilia.com/.
Clark, Frank. *Special Effects in Motion Pictures: Some Methods for Producing Mechanical Special Effects*. New York: SMPTE. 1966.
Clise, Rick. *Special Effects*. Melbourne: Viking. 1986.
Conant, Michael. *Antitrust in the Motion Picture Industry: Economic and Social Analysis*. Berkeley: U of California P. 1960.
Conn, Robert. '*E.T. The Extra-Terrestrial* – review.' *Cinema Papers*, 41, December 1982. 564–65.
Corliss, Richard and Cathy Booth. 'Cinema: Ready, Set, Glow!' *Time*, 26 April 1999. http://www.time.com/time/magazine/article/0,9171,990819-3,00.html.
Cook, David. *A History of Narrative Film*. New York: Norton. 1981.
Coyle, Rebecca. 'The Genesis of Virtual Reality,' in *Future Visions: New Technologies of the Screen*, eds. Philip Hayward and Tana Wollen. London: BFI. 1993. 148–65.
Craig, Robert. 'Establishing New Boundaries for Special Effects: Robert Zemeckis's *Contact* and Computer-Generated Imagery.' *Journal of Popular Film & Television*, 28:4, winter 2001. 158–65.
Cross, Robin. *Movie Magic*. Hempstead: Schuster Young. 1994.
Cross, Robin. *The Big Book of B Movies, or How Low was My Budget*. New York: St. Martin's Press. 1981.
Cubitt, Sean. 'Introduction: *le réel c'est l'impossible* – the Sublime Time of Special Effects.' *Screen*, 40:2, 1996. 123–30.
Cubitt, Sean. 'Phalke, Méliès, and Special Effects Today.' *Wide Angle*, 21:1, January 1999. 114–30.
Cubitt, Sean. *Digital Aesthetics*. London: Thousand Oaks. 1998.
Cubitt, Sean. *The Cinema Effect*. Cambridge: MIT P. 2004.
Culhane, John. *Special Effects in the Movies: How They Do It*. New York: Ballantine. 1981.
Curtin, Michael. 'On Edge: Culture Industries in the Neo-Network Era,' in *Making and Selling Culture*, ed. Richard Ohmann. Hanover: Wesleyan UP. 1996.
Dahan, Yannick. '*Dark City*: l'écran noir de nos rêves.' *Positif*, 449–50, July–August 1998. 144.
Dahan, Yannick. '*Matrix*: le règne de l'hyperréalité.' *Positif*, 461-62, July–August 1999. 138–39.
Darley, Andy. 'From Abstraction to Simulation: Notes on the History of Computer Imaging,'in *Culture, Technology and Creativity in the Late Twentieth Century*, ed. Philip Hayward. London: John Libbey. 1990. 39–64.
Daniels, Bill, David Leedy and Steven Sills. *Movie Money: Understanding Hollywood's (Creative) Accounting Practices*. Beverly Hills: Silman-James P. 1998.
de Baecque, Antoine. 'Canudo, premier cinématophile.' *Cahiers du Cinéma*, no. 498, January 1996. 52–53.
de Sola Pool, Ithiel. *Technologies of Freedom*. Cambridge: Belknap. 1983.
de Solla Price, Derek. *Big Science, Little Science*. New York: Columbia UP. 1963.
de Vany, Arthur. *Hollywood Economics: How Uncertainty Shapes the Film Industry*. London: Routledge. 2004.
'Digital Film Debuts.' *Encore*, 2 July 1999, 17:2. 9.

DreamWorks SKG. http://www.dreamworks.com/.
Dummer, Geoffrey. *Electronic Inventions and Discoveries*. New York: Pergamon. 1978.
Dunlop, Renee, Paul Malcolm and Eric Roth. *The State of Visual Effects in the Entertainment Industry*. Visual Effects Society. http://www.visualeffectssociety.com/documents/VES_StateofVFX_3.pdf.
Earnest, Olden. '*Star Wars*: a Case Study of Motion Picture Marketing.' *Current Research in Film*, 1, 1985. 1–18.
Ebert, Roger. '*Armageddon*.' July 1998. *Chicago Sun-Times Online*. http://rogerebert.suntimes.com/apps/pbcs.dll/article?AID=/19980701/REVIEWS/807010301/1023
Ebert, Roger. '*Final Fantasy: The Spirits Within*.' July 2001. *Chicago Sun-Times Online*. http://rogerebert.suntimes.com/apps/pbcs.dll/article?AID=/20010711/REVIEWS/107110301/1023
Ebert, Roger. '*Independence Day*.' July 1996. *Chicago Sun-Times Online*. http://rogerebert.suntimes.com/apps/pbcs.dll/article?AID=/19960702/REVIEWS/607020301/1023
Ebert, Roger. '*Starship Troopers*.' November 1997. *Chicago Sun-Times Online*. http://rogerebert.suntimes.com/apps/pbcs.dll/article?AID=/19971107/REVIEWS/711070305/1023
Ebert, Roger. '*The Hindenburg*.' January 1975. *Chicago Sun-Times Online*. http://rogerebert.suntimes.com/apps/pbcs.dll/article?AID=/19750101/REVIEWS/501010330/1023
Ebert, Roger. '*The Towering Inferno*.' January 1974. *Chicago Sun-Times Online*. http://rogerebert.suntimes.com/apps/pbcs.dll/article?AID=/19740101/REVIEWS/401010322/1023
'Editorial Note.' *American Cinematographer*, November 1992, 73:11. 6.
Eisenstein, Sergei. *Selected Works*, trans. and ed. Richard Taylor. London: BFI. 1988.
Elliott, Stuart. 'The Media Business: Advertising – The Hype Is With Us; The Lucas Empire Is Invading: Resistance Is Futile.' *The New York Times*, 14 May 1999. http://www.nytimes.com/1999/05/14/business/media-business-advertising-hype-with-us-lucas-empire-invading-resistance-futile.html?pagewanted=2
Ellul, Jacques. *The Technological Society*, trans. John Wilkinson. New York: Knopf. 1964.
Elsaesser, Thomas. 'Film Studies in Search of the Object.' *Film Criticism*, 17:2-3. 40–47.
Elsaesser, Thomas. 'Spectacularity and engulfment: Francis Ford Coppola and *Bram Stoker's Dracula*,' in *Contemporary Hollywood Cinema*, eds. Steve Neale and Murray Smith. London: Routledge. 1998. 191–208.
Elsaesser, Thomas. 'Cinéphilia, or The Uses of Disenchantment,' in *Cinéphilia: Movies, Love And Memory*, eds. Marijke De Valck & Malte Hagener. Amsterdam: Amsterdam UP. 2005. 27–44.
eMarketer. http://www.emarketer.com/estats/sell_eglob.html.
Epstein, Edward J. *The Hollywood Economist: The Hidden Financial Reality Behind the Movies*. Brooklyn: Melvillehouse. 2010.
Essex, Andrew. 'MatrixMania.' *Entertainment Weekly*, no. 485, 14 May 1999. 41.
Everette, Denis. *Reshaping the Media: Mass Communication in an Information Age*. Newbury Park: Sage. 1989.
Feibleman, James. *Technology and Reality*. Boston: Nijhoff. 1982.
Federal Communications Commission (FCC). U.S Congress. http://www.fcc.gov/.
Federal Trade Commission Staff Report. *Economic Report on Corporate Mergers*. Washington: U.S. Government Printing Office. 1969.
Fielding, Raymond. *The Technique of Special Effects Cinematography*. New York: Focal. 1972.
Finch, Christopher. *Special Effects: Creating Movie Magic*. New York: Abbeville. 1984.Fisher, Bob. 'Dawning of the Digital Age.' *American Cinematographer*, 73:4, April 1992. 70–86.
Flanagan, Mary. 'Mobile Identities, Digital Stars, and Post-Cinematic Selves.' *Wide Angle*, 21:1, January 1999. 76–93.
Fry, Ron and Pamela Fourzon. *The Saga of Special Effects*. Englewood Cliffs: Prentice-Hall. 1977.

Fulton, A.R. 'The Machine,' in *The American Film Industry*, ed. Tino Balio. Madison: U of Wisconsin P. 1976. 19–32.

Galison, Peter. *How Experiments End*. Chicago: Chicago UP. 1987.

Gaudreault, André. 'Theatricality, Narrativity and "Trickality": Re-evaluating the Cinema of George Méliès.' *Journal of Popular Film and Television*, 15:3. 110–19.

'General Electric History.' *General Electric*. http://www.ge.com/news/media_history.html.

Genette, Gérard. 'Vraisemblance et Motivation,' in *Figures: essais*. Paris: Seuil. 1966. 71–99.

Gianetti, Louis. *Masters of the American Cinema*. Englewood Cliffs: Prentice-Hall. 1981.

Gibaldi, Joseph. *MLA Handbook for Writers of Research Papers*. New York: The Modern Language Association of America. 1999.

Goldsmith, Ben and Tom O'Regan. *The Film Studio: Film Production in the Global Economy*. Lanham: Rowman & Littlefield. 2005.

Gomery, Douglas. 'Economic and Institutional Analysis: Hollywood as Monopoly Capitalism?' (2005), in *The Contemporary Hollywood Reader*, ed. Toby Miller. New York: Routledge. 2009. 27–36.

Gomery, Douglas. 'Disney's Business History: a Reinterpretation,' in *Disney Discourse: Producing the Magic Kingdom*, ed. Eric Smoodin. London: Routledge. 1994. 71–86.

Gomery, Douglas. *The Hollywood Studio System*. New York: St. Martin's. 1986.

Gomery, Douglas. 'The American Film Industry of the 1970s: Stasis in the "New Hollywood".' *Wide Angle*, 5:4, 1983. 52–59.

Gomery, Douglas. *The Coming of Sound to the American Cinema* (unpub. diss.).Wisconsin U. 1975.

Grant, Barry Keith. 'Introduction,' in *Film Genre Reader III*, ed. Barry Keith Grant. Austin: U of Texas P. 2003. xv–xx.

Grover, Ron. *The Disney Touch – How a Daring Management Team Revived an Entertainment Empire*. Homewood: Richard Irwin. 1991.

Guback, Thomas. 'Theatrical Film,' in Who Owns the Media?, ed. Benjamin Compaine. New York: Knowledge Industry. 1982. 179–250.

Gunning, Tom. 'Now You See It, Now You Don't: The Temporality of the Cinema of Attractions.' *Velvet Light Trap*, 32, fall 1993. 3–12.

Habermas, Jürgen. *Toward a Rational Society*. Boston: Beacon P. 1970.

Hacking, Ian. *Representing and Intervening*. Cambridge: Cambridge UP. 1983.

Haddon, Larry. 'Interactive Games,' in *Future Visions: New Technologies of the Screen*, eds. Philip Hayward and Tana Wollen. London: BFI. 1993. 123–47.

Hafner, Katie and Matthew Lyon. *Where Wizards Stay Up Late: the Origins of the Internet*. New York: Simon. 1996.

Hall, Peter. *Pause: 59 minutes of motion graphics – broadcast design, music video, animation and experimental graphics from around the world*. New York: St. Martin's. 2000.

Hall, Sheldon. 'Tall Revenue Features: The Genealogy of the Modern Blockbuster'. In *Genre and Contemporary Hollywood*, ed. Steve Neale. London: BFI. 2002. 12–26.

Hampton, Benjamin. *History of the American Film Industry From its Beginning to 1931* (1931). New York: Dover. 1970.

Hamus, Réjane. 'Astérix ou l'image numérix.' *Cahiers du Cinéma*, 533, March 1999. 10–11.

Hamus, Réjane. 'Retour vers le passé – images de synthèse et cinéma.' *Cahiers du Cinéma* (hors-série), April 2000. 24–29.

Handel, Samuel. *The Electronic Revolution*. Baltimore: Penguin. 1967.

Harbord, Janet. 'Digital Film and "Late" Capitalism: A Cinema of Heroes?' (2006), in *The Contemporary Hollywood Reader*, ed. Toby Miller. New York: Routledge. 2009. 17–26.

Harbord, Janet. *The Evolution of Film: Rethinking Film Studies*. Cambridge: Polity Press. 2007.

Haskell, Molly. *From Reverence to Rape: the Treatment of Women in the Movies*. New York: Penguin. 1973.

Hayes, R.M. *Trick Cinematography: the Oscar Special Effects Movies*. Jefferson: McFarland. 1986.
Hayward, Philip. 'Industrial Light and Magic – Style, Technology and Special Effects in the Music Video and Music Television,' in *Culture,Technology and Creativity in the Late Twentieth Century*, ed. Philip Hayward. London: John Libbey. 1990. 125–47.
Hayward, Philip. 'Introduction: Technology and the (*Trans*)Formation of Culture,' in *Culture, Technology and Creativity in the Late Twentieth Century*, ed. Philip Hayward. London: John Libbey. 1990. 1–12.
Hayward, Philip and Tana Wollen. 'Introduction: Surpassing the Real,' *in Future Visions – New Technologies of the Screen*, eds. Philip Hayward and Tana Wollen. London: BFI. 1993. 1–9.
Heelan, Patrick. *Space Perception and the Philosophy of Science*. Berkeley: U California P. 1983.
Heidegger, Martin. 'The Question Concerning Technology.' In *The Question Concerning Technology and Other Essays*, trans. William Lovitt. New York: Garland. 1977. 3–35.
Helms, Michael. '*Dark City*: Interview with Andrew Mason and Alex Proyas.' *Cinema Papers*, 124, May 1998. 45.
Henderson, Brian. 'Toward a Non-Bourgeois Camera Style.' *Film Quarterly*, 24:2, winter 1970–1971. 2–14.
Hesse, Mary. *Models and Analogies in Science*. London: Sheed & Ward. 1963.
Hoberman, Jim. '1975–1985: Ten Years That Shook the World.' In *Hollywood: Critical Concepts in Media and Cultural Studies*, ed. Thomas Schatz. 2003. London: Routledge. 315–32.
Hollister, Paul. 'Genius at Work: Walt Disney,' in *Disney Discourse: Producing the Magic Kingdom*, ed. Eric Smoodin. London: Routledge. 1994. 23–41.
Huettig, Mae. *Economic Control of the Motion Picture Industry*. Philadelphia: U of Pennsylvania P. 1944.
Hutchison, David. *Film Magic: the Art and Science of Special Effects*. New York: Prentice-Hall. 1987.
Ihde, Don. *Philosophy of Technology: an Introduction*. New York: Paragon. 1993.
Ihde, Don. *Instrumental Realism*. Bloomington: Indiana UP. 1991.
Ihde, Don. *Technics and Praxis: A Philosophy of Technology*. Dordrecht: Reidel. 1979.
Imes, Jack. *A Guide to Special Effects Cinematography*. New York: Van Nostrand. 1984.
Industrial Light & Magic (ILM). Lucasfilm Inc. LLC. http://www.ilm.com.
'Intel's History of the Microprocessor.' *Intel Museum Online*. Intel Corporation. http://www.intel.com/intel/museum/25anniv/html/hof/hof_main.htm.
'Internet Movie Database (IMdB).' *amazon.com inc*. http://www.us.imb.com.
Jacobs, Lea. 'The B Film and the Problem of Cultural Distinction.' In *Hollywood: Critical Concepts in Media and Cultural Studies*, ed. Thomas Schatz. 2003. London: Routledge. 147–60.
Jacobs, Lewis. *The Rise of the American Film*. New York: Teachers' College. 1967.
Jancovich, Mark. ''A Real Shocker' – Authenticity, Genre and the Struggle for Distinction.' In *The Film Cultures Reader*, ed. Graeme Turner. London: Routledge. 2002. 468–80.
Jenn, Pierre. *George Méliès cinéaste: le montage cinématographique chez George Méliès*. Paris: Albatros. 1984.
Jewell, Richard. 'How Howard Hawks Brought *Baby* Up: An *Apologia* for the Studio System,' in *The Studio System*, ed. Janet Staiger. New Brunswick: Rutgers UP. 1995. 39–49.
Jones, Kent. 'Hollywood et la saga du numérique – A propos de *Matrix* et de *La Menace Fantôme*,' trans. Sylvie Durastanti and Jean Pêcheux. *Cahiers du Cinéma*, 537, July–August 1999. 36–39.
Jones, Steven (ed). *CyberSociety: Computer-Mediated Communication and Community*. Thousand Oaks: Sage. 1995.
Kaplan, E. Ann. 'The State of the Field: Notes Toward an Article.' *Cinema Journal*, 2004, 43:3. 85–88.

Kapsis, Robert. 'Hollywood Genres and the Production of Culture Perspective' (1991), in *The Contemporary Hollywood Reader*, ed. Toby Miller. New York: Routledge. 2009. 3–16.
Keane, Stephen. *CineTech: Film, Convergence and New Media*. London: Palgrave Macmillan. 2007.
Keathley, Christian. *Cinéphilia and History or the Wind in the Trees*. Bloomington: Indiana UP. 2006.
Kerr, Paul. 'My Name is Joseph H. Lewis' (1983), in *The Studio System*, ed. Janet Staiger. New Brunswick: Rutgers UP. 1995. 50–73.
Kittler, Friedrich. *Gramophone, Film, Typewriter*, trans. Geoffrey Winthrop-Young and Michael Wuz. Stanford: California UP. 1986.
Kindem, Gorham. *The American Movie Industry: The Business of Motion Pictures*. Carbondale: Southern Illinois UP. 1982.
Klein, Norman. *The Vatican to Vegas: A History of Special Effects*. New York: The New Press. 2004.
Klingender, Francis and Stuart Legg. *Money Behind the Screen*. London: Lawrence and Wishart. 1937.
Kracauer, Siegfried. *Theory of Film: the Redemption of Physical Reality*. London: Oxford UP. 1960.
Kramer, Peter. 'Post-classical Hollywood,' in *American Cinema and Hollywood – Critical Approaches*, eds. John Hill and Pamela Gibson. New York: Oxford UP. 2000. 63–83.
Krantz, Michael. 'Hollywood gets wired.' *Time*, 149:28, 23 December 1996. 58–60.
Kuhn, Thomas. *The Structure of Scientific Revolutions* (1962). Chicago: U of Chicago P. 1970.
Langford, Barry. *Post-Classical Hollywood: Film Industry, Style and Ideology Since 1945*. Edinburgh : Edinburgh UP. 2010.
Lapsley, Robert and Michael Westlake. *Film Theory: an Introduction*. New York: Manchester UP. 1988.
Latour, Bruno. *Science in Action*. Cambridge: Harvard UP. 1987.
Latour, Bruno. *We Have Never Been Modern*, trans. Catherine Porter. Cambridge: Harvard UP. 1993.
Lee, Nora. 'Motion Control.' *American Cinematographer*, 64:5, May 1983. 60–61.
Lee, Nora. '*Total Recall*: Interplanetary Thriller.' *American Cinematographer*, 71:7, July 1990. 44–52.
Levy, Steven. *Hackers: Heroes of the Computer Revolution*. New York: Dell. 1984.
Lindsay, Vachel. *The Art of the Moving Picture* (1922). New York: Liveright. 1970.
Lounas, Thierry. '*Men In Black*: l'élégance terrestre.' *Cahiers du Cinéma*, 516, September 1997. 70–71.
Lucas Digital Ltd. Lucasfilm Inc. LLC. http://www.lucasdigital.com/.
LucasArts Entertainment Company. Lucasfilm Inc. LLC. http://www.lucasarts.com.
Lucasfilm Inc. Lucasfilm Inc. LLC. http://www.lucasfilm.com/.
Lunenfeld, Peter (ed.). *The Digital Dialectic: New Essays on the New Media*. London: MIT P. 1999.
Magid, Ron. 'Many Hands Make Martian Memories.' *American Cinematographer*, 71:7, July 1990. 54–64.
Maltby, Richard. '"Nobody knows everything": Post-classical historiographies and consolidated entertainment,' in *Contemporary Hollywood Cinema*, eds. Steve Neale and Murray Smith. London: Routledge. 1998. 21–44.
Malthete-Méliès, Madeline. *Méliès et la Naissance du Spectacle Cinématographique*. Paris: Klincksiek. 1984.
Manovich, Lev. 'What is Digital Cinema?'1999. http://www.manovich.net/TEXT/digital-cinema.html.
Marcuse, Herbert. *One Dimensional Man*. Boston: Beacon. 1964.
Maslin, Janet. '*The Phantom Menace*: In the Beginning, the Future.' *The New York Times*, 19 May 1999. http://www.nytimes.com/library/film/051999starwars-film-review.html

Mast, Gerald. *A Short History of the Movies*. Indianapolis: Bobbs-Merrill. 1976.
Mast, Gerald. *Film/Cinema/Movie: A Theory of Experience*. New York: Harper. 1977.
Matthews, M. *Hostile Aliens, Hollywood and Today's News – 1950s Science Fiction Films and 9/11*. New York: Algora. 2007.
Mayer, Michael. *The Film Industries*. New York: Hastings. 1978.
McAlister, Michael. *The Language of Visual Effects*. Los Angeles: Lone Eagle. 1992.
McClean, Shilo. *So what's this all about then? A non-users guide to digital effects in filmmaking*. North Ryde: AFTRS. 1998.
McIntosh, John. 'Bright Lights Ahead for Digital Effects in Hollywood.' *MacWeek*, 11:40, October 1997. 18.
McKenzie, Alan. *Hollywood Tricks of the Trade*. Leicester: Admiral. 1986.
McQuire, Scott. 'Technicalities: Digital Futures.' *Cinema Papers*,121, November 1997.25–40.
McQuire, Scott. *Crossing the Digital Threshold*. Brisbane: AKCCMP/Griffith U. 1997.
Metz, Christian. *Essais sur la signification au cinéma*. Paris: Klincksieck. 1968.
Metz, Christian. *Film Language: a Semiotics of Cinema*, trans. Michael Taylor. London: Oxford UP. 1974.
Metz, Christian. 'Trucage and the Film,' trans. Françoise Meltzer. *Critical Inquiry*, vol.3, summer 1977. 657–75.
Metz, Christian. *The Imaginary Signifier: Psychoanalysis and the Cinema*, trans. Celia Britton, Annwyl Williams, Ben Brewster and Alfred Guzzetti. Bloomington: Indiana UP. 1982.
Millar, Dan. *Cinema Secrets: Special Effects*. London: Chartwell. 1990.
Miller, Toby. 'Introducing Screening Cultural Studies: Sister Morpheme (Clark Kent – Superman's Boyfriend).' *Continuum*, 7:2. 1994. 11–44.
Miller, Toby. 'Introduction,' in *A Companion to Film Theory*, eds. Toby Miller and Robert Stam. Oxford: Blackwell. 1999. 1–8.
Miller, Toby, Nitin Govil, John McMurria, Richard Maxwell and Ting Wang. *Global Hollywood 2*. London: BFI. 2005.
Mitchell, William. *The Reconfigured Eye: Visual Truth in the Post-Photographic Era*. Cambridge: MIT P. 1994.
Mitry, Jean. *Esthétique et psychologie du cinéma (vol.I)*. Paris: Éditions Universitaires. 1963.
Mitry, Jean. *Esthétique et psychologie du cinéma (vol.II)*. Paris: Éditions Universitaires.1965.
Monaco, James. *American Film Now*. New York: Oxford UP. 1979.
Monaco, James. *How to Read a Film – The Art, Technology, Language, History and Theory of Film and Media*. New York: Oxford UP. 1981.
Monaco, James. *How to Read a Film – The World of Movies, Media, and Multimedia*. New York: Oxford UP. 2000.
Mora, Philip. 'Disaster Films.' *Cinema Papers*, March-April 1975. 10–13.
Morris, Meaghan. 'Learning From Bruce Lee: Pedagogy and Political Correctness in Martial Arts Cinema.' *Metro*, 117, 1998. 6–16.
Münsterberg, Hugo. *The Film: a Psychological Study*. New York: Appleton. 1916.
Mulvey, Laura. 'Visual Pleasure and Narrative Cinema,' in *Film Theory and Criticism: Introductory Readings*, eds. Leo Braudy and Marshall Cohen. New York: Oxford UP. 1999. 833–44.
Myers, Edith. 'Oscar computes; the Academy of Motion Picture Arts and Sciences is finding that computers are putting the art and the sciences together.' *Datamation*, 32, 15 April 1986. 68–72.
Nathans, Stephen. 'Beyond *Toy Story*: the products that propel 3D animation into motion.' *CD-ROM Professional*, 9:3, March 1996. 18–20.
Ndalianis, Angela. *Neo-Baroque Aesthetics and Contemporary Entertainment*. Cambridge: MIT P. 2004.
Neale, Steve and Murray Smith. 'Introduction,' in *Contemporary Hollywood Cinema*, eds. Steve Neale and Murray Smith. London: Routledge. 1998. xiv–xxii.

Neale, Steve. 'Hollywood Strikes Back: Special Effects in Recent American Cinema.' *Screen*, 21:3, 1980. 101–5.
Neale, Steve. *Genre and Hollywood*. London: Routledge. 2000.
Neale, Stephen. *Cinema and Technology: Image, Sound, Colour*. Bloomington: Indiana UP. 1985.
Negroponte, Nicholas. *Being Digital*. London: Hodder. 1996.
Neupert, Richard. 'Painting a Plausible World: Disney's Color Prototypes,' in *Disney Discourse: Producing the Magic Kingdom*, ed. Eric Smoodin. London: Routledge. 1994. 106–17.
Nicholls, Peter. *Fantastic Cinema – an Illustrated Survey*. London: Ebury P. 1984.
Nicholson, Dennis. 'Special Effects in *Star Wars*.' *Cinema Papers*, 14, October 1977. 119 & 192.
O'Regan, Tom and Rama Venkatasawmy. 'I Make Films and It's Up to Other People to Pigeon-Hole Them: An Interview with Alex Proyas.' *Metro*, 117, 1998. 29–33.
O'Regan, Tom and Rama Venkatasawmy. 'Only One Day at the Beach: *Dark City* and Australian Filmmaking.' *Metro*, 117, 1998. 16–28.
Ohanian, Thomas and Michael Phillips. *Digital Filmmaking: the Changing Art and Craft of Making Motion Pictures*. Boston: Focal Press. 1996.
Panofsky, Erwin. *Meaning in the Visual Arts*. Hammondsworth: Penguin. 1970.
Panofsky, Erwin. 'Style and Medium in the Motion Pictures' (1934), in *Film Theory and Criticism*, eds. Gerald Mast and Marshall Cohen. New York: Oxford UP.1974. 279–92.
Pennings, Anthony. *History of Information Technology*. Division of Communication and the Arts, Marist C, New York. http://www.academic.marist.edu/pennings/hyprhsty.htm.
Perisic, Zoran. *Special Optical Effects in Film*. London: Focal. 1980.
Phillips, Joseph. 'Film Conglomerate Blockbusters.' *Journal of Communication*, 25, spring 1975. 171–81.
Pierson, Michele. 'CGI Effects in Hollywood Science-Fiction Cinema 1989–1995: The Wonder Years.' *Screen* (special issue), eds. Sean Cubitt and John Caughie, 40:2, 1996. 42–64.
Pierson, Michele. 'No Longer State-of-the-Art: Crafting a Future for CGI.' *Wide Angle*, 21:1, January 1999. 28–47.
Pierson, Michele. *Special Effects: Still in Search of Wonder*. New York: Columbia UP. 2002.
Polsson, Ken. *Chronology of Personal Computers*. http://www.islandnet.com/~kpolsson/comphist.
Polsson, Ken. *Chronology of Video Game Systems*. http://www.islandnet.com/~kpolsson/vidgame.
Polsson, Ken. *Chronology of Walt Disney*. http://www.islandnet.com/~kpolsson/disnehis.
Poster, Mark. *The Second Media Age*. Cambridge: Polity. 1995.
Prince, Stephen, 'True Lies: Perceptual Realism, Digital Images and Film Theory.' *Film Quarterly*, 49:3, Spring 1996. 27–37.
Purves, Barry. 'The Emperor's New Clothes.' *Animation World Magazine*. http://www.awn.com/mag/issue1.1/articles/purves.html.
Pye, Michael and Lynda Myles. *The Movie Brats: How the Film Generation Took Over Hollywood*. New York: Holt. 1979.
Rapp, Friedrich. *An Analytical Philosophy of Technology*, trans. Stanley Carpenter & Theodor Langenbruch. Dordrecht: Reidel 1981.
Renaud, Jean-Luc. 'Towards Higher Definition Television,' in *Future Visions: New Technologies of the Screen*, eds. Philip Hayward and Tana Wollen. London: BFI. 1993. 46–71.
Robins, Kevin. *Into the Image – Culture and Politics in the Field of Vision*. London: Routledge. 1996.
Rodowick, David Norman. *The Virtual Life of Film*. New York: Harvard UP. 2007.

Rose, Frank. 'There's No Business Like Show Business.' *Fortune*, 137:12, 22 June 1998. 52-60.
Rosebush, James. 'Digital Video: Future Predictions.' *CD-ROM Professional*, 7:2, March 1994. 127–32.
Rosenbaum, Jonathan and Adrian Martin (eds). *Movie Mutations: The Changing Face of World Cinephilia*. London: BFI. 2003.
Rosten, Leo. *Hollywood, the Movie Colony, the Movie Makers*. New York: Harcourt. 1941.
Roth, Daniel. 'New-Media Nightmare.' *Fortune*, 137:12, 22 June 1998. 81–33.
Russell, Bertrand. *Authority and the Individual*. London: Allen & Unwin. 1965.
Ryall, Tom. 'Genre and Hollywood'. In *American Cinema and Hollywood: Critical Approaches*, eds. John Hill and Pamela Church Gibson. London: Oxford UP. 2000. 101–15.
Ryan, Roderick. *A History of Motion Picture Color Technology*. New York: Focal. 1978.
Salt, Barry. *Film Style and Technology: History and Analysis*. London: Starword. 1983.
Samuelson, Robert. 'Puzzles of the "New Economy".' *Newsweek*, April 2000. 58.
Schatz, Thomas. '"A Triumph of Bitchery": Warner Bros, Bette Davis and *Jezebel*' (1988), in *The Studio System*, ed. Janet Staiger. New Brunswick: Rutgers UP. 1995. 74–92.
Schatz, Thomas. 'The New Hollywood,' in *Film Theory Goes to the Movies*, eds. Jim Collins, Hilary Radner and Ava Collins. London: Routledge. 1993. 8–36.
Schatz, Thomas. 'New Hollywood, New Millenium,' in *Film Theory and Contemporary Hollywood Movies*, ed. Warren Buckland. New York: Routledge. 2009. 19–46.
Schwam, Stephanie and Jay Cocks (eds.). *The Making of 2001: A Space Odyssey*. New York: Modern Library. 2000.
'Scientific and Technical Awards Database.' Academy of Motion Picture Arts and Sciences (AMPAS). http://www.oscars.org.
Scott, Elaine. *Movie Magic: Behind the Scenes with Special Effects*. New York: Morrow. 1995.
Sergi, Gianluca and Alan Lovell. *Cinema Entertainment: Essays on Audiences, Films and Film makers*. Maidenhead: Open UP. 2009.
Sickels, Robert. *American Film in the Digital Age*. Santa Barbara: Praeger. 2011.
Sjostrom, Ingrid. *Quadratura: Studies in Italian Ceiling Painting*. Stockholm: Almqvist & Wiksell. 1978.
Smith, Murray. 'Theses on the Philosophy of Hollywood History,' in *Contemporary Hollywood Cinema*, eds. Steve Neale and Murray Smith. London: Routledge.1998. 3–20.
Smith, Thomas. *Industrial Light & Magic: the Art of Special Effects*. New York: Ballantine. 1986.
Spielmann, Yvonne. 'Aesthetic Features in Digital Imaging.' *Wide Angle*, 21:1, January 1999. 131–48.
Spufford, Francis and Jenny Uglow (eds). *Cultural Babbage: Technology, Time and Invention*. London: Faber. 1996.
Staiger, Janet. 'Introduction,' in *The Studio System*, ed. Janet Staiger. New Brunswick: Rutgers UP. 1995. 1–14.
'*Star Wars: The Old Republic*.' LucasArts, BioWare, Electronic Arts Inc. http://www.swtor.com/.
Streisand, Brian. 'The patience of Jobs: Apple's cofounder is back on the cutting edge with Pixar.' *U.S. News & World Report*, 119:23, 11 December 1995. 83.
Tahiro, Charles. 'The *Twilight Zone* of Contemporary Hollywood Production?' (2002), in *The Contemporary Hollywood Reader*, ed. Toby Miller. New York: Routledge. 2009. 87–95.
Taylor, Mark and Esa Saarinen. *Imagologies – Media Philosophy*. London: Routledge. 1994.
Telotte, Jay. 'Film and/as Technology: Assessing a Bargain.' *Journal of Popular Film & Television*, 28:4, winter 2001. 146–49.
Tesson, Charles. 'Éloge de l'impureté.' *Cahiers du Cinéma* (hors-série), April 2000. 4–6.

'The Complete Coverage: *Star Wars*.' *American Cinematographer Online*. http://www.theasc.com/magazine/starwars/
'The Visual Effects.'*The Fifth Element*. http://www.gaumont.com/fifth/html/us/movie/fx.htm.
Thompson, Kristin. *Exporting Entertainment: America in the World Film Market 1907–34*. London: BFI. 1985.
Thomson, David. *America in the Dark*. New York: William Morrow. 1977.
'Timeline for Inventing Entertainment: The Motion Pictures and Sound Recordings of the Edison Companies.' *American Memory: Inventing Entertainment*. Library of Congress. http://www.memory.loc.gov/ammem/edhtml/edtime.html.
Todorov, Tzvetan. *Introduction to Poetics*, trans. Richard Howard. Brighton: Harvester. 1981.
Todorov, Tzvetan. *The Poetics of Prose*, trans. Richard Howard. Ithaca: Cornell UP. 1997.
Toy Story. Disney/Pixar Animation Studios. http://www.disney.com/DisneyVideos/ToyStory/about/abfilm.htm.
Turim, Maureen. 'Artisanal Prefigurations of the Digital: Animating Realities, Collage Effects, and Theories of Image Manipulation.' *Wide Angle*, 21:1, January 1999. 48–62.
Tryon, Chuck. *Reinventing Cinema: Movies in the Age of Digital Convergence*. London: Rutgers UP. 2009.
Tzioumakis, Yannis. 'From the Business of Film to the Business of Entertainment: Hollywood in the Age of Digital Technology,' in *American Film in the Digital Age*, Robert Sickels. Santa Barbara: Praeger. 2011. 11–31.
Vasey, Ruth. 'The Hollywood Industry Paradigm,' in *The Sage Handbook of Film Studies*, eds. James Donald and Michael Renov. London: Sage. 2008. 287–311.
Vaz, Mark. *From Star Wars to Indiana Jones: the Best of the Lucasfilm Archives*. San Francisco: Chronicle. 1994.
Vaz, Mark. *Industrial Light & Magic: Into the Digital Realm*. New York: Ballantine. 1996.
Virilio, Paul. *War and Cinema – the Logistic of Perception*, trans. Patrick Camiller.London: Verso. 1989.
'Visual Effects – Seeing is Believing.' Academy of Motion Picture Arts and Sciences (AMPAS).http://www.oscars.org/educationoutreach/teachersguide/visualeffects/activity1.html.
Vogel, Harold. *Entertainment Industry Economics: a Guide for Financial Analysis*. New York: Cambridge UP. 1986.
Wallace, David Foster. 'F/X Porn.' *Waterstone's Magazine*. Winter/Spring 1998. http://www.scribd.com/doc/6447057/David-Foster-Wallace-on-FX-Porn.
Wark, McKenzie. 'Lost in Space: into the digital image labyrinth.' *Continuum: The Australian Journal of Media & Culture*, 7:1, 1993. 140–60.
Wasko, Janet. *Democratic Communications in the Information Age*. Toronto: Garamond. 1992.
Wasko, Janet. *Hollywood in the Age of Information: Beyond the Silver Screen*. Cambridge: Polity. 1994.
Wasko, Janet. *How Hollywood Works*. London: Sage. 2003.
Wasko, Janet. *Movies and Money: Financing the American Film Industry*. Norwood: Ablex. 1982.
'Westinghouse Electric History.' *Westinghouse Electric*. http://www.westinghouse.com/
'What's New – product review.' *American Cinematographer*, July 1990, 71:7. 16–20.
White, Michael. *The Fruits of War – How Military Conflict Accelerates Technology*. London: Simon & Schuster. 2005.
Whitehead, Alfred North. *Science and the Modern World*. New York: Harper & Row. 1972.
Whitney Sr, John. 'Motion Control: An Overview.' *American Cinematographer*, December 1981. 20–52.
Wiener, Norbert. *The Human Use of Human Beings*. London: Sphere. 1968.
Wilkie, Bernard. *Creating Special Effects for TV and Film*. New York: Hastings. 1977.

Williams, Raymond. *Problems in Materialism and Culture: Selected Essays*. London: Verso-New Left Books. 1980.
Williams, Raymond. *Keywords: A Vocabulary of Culture and Society*. New York: Oxford UP. 1976.
Winner, Langdon. *The Whale and the Reactor*. Chicago: U Chicago P. 1986.
Winner, Langdon. *Autonomous Technology:* Technics-out-of-Control as a Theme in Political Thought. Cambridge: MIT P. 1977.
Winston, Brian. *Misunderstanding Media*. London: Routledge. 1986.
Winston, Brian. *Technologies of Seeing: Photography, Cinematography and Television*. London: BFI. 1996.
Winston, Brian. *Media, Technology and Society – A History: From the Telegraph to the Internet*. London: Routledge. 1998.
Wolf, William. *Landmark Films: the Cinema and Our Century*. New York: Paddington. 1979.
Wollen, Peter. *Signs and Meaning in the Cinema*. London: Indiana UP. 1972.
Wollen, Tana. 'The Bigger the Better: From Cinemascope to IMAX,' in *Future Visions: New Technologies of the Screen*, eds. Philip Hayward and Tana Wollen. London: BFI. 1993. 10–30.
Wright, Judith Hess. 'Genre Films and the Status Quo.' In *Film Genre Reader III*, ed. Barry Keith Grant. Austin: U of Texas P. 2003. 42–50.
Wulforst, Harry. *Breakthrough to the Computer Age*. New York: Scribner's. 1982.
Wyatt, Justin. *High Concept: Movies and Marketing in Hollywood*. Austin: U of Texas P. 1994.
Youngblood, Gene. *Expanded Cinema*. New York: Dutton. 1970.
Zucker, Andrew. 'New Media: a Ringside View of Trends in the Industry.' *Animation World Magazine*. http://www.awn.com/mag/issue1.9/articles/zucker 1.9 html/.

Films Cited

[Films are cited in alphabetical order—technical information provided includes: year of release; main producer(s) and director(s); principal provider(s) of VFX]

101 Dalmatians (1961, Disney)—Clyde Geronimi, Hamilton Luske & Wolfgang Reitherman

101 Dalmatians (1996, Disney/Great Oaks)—Stephen Herek

- Industrial Light & Magic; Buena Vista Imaging; Jim Henson's Creature Shop.

102 Dalmatians (2000, Disney/Cruella)—Kevin Lima

- Neal Scanlan Studio; The Secret Lab.

20 Million Miles to Earth (1957, Morningside)—Nathan Juran

- Ray Harryhausen.

2001: A Space Odyssey (1968, MGM)—Stanley Kubrick

- Douglas Trumbull; Wally Veevers; Con Pederson; Tom Howard.

20,000 Leagues Under the Sea (1954, Disney)—Richard Fleischer

- Robert Mattey; John Hench; Joshua Meador; Marcel Delgado; Peter Ellenshaw.

633 Squadron (1964, Mirisch)—Walter Grauman

- Tom Howard; Jimmy Harris; Garth Inns.

7 Faces of Dr Lao (1964, MGM)—George Pal

- Tim Baar; Paul Byrd; Wah Chang; Jim Danforth.

A Beautiful Mind (2001, Universal/DreamWorks SKG)—Ron Howard

- Digital Domain; Keith Vanderlaan's Captive Audience Productions.

A Bug's Life (1998, Disney/Pixar)—John Lasseter & Andrew Stanton

A Few Good Men (1992, Columbia/New Line Cinema/Castle Rock)—Rob Reiner

- Illusion Arts.

A Hard Day's Night (1964, United Artists/Proscenium)—Richard Lester

- Ronnie Wass.

A Night to Remember (1958, Rank Organisation/MacQuitty)—Roy Ward Baker

- Bill Warrington.

A Nightmare on Elm Street (1984, New Line Cinema/Media Home)—Wes Craven

- Tassilo Baur; Charles Belardinelli; Lou Carlucci.

A Nightmare on Elm Street II: Freddy's Revenge (1985, New Line Cinema)—Jack Sholder

- Richard Albain; Paul Boyington; Kevin Yagher.

A Nightmare on Elm Street III: Dream Warriors (1987, New Line Cinema)—Chuck Russell

- DreamQuest Images.

A Nightmare on Elm Street IV: The Dream Master (1988, New Line Cinema)—Renny Harlin

- DreamQuest Images; Reel Efx; Steve Johnson's XFX.

A Nightmare on Elm Street V: The Dream Child (1989, New Line Cinema)—Stephen Hopkins

- K.N.B. Effects Group; Todd Masters Company; Visual Concept Engineering.

A Nightmare on Elm Street VI: Freddy's Dead—The Final Nightmare (1991, New Line
Cinema)—Rachel Talalay

- DreamQuest Images; VIFX.

A Nightmare on Elm Street VII: Wes Craven's New Nightmare (1994, New Line Cinema)—Wes Craven

- Digital Filmworks Inc; Flash Film Works; K.N.B. EFX Group Inc.

A Simple Wish (1997, Universal/Bubble Factory)—Michael Ritchie

- Blue Sky Studios; Matte World Digital; Blue Sky/VIFX; Question Mark FX.

A.I. (2001, Warner/DreamWorks SKG/Amblin Entertainment)—Steven Spielberg

- Industrial Light & Magic; Stan Winston Studio; Pacific Data Images.

Air Force One (1997, Columbia/Beacon/Radiant)—Wolfgang Petersen

- Boss Film Studios; Cinesite Digital Studios.

Airplane (1980, Paramount/Koch)—Jim Abrahams

- Magic Lantern; Motion Pictures Inc; Robert Keith & Company Inc; Visual Concept Engineering.

Airplane II: The Sequel (1982, Paramount)—Ken Finkleman

- Coast Special Effects; Joe Rayner.

Airport (1970, Universal)—George Seaton

- Film Effects of Hollywood.

Airport '77 (1977, Universal)—Jerry Jameson

- Frank Brendel; Albert Whitlock; Bill Taylor; Dennis Glouner.

Akira (1987, ICA/Akira Committee)—Katsuhiro Otomo
Aladdin (1992, Disney)—Ron Clements & John Musker
Ali Baba and the Forty Thieves (1944, Universal)—Arthur Lubin

- John P. Fulton.

Alice Doesn't Live Here Anymore (1974, Warner)—Martin Scorsese
Alice In Wonderland (1933, Paramount)—Norman McLeod

- Farciot Edouart; Gordon Jennings.

Alice's Wonderland (1923, Laugh-O-Gram Films)—Walt Disney
Alien (1979, 20th Century-Fox/Brandywine)—Ridley Scott

- Filmfex Animation Services; System Simulation; Denys Ayling; Brian Johnson; Dick Hewitt; Martin Bower; Bernard Lodge; Bill Pearson; Ray Caple; Nick Allder; Clinton Cavers; Guy Hudson; Carlo Rambaldi.

Aliens (1986, 20th Century-Fox/Brandywine)—James Cameron

- Stan Winston Studio; Peerless Camera Co; L.A. Effects Group; Arkadon; Peter Aston Model Effects.

Alien 3 (1992, 20th Century-Fox/Brandywine)—David Fincher

- Boss Film Studios; Industrial Light & Magic; Video Image; Wildfire Inc.

Alien 4: Resurrection (1997, 20th Century-Fox/Brandywine)—Jean-Pierre Jeunet

- Duboi; Digiscope; Blue Sky Studios; Blue Sky/VIFX; Amalgamated Dynamics; Hunter-Gratzner Industries; Laser Mechanisms; All Effects Company.

Alligator (1980, Group 1/Alligator Associates)—Lewis Teague

- Film Effects of Hollywood.

Alligator II: The Mutation (1991, Golden Hawk)—Jon Hess

- John Criswell.

Altered States (1980, Warner/Gottfried/Melnick)—Ken Russell

- Laser Media Inc; Cinema Research Corporation; The Optical House.

Amblin' (1968)—Steven Spielberg
American Beauty (1999, DreamWorks/Cohen Company)—Sam Mendes

- Computer Film Co; Manex Visual Effects.

American Graffiti (1973, Universal/Lucasfilm/Coppola Company)—George Lucas
American Pie (1999, Universal)—Paul Weitz

- Banned from the Ranch Entertainment.

Amityville II: The Possession (1982, Orion/de Laurentiis)—Damiano Damiani

- Glen Robinson; Gary Zeller; John Caglione Jr.

Amityville III: The Demon (1984, Universal/Orion)—Richard Fleischer

- Jeff Jarvis; Michael Wood; Stuart Ziff; Peter Stolz; Gary Platek; Rick Fichter; David Wood; Al Griswold.

An American in Paris (1951, MGM)—Vincente Minnelli

- Warren Newcombe; Irving G. Ries.

An American Werewolf in London (1981, Polygram/Guber-Peters)—John Landis

- Effects Associates.

Anaconda (1997, Columbia/New Line Cinema)—Luis Llosa

- Sony Pictures ImageWorks; Edge Innovations; Steve Johnson's XFX Inc; Computer Film Co.

Animalympics (1980, Lisberger Studios)—Steve Lisberger

Antz (1998, DreamWorks/Pacific Data Images)—Eric Darnell & Tim Johnson

Apocalypse Now (1979, Zoetrope)—Francis Ford Coppola

- Joe Lombardi; A.D. Flowers; Eddie Ayay; Mario Carmona; John Fraser; Rudy Liszczak; Ted Martin; David St. Ana.

Apocalypse Now Redux (2001, Zoetrope)—Francis Coppola

Apollo 13 (1995, Universal/Imagine)—Ron Howard

- Digital Domain; WonderWorks.

Armageddon (1998, Touchstone/Valhalla/Bruckheimer)—Michael Bay

- DreamQuest Images; Blue Sky Studios/VIFX; Digital Domain; Cinesite Digital Studios; Matte World Digital; Rainmaker Digital Pictures; The Computer Film Co; Buena Vista Imaging; Computer Cafe Inc; Pacific Titles & Optical; Hunter-Gratzner Industries; POP Film; Tippett Studio; Vision Crew Unlimited; WonderWorks; Black Box Digital; Digital Magic Co; 525 Post Productions.

Arrivée d'un Train En Gare de la Ciotat (1896, Lumière)—Auguste Lumière & Louis Lumière

As Good As It Gets (1997, TriStar/Gracie)—James Brooks

Assassins (1995, Warner/Silver Pictures)—Richard Donner

- Dennis Petersen; Casey Pritchett; Jon Belyeu; Chris Cobb; Rick Lupton.

Atlantis—The Lost Continent (1960, MGM)—George Pal

- Projects Unlimited; A. Arnold Gillespie; Jim Danforth; Gene Warren; Matthew Yuricich; Wah Chang; Robert Hoag; Lee LeBlanc.

Attack of the Crab Monsters (1956, Allied Artists)—Roger Corman

- Ed Nelson; Beach Dickerson.

Attack of the Killer Tomatoes (1978, Four Square/NAI)—John de Bello

- Craig Berkos; Robert Matzenauer; Roger Dorney.

Austin Powers: International Man of Mystery (1997, New Line Cinema/Capella)—Jay Roach

- Pacific Vision Productions.

Austin Powers: The Spy Who Shagged Me (1999, New Line Cinema)—Jay Roach

- Pacific Vision Productions; Stan Winston Studio.

Avatar (2009, 20th Century-Fox/Dune/ Lightstorm)—James Cameron

- Stan Winston Studio; Giant Studios; Weta Digital; Industrial Light & Magic; Framestore; Prime Focus; Hybride Technologies; Hydraulx; BUF; Blur Studio; Pixel Liberation Front; Spy Post Digital; Lola Visual Effects; LOOK Effects; Stereo D; Legacy Effects; Camera Control; Lowry Digital Images; The Third Floor.

Back to the Future (1985, Universal/Amblin Entertainment)—Robert Zemeckis

- Industrial Light & Magic.

Back to the Future II (1989, Universal/Amblin Entertainment)—Robert Zemeckis

- Industrial Light & Magic.

Back to the Future III (1990, Universal/Amblin Entertainment)—Robert Zemeckis

- Industrial Light & Magic.

Backdraft (1991, Universal/Trilogy/Imagine)—Ron Howard

- Industrial Light & Magic.

Bad Boys (1995, Columbia/Simpson-Bruckheimer)—Michael Bay

- Richard Lee Jones; Ray Hardesty; Michael Meinardus; Kent Demaine; Steve Rodriguez.

Basic Instinct (1992, TriStar/Carolco/Canal Plus)—Paul Verhoeven

- Cruse & Company; Dreamstate Effects.

Batman (1943, Columbia)—Lambert Hillyer
Batman (1966, 20th Century-Fox/Greenlawn)—Leslie Martinson

- L.B. Abbott.

Batman (1989, Warner/Polygram)—Tim Burton

- The Magic Camera Co.

Batman Returns (1992, Warner/Polygram)—Tim Burton

- Boss Film Studios; Hunter-Gratzner Industries; Matte World Digital; Stan Winston Studio; Video Image.

Batman Forever (1995, Warner/Polygram)—Joel Schumacher

- Pacific Data Images; Rhythm and Hues; Warner Digital Studios.

Batman & Robin (1997, Warner/Polygram)—Joel Schumacher

- Flash Film Works; Pacific Data Images; Warner Digital Studios.

Batteries Not Included (1987, Universal/Amblin Entertainment)—Matthew Robbins

- Industrial Light & Magic.

Battlestar Galactica TV Series (1978–1979, Universal TV/Glen A. Larson)—various

- Apogee Inc.

Battleship Potemkin (1925, Mosfilm/Goskino)—Sergei Eisenstein
Battle Beyond the Stars (1980, New World Pictures)—Jimmy Murakami

- Modern Props Inc; Hal Mann Laboratory.

Battle of Britain (1969, United Artists/Spitfire)—Guy Hamilton

- Glen Robinson; Wally Veevers; Jimmy Harris; Richard Conway; Alf Levy; Ronnie Wass; Martin Body; John Richardson; Ray Caple; Wally Armitage.

Beauty and the Beast (1991, Disney)—Gary Trousdale & Kirk Wise
Becky Sharp (1935, Pioneer Pictures Corp)—Rouben Mamoulian

- Harry Redmond Sr.

Bedknobs and Broomsticks (1971, Disney)—Robert Stevenson

- Danny Lee; Eustace Lycett; Alan Maley.

Ben-Hur: A Tale of Christ (1925, MGM)—Fred Niblo

- Ferdinand P. Earle; Frank Williams; Kenneth Gordon MacLean.

Ben-Hur (1959, MGM)—William Wyler & Andrew Marton

- Arnold Gillespie; Milo Lory; Robert MacDonald.

Beverly Hills Cop (1984, Paramount/Simpson/Bruckheimer)—Martin Brest

- Steve Grumette; Ken Pepiot.

Beverly Hills Cop II (1987, Paramount/Simpson/Bruckheimer/Murphy)—Tony Scott

- Tom Ryba; Tom Tokunaga; Johnny Borgese; David Blitstein.

Beverly Hills Cop III (1994, Paramount)—John Landis

- Global Effects; Illusion Arts Inc.

Beyond the Poseidon Adventure (1979, Warner)—Irwin Allen

- Harold Wellman; Howard Jensen; Gary Zink.

Big Daddy (1999, Giarraputo/Out of the Blue)—Dennis Dugan

Birth of a Nation (1915, David W. Griffith Corp)—D.W. Griffith

Blade (1998, New Line Cinema/Amen Ra/Imaginary Forces)—Stephen Norrington

- Blue Sky Studios/VIFX; Digiscope; Image Savant; Flat Earth Productions; Post Logic; Wildcat Digital Effects; 525 Post Productions; The Production Plant Inc.

Blade Runner (1982, Warner/Ladd Co/Scott)—Ridley Scott

- Douglas Trumbull; Don Baker; Richard Yuricich; David R. Hardberger; Syd Mead; David K. Stewart; David Dryer.

Blade Runner: Director's Cut (1991, Warner/Ladd Co/Scott)—Ridley Scott

Blues Brothers (1980, Universal)—John Landis

- Albert Whitlock; Fred Griggs; Robert Blalack; James Shourt; Bill Taylor.

Blues Brothers 2000 (1998, Universal)—John Landis

- Optical Illusions; Available Light Productions; Steve Johnson's XFX; Models Unlimited SFX.

Bonnie and Clyde (1967, Warner-Seven Arts/Tatira/Hiller)—Arthur Penn

Bound (1996, de Laurentiis/Spelling/Summit)—Andy Wachowski & Larry Wachowski

- FTS EFX Inc.

Bound for Glory (1976, MGM)—Hal Ashby

- Albert Whitlock; Sass Bedig.

Boxcar Bertha (1972, American International Pictures)—Martin Scorsese

Brainscan (1994, Admire/Coral)—John Flynn

- Cineffects Inc.

Brainstorm (1983, MGM)—Douglas Trumbull

- Entertainment Effects Group.

Bram Stoker's Dracula (1992, Columbia/Zoetrope/Osiris)—Francis Ford Coppola

- Fantasy II Film Effects; Matte World Digital; Visual Concept Engineering; Available Light Ltd; 4-Ward Productions.

Braveheart (1995, 20th Century-Fox/Icon/Ladd Co)—Mel Gibson

- R/Greenberg Associates (West)/R/GA Digital Studios; Peerless Camera Co; The Computer Film Co.

Bringing Up Baby (1938, RKO)—Howard Hawks

- Vernon Walker.

Broken Arrow (1996, 20th Century-Fox/WCG)—John Woo

- Anatomorphex; The Chandler Group; Cinesite Digital Studios; Hunter-Gratzner Industries; Matte World Digital; MetroLight Studios; Optical Illusions; Pacific Data Images; Pacific Ocean Post Digital Film Group; Sessums Engineering; VIFX; Video Image; WKR Productions.

Buck Rogers (1939, Universal)—Ford Beebe & Saul Goodkind

- Ed Keyes.

Buck Rogers in the 25th Century TV Series (1979, Universal TV/ Glen A. Larson)—various

- Universal Hartland.

Bug (1975, Paramount)—Jeannot Szwarc

- Phil Cory.

Bullit (1968, Warner/Solar)—Peter Yates
Burnt Offerings (1976, United Artists/PEA)—Dan Curtis

- Cliff Wenger.

Bwana Devil (1952, United Artists)—Arch Oboler

- Henry Maak; Russell Shearman.

Candyman (1992, Columbia/Polygram/Propaganda)—Bernard Rose

- Digital Magic Co; Image Animation; Special Effects Unlimited.

Candyman: Farewell to the Flesh (1996, Columbia/Polygram/Propaganda)—Bill Condon

- Introvision International; Makeup & Effects Laboratories Inc.

Captain Midnight (1942, Columbia)—James W. Horne
Carnosaur (1993, New Horizons)—Adam Simon

- John Buechler; Alan Lasky; Magical Media Industries.

Casablanca (1942, Warner)—Michael Curtiz

- Lawrence Butler; Willard Van Enger.

Casper (1995, Universal/Amblin Entertainment/Harvey)—Brad Silberling

- Industrial Light & Magic.

Cast Away (2000, 20th Century-Fox/DreamWorks SKG/Playtone)—Robert Zemeckis

- Sony Pictures ImageWorks; MetroLight Studios; Travelling Pictures.

Cat Women of the Moon (1953, Three Dimensional Pictures)—Arthur Hilton

- David Commons; Willis Cook; Wah Chang.

Chain Reaction (1996, 20th Century-Fox/Chicago Pacific/Zanuck Co)—Andrew Davis

- Digital Domain.

Chicken Run (2000, DreamWorks/Aardman/Pathé)—Nick Park & Peter Lord

Child's Play (1988, United Artists)—Tom Holland

- Apogee Inc; Ruby-Spears Productions; Cinema Research Corporation.

Child's Play II (1990, Universal)—John Lafia

- Apogee Inc; Cinema Research Corporation; Image Engineering Inc.

Child's Play III (1991, Universal)—Jack Bender

- Kevin Yagher Productions; Mark Rappaport; Thomas Bolland.

Child's Play IV: Bride of Chucky (1998, Universal/Midwinter)—Ronnie Yu

- Kevin Yagher Productions; Gajdecki Visual Effects; MetroLight Studios; Toybox; Nerve Effects Inc; Paul Jones Effects Studio; Perpetual Motion Pictures.

Citizen Kane (1941, RKO)—Orson Welles

- Linwood G. Dunn; Vernon L. Walker; Russell Cully; Bill Leeds; Fitch Fulton; Mario Larrinaga; Douglas Travers.

City of Angels (1998, Warner/Atlas/Taurus/Regency)—Brad Silberling

- Sony Pictures Imageworks.

Clash of the Titans (1981, MGM/Schneer/Harryhausen)—Desmond Davis

- Ray Harryhausen; Camera Effects Co.

Cleopatra (1917, Fox Film Corporation)—J. Gordon Edwards

Cleopatra (1934, Paramount)—Cecil B. DeMille

- Barney Wolff.

Cleopatra (1963, 20th Century-Fox)—Joseph L. Mankiewicz

- L.B. Abbott; William Mittlestedt; Herbert Cheek; Emil Kosa Jr.

Cliffhanger (1993, Carolco/Canal Plus/Pioneer/RCS Video)—Renny Harlin

- Boss Film Studios; Cinema Research Corporation; Cinesite Digital Studios; Digital Magic Co; Pacific Data Images; Video Image.

Close Encounters of the Third Kind (1977, Columbia/EMI)—Steven Spielberg

- Douglas Trumbull; Roy Arbogast; Gregory Jein; Matthew Yuricich; Richard Yuricich; Future General Corporation.

Close Encounters of the Third Kind: Special Edition (1980, Columbia/EMI)—Steven Spielberg

Cocoon (1985, 20th Century-Fox/Zanuck-Brown)—Ron Howard

- Industrial Light & Magic.

Cocoon: The Return (1988, 20th Century-Fox/Zanuck-Brown)—Daniel Petrie

- Industrial Light & Magic; Cannom Creations.

Columbo TV series (1971–1990, Universal TV)—various

Con Air (1997, Buena Vista/Touchstone/Bruckheimer)—Simon West

- Buena Vista Imaging; DreamQuest Images; Matte World Digital; Reelistic F/X; The Burman Studio Inc.

Congo (1995, Paramount/Kennedy/Marshall)—Frank Marshall

- Industrial Light & Magic; Stan Winston Studio; Buena Vista Visual Effects; Banned from the Ranch Entertainment.

Conquest of Space (1955, Paramount)—Byron Haskin

- John P. Fulton; Irmin Roberts; Paul Lerpae, Ivyl Burks; Jan Domella.

Contact (1997, Warner/South Side)—Robert Zemeckis

- Big Sky Studios; Industrial Light & Magic; Pacific Ocean Post Digital; Weta Ltd; Sony Pictures ImageWorks; Warner Digital Studios.

Creature from The Black Lagoon (1954, UIP)—Jack Arnold

- Milicent Patrick.

Crocodile Dundee (1986, Paramount/Rimfire)—Peter Faiman
Crocodile Dundee II (1988, Paramount/Rimfire)—John Cornell

- Chris Murray; Bob Hicks; Steven Kirshoff; Alan Maxwell.

Crocodile Dundee in Los Angeles (2001, Visionview/Bungalow/Silver Lion)—Simon Wincer

- Amalgamated Pixels.

CSI TV Series (2000– , CBS Paramount Network Television/Alliance Atlantis Communications/Jerry Bruckheimer Television)—various

- Stargate Digital; Zoic Studios.

Cyborg Cop (1993, Nu Image/Nu World)—Sam Firstenberg

- Simon Harris.

Dances With Wolves (1990, Majestic/Tig)—Kevin Costner

- K.N.B. Effects Group.

Dante's Peak (1997, Universal/Pacific Western)—Roger Donaldson

- Digital Domain; Cinesite Digital Studios; Grant McCune Design; Illusion Arts Inc; Steve Johnson's XFX Inc; Vision Crew Unlimited; Composite Image Systems; Pacific Titles and Optical; Hollywood Digital.

Dark City (1998, New Line Cinema/Mystery Clock)—Alex Proyas

- DFilm Services.

Daylight (1996, Universal/Davis)—Rob Cohen

- Industrial Light & Magic; MetroLight Studios; VisionArt; Grant McCune Design; Bionics; Illusion Arts Inc.

Dead Poets Society (1989, Warner/Touchstone/Silver Screen)—Peter Weir
Dead Ringers (1988, Morgan Creek)—David Cronenberg

- Randall Balsmeyer; Gordon Smith; Lee Wilson.

Death Becomes Her (1992, Universal)—Robert Zemeckis

- Industrial Light & Magic.

Death Wish (1974, Paramount/de Laurentiis)—Michael Winner
Deep Blue Sea (1999, Warner)—Renny Harlin

- Cinesite Digital Studios; Edge Innovations; Flash Film Works; Grant McCune Design; Hammerhead Productions; Manex Visual Effects; Modern VideoFilm; VisionArt; The Film Factory; Industrial Light & Magic.

Deep Impact (1998, Paramount/DreamWorks)—Mimi Leder

- Industrial Light & Magic; WonderWorks.

Deep Rising (1998, Disney/Hollywood Pictures)—Stephen Sommers

- Buena Vista Imaging; DreamQuest Images; Industrial Light & Magic; Blur Studio; Banned From the Ranch Entertainment.

Dementia 13 (1963, Filmgroup Productions)—Francis Ford Coppola
Die Hard (1988, 20th Century-Fox/Silver Pictures/Gordon Company)—John McTiernan

- Boss Film Studios; Video Image.

Die Hard II: Die Harder (1990, 20th Century-Fox/Silver Pictures)—Renny Harlin

- Industrial Light & Magic; Apogee Inc.

Die Hard With a Vengeance (1995, 20th Century-Fox/Cinergi)—John McTiernan

- Mass.Illusions; Buena Vista Visual Effects; Pacific Title Digital; Sony Pictures ImageWorks.

Dial M for Murder (1954, Warner)—Alfred Hitchcock
Dinosaurus! (1960, Universal)—Irvin Yeaworth

- Tim Baar; Wah Chang; Gene Warren.

Dinosaur (2000, Disney)—Eric Leighton & Ralph Zondag

- Vision Crew Unlimited; The Secret Lab.

Doctor Dolittle (1967, 20th Century-Fox/APJAC)—Richard Fleischer

- L.B. Abbott.

Doctor X (1932, First National/Warner)—Michael Curtiz

- Fred Jackman Jr.

Don Juan (1926, Warner)—Alan Crosland

- Harry Redmond Sr.

Dr Cyclops (1940, Paramount)—Ernest Schoedsack

- Farciot Edouart; Gordon Jennings; Wallace Kelley; Jan Domela; Paul Lerpae.

Dr Dolittle (1998, 20th Century-Fox)—Betty Thomas

- CORE Digital Pictures; Cinesite Digital Studios; Jim Henson's Creature Shop; POP Film; Pacific Title; Mirage; Question Mark FX; VisionArt; Banned From the Ranch Entertainment.

Dr Dolittle 2 (2001, 20th Century-Fox)—Steve Carr

- Animated Engineering; Digital.Art.Media; Hammerhead Productions; Pacific Title Digital; Rhythm and Hues.

Dr Jekyll and Mr Hyde (1920, Famous Players-Lasky Corporation)—John Robertson
Dr Jekyll and Mr Hyde (1931, Paramount)—Rouben Mamoulian
Dr Jekyll and Mr Hyde (1941, MGM)—Victor Fleming

- Warren Newcombe; Peter Ballbusch.

Dr No (1962, United Artists)—Terence Young

- Roy Field; Frank George; Cliff Culley.

Dracula (1931, Universal)—Tod Browning

- Frank H. Booth; John P. Fulton; William Davidson.

Dracula (1979, Universal)—John Badham

- Albert Whitlock; Maurice Binder; Roy Arbogast; Tad Krzanowski.

Dragonheart (1996, Universal)—Rob Cohen

- Industrial Light & Magic; Tippett Studio; Perpetual Motion Pictures; Illusion Arts Inc.

Dragonslayer (1981, Paramount/Disney)—Matthew Robbins

- Industrial Light & Magic; Dreamstate Effects.

Duel (1971, Universal TV)—Steven Spielberg
Earth II (1970, MGM TV)—Tom Gries

- Jim Danforth; Howard A. Anderson; Art Cruickshank; J. McMillan Johnson.

Earth Versus the Flying Saucers (1956, Columbia/Clover)—Fred Sears

- Ray Harryhausen.

Earthquake (1974, Universal/Robson/Filmmakers Group)—Mark Robson

- Albert Whitlock; Glen Robinson; Frank Brendel.

Easy Rider (1969, Columbia)—Dennis Hopper
Edward Scissorhands (1990, 20th Century-Fox)—Tim Burton

- Stan Winston Studio; Dreamstate Effects.

Empire of the Sun (1987, Warner/Amblin Entertainment)—Steven Spielberg

- Industrial Light & Magic.

Equinox (1970, Tonylyn)—Jack Woods & Dennis Muren

- Howard A. Anderson.

Eraser (1996, Warner)—Chuck Russell

- Industrial Light & Magic; Composite Image Systems; Alterian Studios; Manex Visual Effects; Mass.Illusions; MetroLight Studios; Pacific Data Images.

Erin Brockovitch (2000, Jersey Films)—Steven Soderbergh

- Cinesite Digital Studios.

Escamotage d'une Dame au Théâtre Robert-Houdin (1896, Méliès)—Georges Méliès

- Georges Méliès.

Escape to Witch Mountain (1974, Disney)—John Hough

- Art Cruickshank; Danny Lee; David Domeyer; Hans Metz.

E.T. The Extra-Terrestrial (1982, Universal)—Steven Spielberg

- Industral Light & Magic; Carlo Rimbaldi.

Event Horizon (1997, Paramount/Golar/Impact Pictures)—Paul Anderson

- Cinesite Digital Studios; Mass.Illusions; Computer Film Co; AMX Digital.

Evil Dead (1981, New Line Cinema/Renaissance)—Sam Raimi

- Bart Pierce; Sam Raimi.

Evil Dead II: Dead by Dawn (1987, Universal/Renaissance/de Laurentiis)—Sam Raimi

- Acme Effects; Illuminations; Magic Lantern Inc; Anivision; Doug Beswick Productions Inc.

Evil Dead III: Army of Darkness (1993, Universal/Renaissance/de Laurentiis)—Sam Raimi

- Alterian Studios; Introvision International; Perpetual Motion Pictures; K.N.B. EFX Group Inc.

Face/Off (1997, Paramount/Touchstone/WCG)—John Woo

- Video Image; Animal Logic.

Fantasia (1940, Disney)—James Algar & Samuel Armstrong
Fantasia 2000 (1999, Disney)—James Algar & Gaetan Brizzi

Fantastic Voyage (1966, 20th Century-Fox)—Richard Fleischer

- L.B. Abbott; Art Cruickshank; Greg Jensen; Marcel Delgado; Emil Kosa Jr.

Fatal Attraction (1987, Paramount)—Adrian Lyne
Fiddler on the Roof (1971, United Artists/Mirisch)—Norman Jewison
Final Fantasy: The Spirits Within (2001, Square Soft/Bandai/Chris Lee)—Hironobu Sakaguchi & Motonori Sakakibara

- Remo Balcells; Mayumi Arakaki; Takahiko Akiyama; Paolo Costabel; Matthew Davies; Petronella Evers; Yasuko Asakura; Sergio Garcia Abad; Hiroshi Mori; Kerry Shea.

Finian's Rainbow (1968, Warner-Seven Arts)—Francis Ford Coppola
Firefox (1982, Malpaso)—Clint Eastwood

- Apogee Inc.

Fire in the Sky (1993, Paramount)—Robert Lieberman

- Computer Graphics Department; Industrial Light & Magic.

First Blood (1982, Carolco/Anabasis)—Ted Kotcheff

- Thom Noble; Thomas Fisher; George Erschbamer.

First Men in the Moon (1964, Columbia/Ameran)—Nathan Juran

- Ray Harryhausen.

Flash Gordon (1936, Universal)—Frederick Stephani

- Ed Keyes.

Flowers and Trees (1932, Disney)—Burt Gillett
Flubber (1997, Disney/Great Oaks)—Les Mayfield

- DreamQuest Images; Industrial Light & Magic; Computer Café Inc; POP Film; CORE Digital Pictures; Vision Crew Unlimited; Mobility Inc.

Forbidden Planet (1956, MGM)—Fred Wilcox

- Arnold Gillespie; Joshua Meador; Warren Newcombe; Bob Abrams; Joe Alves.

Forrest Gump (1994, Paramount)—Robert Zemeckis

- Industrial Light & Magic.

Frankenstein (1931, Universal)—James Whale

- Jack Pierce; John P. Fulton.

Frankenstein (1994, Columbia/Zoetrope/IndieProd/JSB)—Kenneth Branagh

- Computer Film Co.

Friday the 13th (1980, Paramount/Georgetown/Cunningham)—Sean Cunningham

- Tom Savini.

Friday the 13th, Part II (1981, Paramount/Georgetown/Greengrass)—Steve Miner

- Ross-Gaffney Inc.

Friday the 13th, Part III (1982, Paramount/Georgetown)—Steve Miner

- Makeup & Effects Laboratories Inc.

Friday the 13th, Part IV: The Final Chapter (1984, Paramount/Georgetown)—Joseph Zito

- Tom Savini.

Friday the 13th, Part V: A New Beginning (1985, Paramount)—Danny Steinmann

- Makeup & Effects Laboratories Inc.

Friday the 13th, Part VI: Jason Lives (1986, Paramount)—Tom McLoughlin

- Martin Becker.

Friday the 13th, Part VII: The New Blood (1988, Paramount/Friday Four)—John Buechler

- John Buechler; Lou Carlucci.

Friday the 13th, Part VIII: Jason Takes Manhattan (1989, Paramount)—Rob Hedden

- Cinema Research Corporation; Light & Motion; Reel Efx Inc.

Friday the 13th, Part IX: Jason Goes To Hell—The Final Friday (1993, New Line Cinema)—Adam Marcus

- MetroLight Studios; Bellissimo/Belardinelli Effects Inc.

Friday the 13th, Part X: Jason X (2001, New Line Cinema/Crystal Lake)—James Isaac

- Global Effects Inc; Command Post Toybox.

From Here to Eternity (1953, Columbia)—Fred Zinnemann
From the Earth to the Moon (1958, RKO/Waverly)—Byron Haskin

- Albert Simpson; Lee Zavitz.

Futureworld (1976, American International Pictures)—Richard T. Heffron

- Information International; Evans & Sutherland.

Ghost (1990, Paramount)—Jerry Zucker

- Boss Film Studios; Available Light Productions; Anatomorphex; Illusion Arts; Industrial Light & Magic; Video Image.

Ghost in the Shell (1995, Kodansha/Straight/Bandai)—Mamoru Oshii
Ghostbusters (1984, Columbia/Black Rhino)—Ivan Reitman

- Boss Film Studios; Entertainment Effects Group; Available Light Ltd.

Ghostbusters II (1989, Columbia)—Ivan Reitman

- Industrial Light & Magic; Dreamstate Effects.

Gladiator (2000, Universal/DreamWorks)—Ridley Scott

- Mill Film; Lee Lighting Ltd; AudioMotion Ltd.

Gojira (1954, Toho/Embassy/Transworld/Jewell)—Ishiro Honda

- Eiji Tsuburaya; Sadamasa Arikawa; Fumio Nakadai; Teizo Toshimitsu; Taka Yuki.

Godzilla (1998, TriStar/Centropolis)—Roland Emmerich

- Sony Pictures ImageWorks; Makeup & Effects Laboratories Inc; Anatomorphex; Centropolis Effects LLC; Question Mark FX; VisionArt.

Gone With The Wind (1939, MGM/Selznick)—Victor Fleming

- Jack Cosgrove; Clarence Slifer; F.R. Abbott; Fitch Fulton; Jack Shaw.

Good Will Hunting (1997, Miramax/Lawrence Bender)—Gus Van Zant

- Kavanaugh Special Effects.

Grandma's Reading Glass (1900, Smith)—George Albert Smith
Grease (1978, Paramount)—Randal Kleiser
Gremlins (1984, Warner/Amblin Entertainment)—Joe Dante

- Fantasy II Film Effects; Visual Concept Engineering; Dream-Quest Images; Chris Walas Inc.

Gremlins II: The New Batch (1990, Warner/Amblin Entertainment)—Joe Dante

- Matte World Digital; Makeup & Effects Laboratories Inc; Chris Baker.

Halloween (1978, Falcon)—John Carpenter

- Conrad Rothmann.

Halloween II (1981, Universal/de Laurentiis)—Rick Rosenthal

- Lawrence Cavanaugh; Frank Munoz.

Halloween III: Season of the Witch (1982, Universal/de Laurentiis)—Tommy Wallace

- Charles Moore; William Aldridge; Jon Belyeu.

Halloween IV: The Return of Michael Myers (1988, Trancas Inc)—Dwight Little

- Magical Media Industries.

Halloween V: The Revenge of Michael Myers (1989, Trancas Inc/Magnum)—Dominique Othenin-Girard

- Greg Landerer; Moni Mansano.

Halloween VI: The Curse of Michael Myers (1995, Miramax/Nightfall)—Joe Chappelle

- Frank Ceglia; Brad Hardin; John Cluff.

Halloween H20: 20 Years Later (1998, Dimension/Nightfall)—Steve Miner

- Digiscope; Rainmaker Digital Pictures; Digital Firepower.

Halloween: The Homecoming (2001, Dimension/Nightfall)—Rick Rosenthal

- K.N.B. EFX Group Inc.

Hannibal (2001, Universal/MGM/de Laurentiis)—Ridley Scott

- Mill Films; Visual Effects Co; FB-FX.

Harry Potter and the Chamber of Secrets (2002, Warner/Miracle)—Chris Columbus

- Cinesite(Europe); Framestore CFC; Gentle Giant Studios; Industrial Light & Magic; Lidar Services; Moving Picture Company; Plowman Craven & Associates; Thousand Monkeys; The Visual Effects Company.

Heaven's Gate (1980, United Artists)—Michael Cimino

- Ken Pepiot; Paul Stewart; Kevin Quibell; James Camomile; Stan Parks.

Hell's Angels (1930, United Artists)—Howard Hughes
High Tor (1956, CBS TV)—James Neilson & Franklin J. Schaffner
Highlander (1986, 20th Century-Fox/EMI)—Russell Mulcahy

- Chris Blunden.

Highlander II: The Quickening (1991, Lamb Bear)—Russell Mulcahy

- John Richardson; Allen Gonzales; Sam Nicholson; Jesse Silver.

Highlander III: The Sorcerer (1994, Miramax/Transfilm/Fallingcloud)—Andrew Morahan

- Digital Magic Company; Francois Aubry; Louis Craig.

Highlander IV: Endgame (2000, Mandalay/Dimension/Davis-Panzer)—Douglas Aarniokoski

- Threshold Entertainment Digital Research Lab; Effects Associates Ltd.

Home Alone (1990, 20th Century-Fox)—Chris Colombus
Home Alone II: Lost in New York (1992, 20th Century-Fox)—Chris Colombus
Home Alone III (1997, 20th Century-Fox)—Raja Gosnell
Hook (1991, Columbia)—Steven Spielberg

- Industrial Light & Magic.

House of Wax (1953, Warner)—André de Toth

- Julian Gunzburg.

How Jones Lost His Roll (1905, Edison Co)—Edwin S. Porter
How the Grinch Stole Christmas (2000, Imagine)—Ron Howard

- Digital Domain; MetroLight Studios.

Howard the Duck (1986, Universal/Lucasfilm)—Willard Huyck

- Industrial Light & Magic.

Humorous Phases of Funny Faces (1906, American Vitagraph)—James Stuart Blackton
I Know What You Did Last Summer (1997, Columbia/Mandalay)—Jim Gillespie

- John Milinac; William Purcell; Michael Hudson; Stan Blackwell; Michael Schorr.

Illusions Funambulesques (1903, Méliès)—Georges Méliès

- Georges Méliès.

I, Robot (2004, 20th Century-Fox/Mediastream IV/Davis)—Alex Proyas

- Digital Domain; Weta Digital; Rainmaker Digital Pictures; Pixel Magic; Forum Visual Effects; Film Roman Productions; Motion Analysis Studios; Earthdata Solutions; Patrick Tatopoulos Design; Grand Unified Theories.

I Still Know What You Did Last Summer (1998, Mandalay)—Danny Cannon

- Threshold Digital Research Labs.

Ice Age (2002, 20th Century-Fox)—Chris Wedge & Carlos Saldanha

In The Line of Fire (1993, Columbia/Castle Rock Entertainment)—Wolfgang Petersen

- Cinesite Digital Studios; R/Greenberg Associates; Sony Pictures ImageWorks.

In the Mouth of Madness (1995, New Line Cinema)—John Carpenter

- Industrial Light & Magic; K.N.B. EFX Group Inc.

Inchon (1981, MGM/United Artists)—Terence Young

Independence Day (1996, 20th Century-Fox/Centropolis)—Roland Emmerich

- 20th Century-Fox Digital Unit; Hunter-Gratzner Industries; Digiscope; Matte World Digital; VisionArt; Pacific Ocean Post Digital; Question Mark FX.

Indiana Jones and the Last Crusade (1989, Paramount/Lucasfilm)—Steven Spielberg

- Industrial Light & Magic.

Indiana Jones and the Temple of Doom (1984, Paramount/Lucasfilm)—Steven Spielberg

- Industrial Light & Magic; Modern Film Effects.

Innerspace (1987, Warner/Amblin Entertainment)—Joe Dante

- Industrial Light & Magic.

Intolerance (1916, Triangle Film Corporation/Wark Producing)—D.W. Griffith

Invaders from Mars (1953, National Pictures Corporation)—William Cameron Menzies

- Jack Cosgrove; Howard Lydecker; Jack Rabin; Irving Block.

Ishtar (1987, Columbia)—Elaine May

Island of Lost Souls (1932, Paramount)—Erle C. Kenton

- Gordon Jennings.

It! The Terror from Beyond Space (1958, Vogue)—Edward Cahn

- Paul Blaisdell.

It Came from Beneath the Sea (1955, Clover)—Robert Gordon

- Ray Harryhausen.

It Came from Outer Space (1953, UIP)—Jack Arnold

- David Horsley; Roswell A. Hoffmann.

It's Great to be Alive (1933, 20th Century-Fox)—Alfred L.Werker
Jack the Giant Killer (1962, United Artists)—Nathan Juran

- Jim Danforth; Howard A. Anderson; Lloyd Vaughan; Tim Baar; Wah Chang; Gene Warren; Augie Lohman.

Jason and the Argonauts (1963, Columbia/Morningside)—Don Chaffey

- Ray Harryhausen.

Jaws (1975, Universal/Zanuck-Brown)—Steven Spielberg

- Roy Arbogast; Robert Mattey; Richard Stutsman; Eddie Surkin.

Jaws II (1978, Universal)—Jeannot Szwarc

- Roy Arbogast; Robert Mattey; Eddie Surkin.

Jaws III (1983, Universal/Landsburg)—Joe Alves

- Praxis Film Works; Private Stock Effects Inc.

Jaws IV: The Revenge (1987, Universal)—Joseph Sargent

- Henry Millar; Mike Millar; Doug Hubbard; Dave Hubbard.

Jerry Maguire (1996, TriStar/Gracie)—Cameron Crowe
JFK (1991, Warner/Canal Plus/Regency)—Oliver Stone

- Miller Drake; Margaret Johnson; Bob Shelley; Randy Moore.

Joe's Apartment (1996, Warner/Geffen/MTV)—John Payson

- Blue Sky Studios; VIFX.

Johnny Mnemonic (1995, TriStar/Cinévision/Alliance)—Robert Longo

- Sony Pictures ImageWorks; Fantasy II Film Effects; CORE Digital Pictures; Gajdecki Visual Effects; FXSmith Inc.

Journey to the Center of the Earth (1959, 20th Century-Fox)—Henry Levin

- L.B. Abbott; James B. Gordon; Emil Kosa Jr.

Judge Dredd (1995, Hollywood Pictures/Cinergi)—Danny Cannon

- Mass.Illusions; Kleiser-Walczak Construction Co; Digital Fauxtography; Digital Magic Company; Manex Visual Effects; Sony Pictures ImageWorks.

Jumanji (1995, TriStar/Interscope/Teitler)—Joe Johnston

- Industrial Light & Magic; Amalgamated Dynamics.

Jurassic Park (1993, Universal/Amblin Entertainment)—Steven Spielberg

- Industrial Light & Magic; Stan Winston Studio.

Jurassic Park II: The Lost World (1997, Universal/Amblin Entertainment)—Steven Spielberg

- Industrial Light & Magic; Stan Winston Studio.

Jurassic Park III (2001, Universal/Amblin Entertainment)—Joe Johnston

- Industrial Light & Magic; Stan Winston Studio.

Just Imagine (1930, Fox Pictures)—David Butler
King Kong (1933, RKO)—Merian Cooper & Ernest Schoedsack

- Willis O'Brien.

King Kong (1976, Paramount/de Laurentiis)—John Guillermin

- Carlo Rimbaldi; Glen Robinson; Joe Day; Louis Lichtenfield; Aldo Puccini.

Knick Knack (1989, Pixar)—John Lasseter
LA Confidential (1997, Warner/Regency)—Curtis Hanson

- Computer Film Co.

La Cucaracha (1934, Pioneer Pictures Corp)—Lloyd Corrigan
La Lune à Un Mètre (1898, Méliès)—Georges Méliès

- Georges Méliès.

La Sortie Des Usines Lumière (1895, Lumière)—Louis Lumière
Ladri di Biciclette (1948, PDS-ENIC)—Vittorio de Sica
Lara Croft: Tomb Raider (2001, Paramount/Core Design/Tele-Munchen)—Simon West

- Digital Film; Mill Film; Cinesite (Europe); Computer Film Co; The Moving Picture Company; Peerless Camera Co.

Last Action Hero (1993, Columbia)—John McTiernan

- Industrial Light & Magic; Fantasy II Film Effects; Sony Pictures ImageWorks; Visual Concept Engineering; Dreamstate Effects; Composite Image Systems; Cinesite Digital Studios; Boss Film Studios; Alterian Studios; Baer Animation; Image G; Playhouse Pictures; Praxis Film Works; Available Light Ltd.

Le Grand Canal á Venise (1896, Lumière)—Alexandre Promio
Le Puits Fantastique (1903, Méliès)—Georges Méliès

- Georges Méliès.

Le Voyage Dans La Lune (1902, Méliès)—Georges Méliès

- Georges Méliès.

Le Voyage de Gulliver à Lilliput et Chez les Géants (1902, Méliès)—Georges Méliès

- Georges Méliès.

Lethal Weapon (1987, Warner/Silver Pictures)—Richard Donner

- Chuck Gaspar; Greg Callas; Elmer Hui; Thomas Mertz; Harold Selig; Bruce Robles.

Lethal Weapon 2 (1989, Warner/Silver Pictures)—Richard Donner

- Calvin Acord; Chris Burton; Don Hathaway; Steve Luport; David Peterson; Richard Hill; Larry Fuentes.

Lethal Weapon 3 (1992, Warner/Silver Pictures)—Richard Donner

- Matt Sweeney; Jon Belyeu; William Aldridge; Don Hathaway; Larry Fuentes;Ken Ebert; Arnold Peterson; Greg Jensen.

Lethal Weapon 4 (1998, Warner/Silver Pictures)—Richard Donner

- CinesiteDigital Studios; Mirage; Pacific Titles & Optical.

Leviathan (1989, MGM)—George P. Cosmatos

- Industrial Light & Magic; Cinema Research Corporation; Perpetual Motion Pictures.

L'homme à la Tête de Caoutchouc (1902, Méliès)—Georges Méliès

- Georges Méliès.

Liar Liar (1997, Universal/Imagine)—Tom Shadyac
Light & Heavy (1991, Pixar)—John Lasseter
Lights of New York (1928, Warner)—Bryan Foy
Logan's Run (1976, MGM)—Michael Anderson

- L.B. Abbott; James Liles; Frank Van der Veer; Matthew Yuricich; Glen Robinson.

London After Midnight (1927, MGM)—Tod Browning
Look Who's Talking (1989, TriStar/MCEG)—Amy Heckerling
Looker (1981, Warner/Ladd Co)—Michael Crichton

- Information International; Gary Demos; Doug Hubbard; Joe Day; Michael Lawler; Richard Taylor; Bert Terreri; John H. Whitney Jr.

Lost City of the Jungle (1946, Universal)—Lewis D. Collins& Ray Taylor
Lost in Space (1998, New Line Cinema/Prelude)—Stephen Hopkins

- Cinesite (Europe); Uli Meyer Studios; Mattes and Miniatures Atlantic Digital.

Luxo Jr. (1986, Pixar)—John Lasseter
M (1931, Nero Film)—Fritz Lang
Mandrake the Magician (1939, Columbia)—Norman Deming & Sam Nelson
Mark of the Vampire (1935, MGM)—Tod Browning

- Warren Newcombe; Tom Tutwiler.

Marooned (1969, Columbia)—John Sturges

- Lawrence Butler; Donald Glouner; Robie Robinson; Chuck Gaspar.

Mars Attacks! (1996, Warner)—Tim Burton

- Industrial Light & Magic; Warner Digital Studios; Acme Models.

Mary Poppins (1964, Disney)—Robert Stevenson

- Eustace Lycett; Peter Ellenshaw; Robert Mattey.

Max Payne (2008, Dune/Abandon/Collision/Foxtor)—John Moore

- SPIN VFX; Soho VFX; Mr. X; Modern VideoFilm; Acme F/X; HimAnI Productions.

Memoirs of an Invisible Man (1992, Warner/Canal Plus/Alcor/Regency)—John Carpenter

- Industrial Light & Magic; Effects Associates Inc.

Men In Black (1997, Columbia/Amblin Entertainment)—Barry Sonnenfeld

- Autumn Light Entertainment; Cinovation Studios; Industrial Light & Magic; Visual Concept Engineering; K.N.B. EFX Group Inc; Steve Johnson's XFX Inc; VisionArt; Banned From the Ranch Entertainment; ME FX; Question Mark FX; Mo42 Special Effects; Storyboard Inc.

Men In Black II (2002, Columbia/Amblin Entertainment)—Barry Sonnenfeld

- Industrial Light & Magic; Cinovation Studios.

Meteor (1979, AIP/Palladium)—Ronald Neame

- Margo Anderson; Glen Robinson; Robert Staples; Allen Blaisdell; Sam DiMaggio; Dave Cornell; Byron Bauer; Jim Doyle; Marcia Dripchak; John Greenleigh.

Metropolis (1926, UFA)—Fritz Lang

- Eugene Schufftan.

Mighty Joe Young (1949, RKO)—Ernest Schoedsack

- Willis O'Brien; Ray Harryhausen; Linwood Dunn; Harold Stine; Jack Shaw; Bert Willis; Marcel Delgado.

Mighty Joe Young (1998, Disney/RKO)—Ron Underwood

- Industrial Light & Magic; DreamQuest Images; Matte World Digital.

Mighty Morphin Power Rangers: The Movie (1995, 20th Century-Fox/Toei)—Brian Spicer

- Makeup Effects Laboratories; Video Image.

Mission: Impossible (1996, Paramount)—Brian de Palma

- Industrial Light & Magic; Cinesite Digital Studios; Computer Film Co.

Mission: Impossible II (2000, Paramount)—John Woo

- Cinesite Digital Studios; Cinesite (Europe); Matte World Digital; Pacific Data Images; Manex Visual Effects; Computer Film Co.

Mortal Kombat (1995, New Line Cinema/Threshold)—Paul Anderson

- Hunter-Gratzner Industries; Buena Vista Visual Effects; Metro-Light Studios.

Mortal Kombat II: Annihilation (1997, New Line Cinema/Threshold)—John Leonetti

- Kleiser-Walczak Construction Co; Digital Magic Co; Alterian Studios; Metrolight Studios; Optical Illusions Inc; Rainmaker Digital Pictures; Vision Crew Unlimited; Wildcat Digital Effects.

Mrs. Doubtfire (1993, 20th Century-Fox/Blue Wolf)—Chris Colombus

Mouse Hunt (1997, DreamWorks)—Gore Verbinski

- Animal Logic; Stan Winston Studio; Rhythm & Hues.

Multiplicity (1996, Columbia)—Harold Ramis

- Boss Film Studios.

Mutiny on the Bounty (1935, MGM)—Frank Lloyd
My Best Friend's Wedding (1997, TriStar/Predawn)—P.J. Hogan

- Alterian.

National Lampoon's Animal House (1978, Universal)—John Landis
Night Gallery TV series (1970–1973, Universal TV)—various
Nikita (1990, Gaumont/Tiger)—Luc Besson
Nosferatu, eine Symphonie des Grauens (1921, Prana-Film/Jofa-Atelier Berlin)—F.W. Murnau
Obsession (1976, Columbia)—Brian De Palma

- Joe Lombardi.

One Million BC (1940, United Artists)—Hal Roach & Hal Roach Jr.

- Roy Seawright; Jack Shaw.

One Million Years BC (1966, Hammer/Seven Arts)—Don Chaffey

- Ray Harryhausen.

Only Angels Have Wings (1939, Columbia)—Howard Hawks

- Roy Davidson; Edwin Hahn; Harry Redmond Jr.

Out of the Inkwell (1918, Bray Studios)—Dave Fleischer & Max Fleischer
Partie de Cartes (1895, Lumière)—Louis Lumière
Patton (1970, 20th Century-Fox)—Franklin J. Schaffner

- L.B. Abbott; Art Cruickshank; Alex Weldon.

Paulie (1998, DreamWorks/Mutual Film)—John Roberts

- Sony Pictures ImageWorks; Santa Barbara Studios; Light Matters Inc; Banned From the Ranch Entertainment; Hammerhead Productions; Pixel Envy.

Phenomenon (1996, Touchstone)—Jon Turteltaub

- Sony Pictures ImageWorks.

Philadelphia (1993, TriStar/Clinca Estetico)—Jonathan Demme
Plane Crazy (1928, Disney)—Ub Iwerks
Planet of the Apes (1968, 20th Century-Fox/Apjac)—Franklin Schaffner

- John Chambers; L.B. Abbott; Art Cruickshank; Emil Kosa Jr.

Planet of the Apes (2001, 20th Century-Fox/Zanuck Co)—Tim Burton

- Industrial Light & Magic; Sphere FX Ltd.

Platoon (1986, Hemdale/Kopelson)—Oliver Stone

- Yves De Bono; Andrew Wilson.

Pocahontas (1995, Disney)—Mike Gabriel & Eric Goldberg
Point of No Return (1993, Warner)—John Badham
Pokémon: The First Movie (1999, Nintendo/Pikachu Project 99/ Shogakukan)—Michael Haigney & Kunihiko Yuyama
Police Academy (1984, Warner)—Hugh Wilson
Poltergeist (1982, MGM)—Tobe Hooper

- Industrial Light & Magic.

Poltergeist II: The Other Side (1986, MGM)—Brian Gibson

- Boss Film Studios.

Poltergeist III (1988, MGM)—Gary Sherman

- Gary Sherman; Calvin Acord.

Practical Magic (1998, Warner/Village Roadshow/Fortis)—Griffin Dunne

- Cinesite Digital Studios; Composite Image Systems; Hammerhead Productions; Tippett Studio.

Predator (1987, 20th Century-Fox/Amercent)—John McTiernan

- Stan Winston Studio; DreamQuest Images; Howard A. Anderson Company; Video Image.

Predator II (1990, 20th Century-Fox)—Stephen Hopkins

- Stan Winston Studio; Visual Concept Engineering; Howard A. Anderson Company; Pacific Titles & Optical; Video Image.

Pretty Woman (1990, Touchstone/Silver Screen)—Garry Marshall
Rainbow (1995, Filmline/Winchester)—Bob Hoskins

- Sony Pictures Imageworks.

Raiders of the Lost Ark (1981, Paramount/Lucasfilm)—Steven Spielberg

- Industrial Light & Magic; Modern Film Effects.

Rain Man (1988, United Artists/Mirage/Star Partners)—Barry Levinson
Rambo: First Blood, Part II (1985, Carolco/Anabasis)—George P. Cosmatos

- Thomas Fisher; Jay King; William Purcell; Cliff Wenger Jr.

Rambo III (1988, Carolco)—Peter MacDonald

- Introvision International.

Ransom (1996, Touchstone/Imagine/Grazer)—Ron Howard

- John Alagna; Michael Curtis; Ed Hanson; Eli Jarra.

Red's Dream (1987, Pixar)—John Lasseter
Redskin (1928, Paramount Famous Lasky Corporation)—Victor Schertzinger
Red Sonja (1985, de Laurentiis/Famous)—Richard Fleischer

- Van der Veer Photo Effects; Germano Natali.

Repas de Bébé (1895, Lumière)—Louis Lumière
Revenge of the Shogun Women (1977, Eastern Media)—Mei Chung Chang
Revolution (1985, Warner/Goldcrest Films International)—Hugh Hudson

- Alan Whibley; Nick Middleton; Dave Eltham.

Robin Hood: Prince of Thieves (1991, Warner/Morgan Creek)—Kevin Reynolds

- Matte World Digital.

Robinson Crusoe On Mars (1964, Paramount)—Byron Haskin

- Albert Whitlock; Lawrence Butler; Farciot Edouart.

RoboCop (1987, Orion)—Paul Verhoeven

- Robert Blalack; Phil Tippett; Rocco Gioffre; Tom St. Amand; Rob Bottin.

RoboCop II (1990, Orion/Tobor)—Irvin Kershner

- Tippett Studio; Matte World Digital.

RoboCop III (1993, Orion)—Fred Dekker

- Tippett Studio; Dreamstate Effects; Pacific Data Images; Light Visions; Visual Concept Engeering.

Robots (2005, 20th Century-Fox/Blue Sky Studios)—Chris Wedge & Carlos Saldanha
Rocky (1976,United Artists/Chartoff-Winkler)—John G. Avildsen
Rocky II (1979, Chartoff-Winkler)—Sylvester Stallone
Rocky III (1982, MGM/United Artists)—Sylvester Stallone
Rocky IV (1985, MGM)—Sylvester Stallone
Rocky V (1990, United Artists/Chartoff-Winkler)—John G. Avildsen
Runaway Bride (1999, Touchstone/Interscope/Lakeshore)—Garry Marshall
Rush Hour (1998, New Line Cinema/Birnbaum)—Brett Ratner

- Wildcat Digital Effects; Composite Image Systems; The Production Plant; Optical Illusions Inc.

Sangaree (1953, Paramount)—Edward Ludwig

- John P. Fulton; Farciot Edouart; Paul Lerpae.

Saving Private Ryan (1998, Paramount/ DreamWorks)—Steven Spielberg

- Industrial Light & Magic.

Saw (2004, Evolution Entertainment/Twisted Pictures)—James Wan

- Title House Digital.

Saw II (2005, Evolution Entertainment/Twisted Pictures/Lions Gate)—Darren Lynn Bousman

- C.O.R.E. Digital Pictures.

Saw III (2006, Evolution Entertainment/Twisted Pictures/Lions Gate)— Darren Lynn Bousman

- Switch VFX; Good Enterprises; Grand Unified Theories.

Saw IV (2007, Twisted Pictures/Lions Gate)—Darren Lynn Bousman

- Switch VFX; Acme F/X.

Saw V (2008, Twisted Pictures)—David Hackl

- Switch VFX; Acme F/X.

Saw VI (2009, Twisted Pictures)—Kevin Greutert

- Switch VFX; Acme F/X.

Saw 3D: The Final Chapter (2010, Twisted Pictures)—Kevin Greutert

- Switch VFX; Acme F/X.

Scary Movie (2000, Dimension/Wayans Bros/Brillstein-Grey)—Keenen Wayans

Schindler's List (1993, Universal/Amblin Entertainment)—Steven Spielberg

- Industrial Light & Magic.

Scott of the Antarctic (1948, Eiling)—Charles Frend

- Jim Morahan; Richard Dendy; Geoffrey Dickinson; Norman Ough; Sydney Pearson.

Sculpture Moderne (1906, Pathé)—Segundo de Chomòn

Se7en (1995, New Line Cinema)—David Fincher

- Greg Kimble; Peter Frankfurt; Steven Puri; Tim Thompson; Danny Cangemi; Peter Albiez.

Sextone for President (1988, Kleiser-Walczak)—Jeff Kleiser & Diana Walczak

- Kleiser-Walczak Construction Co.

Shakespeare in Love (1998, Universal/Miramax/Bedford Falls)—John Madden

- The Magic Camera Co; United Special Effects.

Shane (1953, Paramount)—George Stevens

- Farciot Edouart; Gordon Jennings.

Sherlock Holmes: A Game of Shadows (2011, Warner/Village Roadshow)—Guy Ritchie

- Plowman Craven & Associates; BlueBolt; Framestore; Mark Roberts Motion Control; Moving Picture Company; ReelEye Company; The Visual Effects Company.

Shrek (2001, DreamWorks/Pacific Data Images)—Andrew Adamson & Vicky Jenson
Shrek 2 (2004, DreamWorks/Pacific Data Images)—Andrew Adamson & Kelly Asbury
Shrunken Heads (1994, Paramount/Full Moon)—Richard Elfman

- Alchemy FX; Paul Gentry; Brian Hanable; Candice Scott.

Silence of the Lambs (1991, Orion/Rank/Demme)—Jonathan Demme

- Dwight Benjamin-Creel.

Silent Running (1972, Universal)—Douglas Trumbull

- John Dykstra; Vernon Archer; Gerald Endler; Marlin Jones; James Rugg; Richard Helmer; William Shourt; Richard Alexander; Wayne Smith; Richard Yuricich; Douglas Trumbull.

Sinbad the Sailor (1947, RKO)—Richard Wallace

- Vernon Walker; Harold Wellman.

Sinbad and the Eye of the Tiger (1977, Columbia)—Sam Wanamaker

- Ray Harryhausen.

Singin' in the Rain (1952, MGM)—Stanley Donen & Gene Kelly

- Warren Newcombe; Irving Ries; Mark Davis.

Sister Act (1992, Touchstone)—Emile Ardolino

Sleepers (1996, Warner/Polygram/Propaganda/Baltimore)—Barry Levinson

- Industrial Light & Magic.

Small Soldiers (1998, Universal/ DreamWorks/Amblin Entertainment)—Joe Dante

- Industrial Light & Magic; Stan Winston Studio; Digiscope; Optical Illusions Inc.

Snake Eyes (1998, Paramount/Touchstone)—Brian de Palma.

- Industrial Light & Magic; Texa FX Group; Computer Film Co.

Snow White and the Seven Dwarfs (1937, Disney)—David Hand
Something Evil (1972, CBS Entertainment/Belford)—Steven Spielberg
Son of Kong (1933, RKO)—Ernest Schoedsack

- Willis O'Brien.

Soylent Green (1973, MGM)—Richard Fleischer

- Braverman Productions; Robert Hoag; Augie Lohman; Matthew Yuricich.

Space Jam (1996, Warner/Northern Lights)—Joe Pytka

- Cinesite Digital Studios; Vision Crew Unlimited; Charles Gammage Animation; Effects Associates Ltd; Uli Meyer Productions; Exceptional Opticals Inc; Spaff Animation; Rees/Leiva Productions; Warner Bros Feature Animation; Stardust Pictures; Premier Films Ltd.

Space Camp (1986, ABC)—Harry Winer

- Van der Veer Photo Effects; Video Image.

Spawn (1997, New Line Cinema/Juno)—Mark Dippé

- CORE Digital Pictures; Industrial Light & Magic; K.N.B. EFX Group Inc; Western Images; Rhythm & Hues; EFX Unlimited Inc.

Spawn of the North (1938, Paramount)—Henry Hathaway

- Gordon Jennings; Farciot Edouart; Barney Wolff; Jan Domela; Paul Lerpae.

Special Effects—IMAX (1996, Nova/WGBH Boston/ILM)—Ben Burtt

- Industrial Light & Magic; New Wave International; Imagica Corporation.

Species (1995, MGM)—Roger Donaldson

- H.R. Geiger; Richard Edlund; Boss Film Studios; Steve Johnson's XFX Inc.

Species II (1998, MGM/FGM)—Peter Medak

- Steve Johnson's XFX Inc; Digital Magic Co; Vision Crew Unlimited.

Speed (1994, 20th Century-Fox)—Jan de Bont

- Sony Pictures ImageWorks; Alterian Studios; Video Image; Grant McCune Design; Sessums Engineering; The Burman Studio Inc.

Speed II: Cruise Control (1997, 20th Century-Fox)—Jan de Bont

- Industrial Light & Magic; Rhythm and Hues.

Star Trek TV Series (1966–1969, Paramount TV)—various
Star Trek: The Motion Picture (1979, Paramount)—Robert Wise

- Douglas Trumbull; John Dykstra; Apogee Inc; Foundation Imaging; Robert Abel & Associates.

Star Trek II: The Wrath of Khan (1982, Paramount)—Nicholas Meyer

- Industrial Light & Magic.

Star Trek III: The Search for Spock (1984, Paramount)—Leonard Nimoy

- Industrial Light & Magic.

Star Trek IV: The Voyage Home (1986, Paramount)—Leonard Nimoy

- Industrial Light & Magic; Westheimer Co; Video Image; Novocom Inc.

Star Trek V: The Final Frontier (1989, Paramount)—William Shatner

- Makeup & Effects Laboratories Inc; Associates & Ferren; Novocom Inc; Illusion Arts Inc.

Star Trek VI: The Undiscovered Country (1991, Paramount)—Nicholas Meyer

- Industrial Light & Magic; Matte World Digital.

Star Trek VII: Generations (1994, Paramount)—David Carson

- Industrial Light & Magic; Santa Barbara Studios; Special Effects Unlimited; Composite Image Systems; Illusion Arts Inc.

Star Trek VIII: First Contact (1996, Paramount)—Jonathan Frakes

- Industrial Light & Magic; Matte World Digital; Pacific Ocean Post Digital; Todd Masters Company; Makeup & Effects Laboratories Inc; VisionArt; Question Mark FX; Illusion Arts Inc.

Star Trek IX: Insurrection (1998, Paramount)—Jonathan Frakes

- Sony Pictures ImageWorks; Blue Sky Studios; Composite Image Systems; Fulcrum Studios LLC; Hunter-Gratzner Industries; Pacific Ocean Post Digital; Makeup & Effects Laboratories Inc; Video Image; Santa Barbara Studios.

Star Wars (1977, 20th Century-Fox/Lucasfilm)—George Lucas

- Industrial Light & Magic; DePatie-Freleng Enterprises; Master Film Effects; Ray Mercer & Co; Image West Ltd; Modern Film Effects.

Star Wars: The Empire Strikes Back (1980, Lucasfilm)—Irvin Kershner

- Industrial Light & Magic; Modern Film Effects; Ray Mercer & Co; Westheimer Co.

Star Wars: Return of the Jedi (1983, Lucasfilm)—Richard Marquand

- Industrial Light & Magic; Dreamstate Effects; California Film; Movie Magic.

Star Wars—Special Edition (1997, Lucasfilm)—George Lucas
Star Wars: The Empire Strikes Back—Special Edition (1997, Lucasfilm)—Irvin Kershner
Star Wars: Return of the Jedi—Special Edition (1997, Lucasfilm)—Richard Marquand
Star Wars: Episode I—The Phantom Menace (1999, Lucasfilm)—George Lucas

- Industrial Light & Magic; Gentle Giant Studios.

Star Wars: Episode II—Attack of the Clones (2002, Lucasfilm)—George Lucas

- Industrial Light & Magic; Any Effects; FB-FX; Gentle Giant Studios.

Star Wars: Episode III—Revenge of the Sith (2005, Lucasfilm)—George Lucas

- Industrial Light & Magic; Gentle Giant Studios; The Third Floor.

Star Wars: Episode I—The Phantom Menace 3D (2012, Lucasfilm)—George Lucas
Stargate (1994, Carolco/Canal Plus/Centropolis)—Roland Emmerich

- Hunter-Gratzner Industries; Kleiser-Walczak Construction Co; Cinema Research Corporation; Available Light Ltd; Cobitt Design Inc; SPFX Inc.

Starship Troopers (1997, Columbia/TriStar/Touchstone)—Paul Verhoeven

- Amalgamated Dynamics; Industrial Light & Magic; Sony Pictures ImageWorks; Thunderstone; Kevin Yagher Productions; Visual Concept Engineering; Video Image; Boss Film Studios; Banned From the Ranch Entertainment; Dreamstate Effects; Manex Visual Effects; Tippett Studio.

Stay Tuned (1992, Warner/Morgan Creek)—Peter Hyams

- Rhythm and Hues.

Steamboat Willie (1928, Disney)—Ub Iwerks & Walt Disney
Strange Days (1995, 20th Century-Fox/Lightstorm)—Kathryn Bigelow

- Digital Domain; Fantasy II Film Effects; Video Image; Light Matters Inc; Banned From the Ranch Entertainment; Computer Film Co.

Street Fighter (1994, TriStar/Capcom)—Steven de Souza

- International Creative Effects; Photon Stockman.

Stuart Little (1999, Columbia/Global Medien)—Rob Minkoff

- Sony Pictures ImageWorks; Rhythm & Hues; Centropolis Effects LLC; Patrick Tatopoulos Design Inc.

Super Mario Bros. (1993, Hollywood Pictures/Nintendo/Cinergi)—Rocky Morton & Annabel Jankel

- Animated Engineering; Cinesite Digital Studios.

Superman (1948, Columbia)—Spencer Gordon Bennet & Thomas Carr

- Howard Swift.

Superman (1978, Salkind/Dovemead/Film Export)—Richard Donner

- Cineflex Ltd; Cinema Research Corporation; Van der Veer Photo Effects; Howard A. Anderson Co.

Superman II (1980, Warner/Salkind/Film Export)—Richard Lester

- Cinema Research Corporation; Howard A. Anderson Co; Zoptic.

Superman III (1983, Cantharaus/Dovemead/Film Export)—Richard Lester

- Effects Associates Ltd.

Superman IV: The Quest For Peace (1987, Warner/Cannon/Golan-Globus)—Sidney Furie

- Westheimer Company; Visual Concept Engineering; Ray Mercer & Co; Acme Effects; Howard A. Anderson Co; Disney Photographic Effects; Cinema Research Corporation; Buena Vista Visual Effects; Available Light Ltd.

Surprise (1991, Pixar)—John Lasseter

Tarantula (1955, UIP)—Jack Arnold

- David Horsley.

Tarzan (1999, Disney/Edgar R. Burroughs Inc)—Chris Buck & Kevin Lima

Taxi Driver (1976, Columbia)—Martin Scorsese

Teenage Mutant Ninja Turtles (1990, New Line Cinema/Golden Harvest)—Steve Barron

- Jim Henson's Creature Shop.

Teenage Mutant Ninja Turtles II: The Secret of the Ooze (1991, New Line Cinema/Golden Harvest/Northshore)—Michael Pressman

- Jim Henson's Creature Shop.

Teenage Mutant Ninja Turtles III (1993, Golden Harvest/Clearwater)—Stuart Gillard

- Eric Allard; Joseph Mercurio; Beecher Tomlinson; Jesse Silver.

Terminator (1984, Pacific Western/Hemdale/Cinema 84/EuroFilm)—James Cameron

- Fantasy II Film Effects; Stan Winston Studio; Ray Mercer & Co.

Terminator 2: Judgment Day (1991, Carolco/Pacific Western/Lightstorm)—James Cameron

- Industrial Light & Magic; Fantasy II Film Effects; Stan Winston Studio; Video Image; Pacific Data Images; The Artificial Lighting Company; 4-Ward Productions; Pacific Titles & Opticals; Make-up Effects Unlimited; Electric Image Inc; Genesis Optical Efx.

Terminator 3 (2002, Intermedia/Toho-Towa/VCL/C-2)—Jonathan Mostow

- Industrial Light & Magic; Blue Sky Studios/VIFX; Cinesite Digital Studios; Sphere FX Ltd; POP Film; Matte World Digital; Pacific Data Images; Sony Pictures ImageWorks; Vision Crew Unlimited; Stan Winston Studio; Banned From the Ranch Entertainment.

Terry and the Pirates (1940, Columbia)—James W. Horne

The Abyss (1989, 20th Century-Fox/Pacific Western/Lightstorm)—James Cameron

- Industrial Light & Magic; DreamQuest Images; XFX Images; WonderWorks; Stetson Visual Services; Fantasy II Film Effects; Video Image; The Design Setters Corporation; Steve Johnson's XFX Inc.

The Ace of Scotland Yard (1929, Universal)—Ray Taylor

The Adventures of André and Wally B. (1984, Lucasfilm)—Alvy Ray Smith

The Adventures of Baron Munchausen (1988, Columbia)—Terry Gilliam

- Digital Pictures Ltd; The Magic Camera Co; Peerless Camera Co.

The Adventures of Robin Hood (1938, Warner)—Michael Curtiz & William Keighley

The Adventures of Sir Galahad (1949, Columbia)—Spencer Gordon Bennet

The Agony and the Ecstasy (1965, 20th Century-Fox)—Carol Reed

- L.B. Abbott; Art Cruickshank; Emil Kosa Jr.

The Amityville Horror (1979, AIP/Cinema 77)—Stuart Rosenberg

- Dell Rheaume; William Cruse; Allen Blaisdell.

The Andromeda Strain (1971, Universal)—Robert Wise

- James Shourt.

The Avengers (1998, Warner)—Jeremiah Chechik

- Cinesite (Europe); The Magic Camera Co; Computer Film Co.

The Battle of Santiago (1899, Lubin)—James Stuart Blackton & Albert Smith

- James Stuart Blackton; Albert Smith.

The Beast from 20,000 Fathoms (1953, Mutual)—Eugène Lourié

- Ray Harryhausen; Willis Cook; George Lofgren; Eugène Lourié.

The Beast of Hollow Mountain (1956, United Artists)—Edward Nassour & Ismael Rodriguez

- Louis DeWitt; Jack Rabin.

The Beastmaster (1982, MGM/Ecta)—Don Coscarelli

- Cruse & Co.

The Bible: In the Beginning . . . (1966, 20th Century-Fox/Seven Arts)—John Huston

- Linwood G. Dunn; Augie Lohman; Zeus LaNero.

The Birds (1963, Universal)—Alfred Hitchcock

- Ub Iwerks; Albert Whitlock; Dave Fleischer; Larry Hampton; Bob Broughton; Roswell Hoffmann; Chuck Gaspar.

The Birth of a Nation (1915, Griffith/Epoch Corp)—D.W. Griffith
The Black Hole (1979, Disney)—Gary Nelson

- Art Cruickshank; Harrison Ellenshaw; Peter Ellenshaw; Eustace Lycett.

The Black Scorpion (1957, Warner/Melford-Dietz)—Edward Ludwig

- Willis O'Brien; Pete Peterson.

The Blair Witch Project (1999, Haxan)—Daniel Myrick & Eduardo Sanchez
The Blob (1958, Tonylyn/Fairview)—Irvin Yeaworth Jr.

- Bart Sloane.

The Blob (1988, TriStar/Palisades)—Chuck Russell

- Alterian Studios; Makeup & Effects Laboratories Inc.

The Bride of Frankenstein (1935, Universal)—James Whale

- John P. Fulton; David Horsley; Ken Strickfaden.

The Bridge on the River Kwai (1957, Columbia/Horizon)—David Lean
The Cabinet of Dr. Caligari (1919, Decla-Bioscop AG)—Robert Wiene
The Cemetery Club (1993, Touchstone)—Bill Duke

- Sam Barkin.

The Company of Wolves (1984, ITC/Palace)—Neil Jordan

- ReelEye Company.

The Conversation (1974, Paramount/American Zoetrope)—Francis Ford Coppola
The Conquest of Space (1955, Paramount)—Byron Haskin

- John P. Fulton; Farciot Edouart; Jan Domela; Ivyl Burks; Paul Lerpae.

The Crow (1994, Pressman/Entertainment Media/Most)—Alex Proyas

- DreamQuest Images; Ultimate Effects.

The Crow II: City of Angels (1996, Miramax/Dimension/Most)—Tim Pope

- Digiscope; Buzz Image Group.

The Crow III: Salvation (2000, Pressman/Most)—Bharat Nalluri

- Thomas Rainone; Roger Nall; Doyle Rockwell; Michael Shelton.

The Dam Busters (1955, ABPC)—Michael Anderson

- George Blackwell; Gilbert Taylor; Ronnie Wass

The Day the Earth Stood Still (1951, 20th Century-Fox)—Robert Wise

- L.B. Abbott; Fred Sersen; Emil Kosa.

The Day the Earth Stood Still (2008, 20th Century-Fox/Dune Entertainment III/Earth Canada/Hammerhead)—Scott Derrickson

- The Aaron Sims Company; MastersFX; Pacific Title; Weta Digital; Cinesite; Hydraulx; Flash Film Works; Cos Fx Films; Digital Dimension; At The Post; Image Engine Design; Persistence of Vision; Pixel Liberation Front.

The Deep (1977, EMI/Casablanca)—Peter Yates

- Ira Anderson Jr.

The Devil-Doll (1936, MGM)—Tod Browning
The Execution of Mary, Queen of Scots (1895, Edison)—Alfred Clark
The Exorcist (1973, Warner/Hoya)—William Friedkin

- Marcel Vercoutere; Marv Ystrom; Dick Smith.

The Exorcist II: The Heretic (1977, Warner)—John Boorman

- Van der Veer Photo Effects.

The Exorcist III (1990, Morgan Creek)—William Blatty

- DreamQuest Images.

The Exorcist: Special Edition (1998, Warner)—William Friedkin

- Jennifer Law-Stump; Pacific Title Digital.

The Faculty (1998, Dimension/Los Hooligans)—Roberto Rodriguez

- K.N.B. EFX Group Inc; Centropolis Effects; Rhythm & Hues; XFX Images; Threshold Entertainment Digital Research Lab; Hybride Technologies; Digiscope; Cinema Production Services.

The Fifth Element (1997, TriStar/Gaumont)—Luc Besson

- Digital Domain; Vision Crew Unlimited; The Magic Camera Co.

The Firm (1993, Paramount)—Sydney Pollack
The Flintstones (1994, Universal/Hanna-Barbera/Amblin Entertainment)—Brian Levant

- Industrial Light & Magic; Makeup & Effects Laboratories Inc.

The Flintstones in Viva Rock Vegas (2000, Universal/Hanna-Barbera)—Brian Levant

- Jim Henson's Creature Shop; Cinesite Digital Studios; Rhythm & Hues; MetroLight Studios.

The Fly (1958, 20th Century-Fox)—Kurt Neumann

- L.B. Abbott; James B. Gordon.

The Fly (1986, 20th Century-Fox/Brooksfilms)—David Cronenberg

- Available Light Productions; Intrigue; Chris Walas.

The Fly II (1989, 20th Century-Fox)—Chris Walas

- Available Light Productions.

The Fugitive (1993, Warner)—Andrew Davis

- Ultimate Effects; Introvision International.

The Gallopin' Gaucho (1928, Disney)—Ub Iwerks
The Godfather (1972, Paramount)—Francis Coppola

- Paul J. Lombardi; Sass Bedig; A.D. Flowers.

The Godfather, Part II (1974, Paramount/Coppola Co)—Francis Coppola

- Paul J. Lombardi; A.D. Flowers.

The Godfather, Part III (1990, Paramount/Zoetrope)—Francis Coppola

- Industrial Light & Magic.

The Golden Voyage of Sinbad (1974, Columbia/Morningside)—Gordon Hessler

- Ray Harryhausen.

The Great Gatsby (1974, Paramount/Newdon)—Jack Clayton
The Great Train Robbery (1903, Edison Co)—Edwin S. Porter
The Greatest Show on Earth (1952, Paramount)—Cecil B. de Mille

- Gordon Jennings; Paul Lerpae; Barney Wolff; Devereaux Jennings.

The Green Mile (1999, Warner/Castle Rock)—Frank Darabont

- Industrial Light & Magic; Matte World Digital; Rhythm & Hues; Pacific Ocean Post Digital; K.N.B EFX Group Inc.

The Hindenburg (1975, Universal)—Robert Wise

- Albert Whitlock; Clifford Stine; Bill Taylor; Glen Robinson; Andy Evans; Frank Brendel; Robert Beck.

The Hurricane (1937, United Artists/Goldwyn)—John Ford

- James Basevi; Ray Binger; R.T. Layton; Lee Zavitz.

The Hunt For Red October (1990, Paramount)—John McTiernan

- Industrial Light & Magic; Boss Film Studios; Slagle Minimotion Inc; Video Image; Chandler Group Visual Effects.

The Incredible Shrinking Man (1957, UIP)—Jack Arnold

- Everett Broussard; Roswell Hoffmann; Charles Baker; Fred Knoth.

The Indian in the Cupboard (1995, Columbia/Paramount/Kennedy-Marshall)—Frank Oz

- Industrial Light & Magic; Pacific Titles & Optical.

The Invisible Boy (1957, MGM)—Herman Hoffman

- Irving Block; Louis DeWitt; Jack Rabin.

The Invisible Man (1933, Universal)—James Whale

- John P. Fulton; John Mescall; Frank Williams.

The Invisible Ray (1936, Universal)—Lambert Hillyer

- John P. Fulton; Raymond Lindsay.

The Iron Horse (1924, Fox Film Corporation)—John Ford
The Island at the Top of the World (1974, Disney)—Robert Stevenson

- Art Cruickshank; Danny Lee; Peter Ellenshaw; Allan Maley; Mike Reedy.

The Jazz Singer (1927, Warner)—Alan Crosland

- Nugent Slaughter.

The Karaté Kid (1984, Columbia/Delphi)—John G. Avildsen
The Karaté Kid II (1986, Columbia/Delphi)—John G. Avildsen
The Karaté Kid III (1989, Columbia/Weintraub)—John G. Avildsen
The King and I (1956, 20th Century-Fox)—Walter Lang

- Ray Kellogg; Doug Hubbard.

The Last Starfighter (1984, Universal/Lorimar)—Nick Castle

- Digital Productions Inc; Industrial Light & Magic; Van der Veer Photo Effects; Apogee Inc.

The Lawnmower Man (1992, New Line Cinema/Allied Vision/Fuji Eight)—Brett Leonard

- Angel Studios; David Stipes Productions; Homer & Associates; Reel Efx Inc; Western Images; XAOS; EFX Unlimited Inc.

The Lawnmower Man II: Beyond Cyberspace (1996, New Line Cinema)—Farhad Mann

- Cinesite Digital Studios.

The Lion King (1994, Disney)—Roger Allers & Rob Minkoff
The Lion King 3D (2011, Disney)—Roger Allers & Rob Minkoff
The Lone Ranger and the Lost City of Gold (1958, United Artists)—Lesley Selander
The Longest Day (1962, 20th Century-Fox)—Ken Annakin, Andrew Marton, Bernhard Wicki & Darryl F. Zanuck

- Jean Fouchet; David Horsley; Wally Veevers; Bob Cuff; Karl Baumgartner; Karl Helmer; Alex Weldon.

The Lord of the Rings: The Fellowship of the Ring (2001, New Line Cinema/WingNut)—Peter Jackson

- Animal Logic; Digital Domain; EYETECH Optics; Gentle Giant Studios; Hatch Production; Oktobor; Rhythm and Hues; Weta Digital.

The Lord of the Rings: The Two Towers (2002, New Line Cinema/WingNut)—Peter Jackson

- Animal Logic; Digital Domain; EYETECH Optics; Gentle Giant Studios; Hatch Production; Oktobor; Rhythm and Hues; Sony Pictures ImageWorks; Weta Digital.

The Lord of the Rings: The Return of the King (2003, New Line Cinema/WingNut)—Peter Jackson

- Weta Digital; Tweak Films; Oktobor; Sandbox F/X; Rhythm and Hues; Rising Sun Pictures; Gentle Giant Studios; Hybrid Enterprises; Motion Works.

The Lost World (1925, First National)—Harry Hoyt

- Willis O'Brien; Ralph Hammeras; Fred Jackman.

The Lost World (1960, 20th Century-Fox/Saratoga)—Irwin Allen

- Willis O'Brien; L.B. Abbott.

The Lost World (1992, Worldvision)—Timothy Bond

- Philip Sharpe.

The Lost World (1998, Trimark)—Bob Keen

- Intrigue.

The Maltese Falcon (1941, Warner/First National)—John Huston
The Man Who Wasn't There (1983, Paramount)—Bruce Malmuth

- Eric Brevig; Robb King; Martin Becker.

The Mask (1994, New Line Cinema/Dark Horse)—Chuck Russell

- Industrial Light & Magic; DreamQuest Images.

The Mask of Fu Manchu (1932, MGM)—Charles Brabin & Charles Vidor

- Warren Newcombe.

The Mask of Zorro (1998, TriStar/Amblin Entertainment)—Martin Campbell

- Illusion Arts; Digital Film; Moving Picture Co.

The Matrix (1999, Warner/Silver Pictures)—Andy Wachowski & Larry Wachowski

- Mass.Illusions; Manex Visual Effects; Amalgamated Pixels; Animal Logic; DFilm Services; Makeup Effects Group Studio; Bullet Time.

The Matrix Reloaded (2003, Warner/Silver Pictures)—Andy Wachowski & LarryWachowski

- Industrial Light & Magic; Manex Visual Effects; Amalgamated Pixels; Bullet Time.

The Matrix Revolutions (2003, Warner/Silver Pictures)—Andy Wachowski & Larry Wachowski

- Sony Pictures Imageworks; BUF; CIS Hollywood; ESC Entertainment; Pacific Title; Tippett Studio.

The Monolith Monsters (1957, UIP)—John Sherwood

- Clifford Stine.

The Motorist (1905, Robert W. Paul)—Walter R. Booth

- Robert W. Paul.

The Mummy (1932, Universal)—Karl Freund

- John Fulton; Jack Pierce.

The Mummy (1999, Universal/Alphaville)—Stephen Sommers

- Industrial Light & Magic; Cinesite Digital Studios; Vision Crew Unlimited.

The Mummy Returns (2001, Universal/Alphaville)—Stephen Sommers

- Industrial Light & Magic; Digiscope; Manex Visual Effects; Computer Film Co.

The Mysterious Island (1929, MGM)—Lucien Hubbard

- J. Ernest Williamson.

The Mysterious Island (1961, Columbia)—Cy Endfield

- Ray Harryhausen.

The Next Karaté Kid (1994, Columbia)—Christopher Cain
The Nutty Professor (1963, Paramount/Lewis)—Jerry Lewis

- Paul Lerpae.

The Nutty Professor (1996, Universal/Imagine)—Tom Shadyac

- Rhythm & Hues; Computer Film Co; Light Matters Inc.

The Nutty Professor II: Meet the Klumps (2000, Universal/Imagine)—Peter Segal

- CORE Digital Pictures; Double Negative Ltd; Pacific Title and Art Studio.

The Peacemaker (1997, DreamWorks)—Mimi Leder

- Pacific Data Images; Effects Associate Ltd.

The Perfect Storm (2000, Warner/Radiant/Baltimore Spring)—Wolfgang Petersen

- Industrial Light & Magic.

The Phantom (1943, Columbia)—B. Reeves Eason
The Phantom (1996, Paramount/Village Roadshow/Ladd Co)—Simon Wincer

- Buena Vista Visual Effects; Cinema Research Corporation; D. Rez Hollywood; Chris Walas Inc; Jim Henson's Creature Shop; Stirber Visual Effects Network Inc.

The Phantom of the Opera (1925, Universal)—Rupert Julian

- Jerome Ash; Trey Freeman.

The Polar Express (2004, Warner/Universal CGI/Castle Rock)—Robert Zemeckis

- Sony Pictures Imageworks; Gentle Giant Studios; NAC Co; Effects & Prop Animation; W.M. Creations.

The Poseidon Adventure (1972, Kent)—Ronald Neame

- L.B. Abbott; A.D. Flowers.

The Prince of Egypt (1998, DreamWorks)—Brenda Chapman, Steve Hickner & Simon Wells

The Private Lives of Elizabeth and Essex (1939, Warner)—Michael Curtiz

- Byron Haskin; Hans F. Koenekamp.

The Rainmaker (1997, Zoetrope/Constellation)—Francis Coppola

- Guy Clayton Jr.; Pierre Maurer; Reuben Goldberg.

The Rain People (1969, American Zoetrope)—Francis Ford Coppola

The Rains Came (1939, 20th Century-Fox)—Clarence Brown

- Edmund Hansen; Fred Sersen.

The Relic (1997, Paramount/Universal/BBC/Polygram/Toho-Towa)—Peter Hyams

- Stan Winston Studio; Video Image; Blue Sky Studios/VIFX.

The Robe (1953, 20th Century-Fox)—Henry Koster

- Ray Kellogg; James Gordon; Matthew Yuricich.

The Rock (1996, Hollywood Pictures/Simpson/Bruckheimer)—Michael Bay

- DreamQuest Images; Alterian Studios; Buena Vista Imaging; Video Image.

The Rocketeer (1991, Disney)—Joe Johnston

- Industrial Light & Magic.

The Saint (1997, Paramount)—Phil Noyce

- Digital Film; Bionics; Cinesite (Europe); Computer Film Co.

The Santa Clause (1994, Disney/Hollywood Pictures)—John Pasquin

- Buena Vista Visual Effects; Amalgamated Dynamics; Lebensfeld Productions; Global Effects Inc.

The Seventh Voyage of Sinbad (1958, Morningside)—Nathan Juran

- Ray Harryhausen.

The Searchers (1956, Warner)—John Ford

- George Brown.

The Shadow (1940, Columbia)—James W. Horne

- Ken Strickfaden.

The Shadow (1994, Universal/Bregman)—Russell Mulcahy

- Fantasy II Film Effects; Chandler Group Visual Effects; Dreamstate Effects; Composite Components Co; Howard A. Anderson Co; Illusion Arts Inc; Hunter-Gratzner Industries; Matte World Digital; Lumeni Productions Inc; Pacific Titles & Optical; Stetson Visual Services.

The Shawshank Redemption (1994, Columbia/Castle Rock)—Frank Darabont

- Melissa Taylor; Bob Williams.

The Sixth Sense (1999, Hollywood Pictures/Spyglass/Kennedy-Marshall)—Night Shyamalan

- DreamQuest Images; Stan Winston Studio.

The Son of Kong (1933, RKO) - Ernest B. Schoedsack

- Willis O'Brien; Harry Redmond Jr; Juan Larrinaga.

The Sound of Music (1965, 20th Century-Fox)—Robert Wise

- L.B. Abbott; Emil Kosa Jr.

The Space Children (1958, Paramount)—Jack Arnold

- John P. Fulton; Farciot Edouart.

The Sting (1973, Universal)—George Roy Hill
The Story of the Kelly Gang (1906, Tait/Johnson/Gibson)—Charles Tait
The Sugarland Express (1974, Universal/Zanuck-Brown)—Steven Spielberg

- Frank Brendel.

The Ten Commandments (1923, Famous Players-Lasky Corp)—Cecil B. De Mille
The Ten Commandments (1956, Paramount)—Cecil B. De Mille

- John P. Fulton; Jan Domela; Paul Lerpae; David Horsley.

The Thief of Bagdad (1924, Douglas Fairbanks Pictures)—Raoul Walsh

- Hampton Del Ruth; Coy Watson Sr.

The Thief of Bagdad (1940, London Film Productions/United Artists)—Ludwig Berger & Michael Powell

- Lawrence W. Butler; Peter Ellenshaw; Wally Veevers; Tom Howard.

The Thing (1982, Universal)—John Carpenter

- Albert Whitlock; Roy Arbogast; Jim Danforth; Henry Schoessler; Peter Kuran; Andrew Miller; Susan Turner; James Belohovek.

The Thing from Another World (1951, Winchester)—Christian Nyby

- Linwood G. Dunn; Donald Steward; Harold E. Stine.

The Three Caballeros (1944, Disney)—Norman Ferguson

- Ub Iwerks; Richard C. Jones; Edwin Aardal; John McManus; Joshua Meador.

The Towering Inferno (1974, 20th Century-Fox/Warner)—John Guillermin & Irwin Allen

- Douglas Trumbull; L.B. Abbott; A.D. Flowers; Van der Veer Photo Effects.

The Truman Show (1998, Paramount)—Peter Weir

- Available Light Productions; Cinesite Digital Studios; The Computer Film Company; EDS Digital Studios; Matte World Digital; Stirber Visual Effects Network.

The Twentieth Century Tramp (1902, Edison Co)—Edwin S. Porter

- Edwin S. Porter.

The Untouchables (1987, Paramount)—Brian de Palma

- Allen Hall; Charles Stewart; Janos Pilenyi; Phil Gosiewski.

The Usual Suspects (1995, Polygram/Spelling)—Bryan Singer

- Roy Downey; David Long.

The Valley of Gwangi (1969, Warner)—Jim O'Connolly

- Ray Harryhausen.

The Vanishing Shadow (1934, Universal)—Lew Landers

- Elmer A. Johnson; Raymond Lindsay; Ken Strickfaden.

The War of the Worlds (1953, Paramount)—Byron Haskin

- Gordon Jennings; Aubrey Law; Irmin Roberts; Walter Hoffman; Jan Domela; Jack Caldwell; Ivyl Burks; Paul Lerpae; Bob Springfield; Chester Pate.

The Waterboy (1998, Touchstone)—Frank Coraci

- Flash Film Works.

The Whole Dam Family and the Dam Dog (1905, Edison Co)—Edwin S. Porter

The Witches of Eastwick (1987, Warner/Kennedy-Miller/Guber-Peters)—George Miller

- Industrial Light & Magic.

The Wizard of Oz (1939, MGM)—Victor Fleming, King Vidor & Richard Thorpe

- Arnold Gillespie.

The Wolf Man (1941, Universal)—George Waggner

- John P. Fulton; Ellis Burman; Jack P. Pierce.

The X-Files (1998, 20th Century-Fox/Ten Thirteen)—Rob Bowman

- Hunter-Gratzner Industries; Amalgamated Dynamics; Blue Sky Studios/VIFX; Todd-AO Digital Images; K.N.B. EFX Group Inc; Hollywood Digital; Trans FX; Digital Filmworks Inc; Full Scale Effects Inc; O'Connor FX; Pixel Envy; POP Film; Mann Consulting; Light Matters Inc.

The Young Indiana Jones Chronicles TV Series (1992-1993, Paramount TV/Lucasfilm/Amblin Entertainment)—Jim O'Brien

- Industrial Light & Magic; Digital Magic Co.

Thelma & Louise (1991, MGM/Pathé)—Ridley Scott

- Stan Parks; Kevin Quibell; Paul Stewart.

There's Something About Mary (1998, 20th Century-Fox)—Bobby Farrelly & Peter Farrelly

Things to Come (1936, London Film Productions/United Artists) –William Cameron Menzies

- Ned Mann; Wally Veevers; Lawrence Butler; Edward Cohen; W. Percy Day; Peter Ellenshaw.

Thirty Seconds Over Tokyo (1944, MGM)—Mervyn LeRoy

- Arnold Gillespie; Donald Jahraus; Warren Newcombe.

This Is Cinerama (1952, Cinerama Corp)—Merian Cooper

This Island Earth (1955, UIP)—Joseph Newman

- David Horsley; Clifford Stine; Charles Baker; Roswell Hoffmann; Frank Tipper.

Three Men and a Baby (1987, Touchstone/Interscope/Silver Screen)—Leonard Nimoy

THX 1138 (1971, Warner/American Zoetrope)—George Lucas

- Hal Barwood.

Tin Toy (1988, Pixar)—John Lasseter
Titan A.E (2000, 20th Century-Fox)—Don Bluth & Gary Goldman
Titanic (1953, 20th Century-Fox)—Jean Negulesco

- Ray Kellogg.

Titanic (1997, 20th Century-Fox/Paramount/Lightstorm)—James Cameron

- Industrial Light & Magic; Digital Domain; Blue Sky Studios/ VIFX; Pixel Envy; Digiscope; Rainmaker Digital Pictures; Vision Crew Unlimited; 4-Ward Productions; POP Film; Banned From the Ranch Entertainment; Composite Image Systems; Video Image; Cinesite Digital Studios; Donald Pennington Inc; Hammerhead Productions; Perpetual Motion Pictures; Matte World Digital; Light Matters Inc; Title House Inc.

Titanic 3D (2012, 20th Century-Fox/Paramount/Lightstorm)—James Cameron
Tootsie (1982, Columbia/Delphi/Mirage/Punch)—Sydney Pollack
Top Gun (1986, Paramount)—Tony Scott

- USFX; Colossal Pictures; Intervideo.

Topper Takes a Trip (1939, Roach)—Norman McLeod

- Roy Seawright.

Total Recall (1990, Columbia/Carolco)—Paul Verhoeven

- Industrial Light & Magic; DreamQuest Images; Hunter-Gratzner Industries; Metrolight Studios; Stetson Visual Services; Cinema Research Corporation.

Toy Story (1995, Disney/Pixar)—John Lasseter
Toy Story 2 (1999, Disney/Pixar)—John Lasseter & Ash Brannon
Trois Hommes et un Couffin (1985, TF1/Flach/Soprofilms)—Coline Serreau
Tron (1982, Disney/Lisberger-Kushner)—Steven Lisberger

- Digital Effects Inc; MAGI-Synthavision; Robert Abel & Associates; Stargarte Films Inc.

Tron: Legacy (2010, Disney/LivePlanet)—Joseph Kosinski

- Digital Domain; Mova; Mr. X; Prana Studios; Prime Focus; Prologue Films; Quantum Creation FX; Whiskytree.

Troy (2004, Warner/Radiant)—Wolfgang Petersen

- Moving Picture Company; Framestore CFC; Cinesite; Lola; Artem Digital Ltd; Cine Image Film Opticals; FB-FX; Plowman Craven & Associates.

True Lies (1994, 20th Century-Fox/Universal/Lightstorm)—James Cameron

- Digital Domain; Makeup & Effects Lab; Pacific Data Images; Boss Film Studios; Stetson Visual Services; Pacific Titles & Optical; Light Matters Inc; Sturm Special Effects International; Fantasy II Film Effects; Cinesite Digital Studios; Shockwave Entertainment.

Twelve Monkeys (1995, Universal/Atlas)—Terry Gilliam

- Hunter-Gratzner Industries; Mill Film; Peerless Camera Co.

Twister (1996, Universal/Warner/Amblin Entertainment)—Jan de Bont

- Industrial Light & Magic; Banned From the Ranch Entertainment.

Union Pacific (1939, Paramount)—Cecil B. DeMille

- Gordon Jennings; Farciot Edouart; Jan Domela; Paul Lerpae; Barney Wolff.

Universal Soldier (1992, Carolco/Centropolis/IndieProd)—Roland Emmerich

- Matt Kutcher; Kit West; Volker Engel.

Universal Soldier: The Return (1999, IndieProd/Baumgarten-Prophet)—Mick Rodgers

- K.N.B. EFX Group Inc.; E=mc2; Cinema Research Corporation.

Urban Legend (1998, Canal Plus/Original/Phoenix)—Jamie Blank

- Digiscope; Pacific Title; Mirage.

Vertigo (1958, Paramount)—Alfred Hitchcock

- Farciot Edouart; John P. Fulton; Wallace Kelley; John Whitney Sr.

Volcano (1997, 20th Century-Fox/Donner/Moritz)—Mick Jackson

- Digiscope; Digital Magic Co; POP Film; Pixel Envy; Video Image; Stirber Visual Effects Network Inc; Light Matters Inc.

Voyage to the Bottom of the Sea (1961, 20th Century-Fox)—Irwin Allen

- L.B Abbott.

WarGames (1983, MGM/Sherwood)—John Badham

- Joe Digaetano; Michael L. Fink; Colin Cantwell; Linda Fleisher; Marcia Dripchak; Geoffrey Kirkland.

War and Peace (1956, Paramount)—King Vidor

- Costel Grozea.

War of the Worlds (2005, Paramount/Dreamworks/Amblin)—Steven Spielberg

- Industrial Light & Magic; Stan Winston Studio; Pacific Title; Gentle Giant Studios.

Waterworld (1995, Universal)—Kevin Reynolds

- Boss Film Studios; Cinesite Digital Studios; Rhythm & Hues; Stetson Visual Services; Out of the Blue Visual Effects; EFilm; DreamQuest Images; Editel; Composite Image Systems; 525 Post Production; Matte World Digital; Pacific Ocean Post Digital; OCS/Freeze Frame/Pixel Magic; Hunter-Gratzner Industries.

We Will Rock You: Queen Live in Concert (1982, Mobilevision/Yellowbill)—Saul Swimmer

Werewolf of London (1935, Universal)—Stuart Walker

- John P. Fulton; David Horsley.

Westworld (1973, MGM)—Michael Crichton

- Information International.

What Lies Beneath (2000, DreamWorks/20th Century-Fox)—Robert Zemeckis

- Sony Pictures ImageWorks; Station X Studios; Computer Film Co.

What Price Glory? (1926, Fox)—Raoul Walsh

- L.B Abbott.

When Worlds Collide (1951, Paramount)—Rudolph Maté

- Tim Baar; Farciot Edouart; Gordon Jennings; Harry Barndollar; Barney Wolff; Jan Domela; Paul Lerpae.

Who Framed Roger Rabbit? (1988, Touchstone/Amblin Entertainment)—Robert Zemeckis

- Industrial Light & Magic.

Wild Wild West (1999, Warner)—Barry Sonnenfeld

- Cinesite Digital Studios; Industrial Light & Magic.

Willow (1988, Lucasfilm/MGM)—Ron Howard

- Industrial Light & Magic.

Wings (1927, Paramount)—William A. Wellman

- L.B. Abbott; Roy Pomeroy; Barney Wolff

X-Men (2000, 20th Century-Fox/Marvel/Schuler-Donner)—Bryan Singer

- Digital Domain; CORE Digital Pictures; Hammerhead Productions; FXSmith Inc; Cinesite Digital Studios; Kleiser-Walczak Construction Co; Matte World Digital; Pacific Ocean Post Digital.

Young Sherlock Holmes (1985, Paramount/ Amblin Entertainment/ ILM)—Barry Levinson

Index